THE SILICON WIZARDS

By

Robert L Skidmore

ISBN: 0-7596-8508-8

This book is printed on acid free paper.

1stBooks - rev. 12/13/01

Other Books by Robert L Skidmore:

The Satterfield Saga Series

If Genes Could Talk
Yankee Doodles in Black Hats
Green Eyes, Red Skies, Pale Ales, White Sails
Point of Rocks
The Drums, the Fog, the Terrible Shadows of War
Did You Lose Your Elephant
The Ballbreaker
The Sorry World

Inspector Richard Thatcher Series

The City of Lost Dreams
Cluster of Spies

Dedication

Again, For Margaret and Tad

Chapter 1

Charles Swift stared at his image in the mirror and shook his head in disgust. He looked like a damned twelve year old, a geek with unruly sandy hair, horn rimmed glasses that kept sliding down a prominent nose, thin lips, square chin. No wonder, Charles thought, that girls avoided him. Charles retreated from the bathroom and went to the window where he scraped a hole in the ice and peered into the gray gloom. Two feet of snow blanketed the landscape, rendering the unplowed roads and unshoveled walks formidable. Charles smiled at the sight of four intrepid students who ignored the heavy flakes floating from the sullen sky as they pelted each other with large, soggy snowballs. Charles, amused by his housemates' antics, turned and leaped to touch the ceiling.

"For Christ sakes Charles," his roommate complained. "Perch somewhere. I have to study for this exam even if you don't."

"You need a break," Charles countered.

"You need a life."

Undaunted, Charles leaped again. This time his heavy ski boots landed with a thud. Before Bill Oldham could complain, the student in the room below grabbed the broomstick that he kept next to his desk just for this purpose and thumped against the ceiling.

"We would all be happy if you went out for the ski team," Bill said.

Charles smiled at the gibe. It was no secret that the boots were an affectation. Charles did not ski nor engage in any type of athletic activity. He doubted that jumping and touching the ceiling counted. At five ten and a meager hundred and thirty pounds, he had no chance at passing as a jock. When younger, he had often yearned for an athlete's body, not because he wanted to compete on the court or field, but because girls seemed to be attracted to the type. Such concerns no longer troubled him, much, not all the time. During his sophomore year in high school, he had settled on his niche. Fortunately, he was blessed with a sharp analytical mind. The jocks made the marks on the playing fields while Charles distinguished himself in the classroom. At least he did for a few days every year until he grew bored with the academic routine.

1

If it had not been for his discovery of the wonderful world of computers, Charles' adolescence would have been terrible. Despite his fine mind, his high school grades had not been particularly good. He crunched the tests but failed to maintain a sufficient interest in his academic courses to perform adequately on the day to day work. Consequently, the C's, D's, and Incompletes for daily work more than offset the A's of his brilliant examination papers. If it had not been for his mother's contacts—she was a prominent board member of several important corporations—and the fact he had aced his SATs, Charles would not have been accepted by Harvard's Overseers. If the truth were known, he ranked below borderline for applicants.

Charles had arrived at Cambridge full of confidence, expecting to conquer Harvard. Unfortunately, things had not turned out as he had expected. It had not taken long to learn that there were two types of students at Harvard, the aristocratic elite and the extremely bright scholarship boys. To his disappointment, Charles had not meshed with either group. He could do nothing about the hard fact that his father was merely a successful Seattle lawyer. Charles had always assumed that his family ranked in the upper class. They belonged to the country club and lived in one of the better domiciles in their exclusive suburb. Their home even abutted Seattle's fabulous Lake Washington. It had been a shock to find Harvard's aristocrats unimpressed. This was not the worst, though. Charles had expected to dominate Harvard with his brilliant brain. He had arrived with no doubts about his ability to compete academically with the best of the scholarship boys. Charles' first months had rudely introduced him to the cold fact that there were others with even finer minds, and his bad study habits, acquired against the lesser competition of his protected suburb, rendered him an also ran. He found it difficult adjusting to his place in the second and third files. Now, in his second year at Harvard, he was adrift.

Charles smiled at Bill, then decided to give him a break.

"Think I'll check the game out."

"Thank God."

Charles was a regular at the non-stop poker game on the second floor.

"And he entered the room with a clatter," Charles shouted as he flung open the closed door.

"Oh shit," the players chorused.

The familiar greeting delighted Charles because he knew the other players sincerely did not welcome his presence. His appearance changed the nature of the game. Charles, a methodical player, carefully calculated the odds on every card. Charles won steadily, not big on any one day, but consistently. Charles knew to a penny exactly how much he had won over the course of the year, but he kept his own counsel. The others could only estimate the size of Charles' winnings by calculating their own losses. They did not know, however, that Charles was indifferent to the money. Winning was important.

"What's the game?" Charles asked as he took the last empty chair.

"Cut the shit, Charles," the dealer said. The portly boy whose father ran a trucking company in Boise waited for Charles to ante a dollar. They used dollar bills instead of chips, and each boy had his stash of Washingtons piled in front of him, some much smaller than others. Charles deliberately pulled a modest roll of ones from his pocket.

"You're going to need more than that," the dealer jibed.

"Probably," Charles agreed, then leaped and touched the ceiling for luck before sitting down.

Back in their room, Bill Oldham attempted to focus on his notes. He had memorized the answers to all the questions he anticipated would appear on their six weeks exam in English Lit. He knew the answers cold and now scanned for the teaser questions that would make the difference between a B and an A grade. A scholarship student, who had to maintain his high grade point average in order to continue financially at Harvard, Bill envied his roommate and their peers whose wealthy parents granted them the freedom to be indifferent to academic ranking.

Bill took off his reading glasses and rubbed his burning eyes. To Hell with it, he decided. He dropped the glasses on his notes and went to the window where he peered out of the small hole in the frost that Charles had rubbed. The heavy wet snowflakes, some as large as silver dollars, continued to fall. Bill had never seen such snow, not even in his hometown of Clarksburg located high in the West Virginia mountains. While he watched, a snow plow mounted on a sand and salt filled dump truck turned the corner and roared past the house depositing a three foot ridge of snow beside the line of parked cars, effectively blocking them in place. Bill chuckled. He did not own a car, but he could imagine the reaction of those who did when they discovered the work that awaited if they hoped to free their chariots. As Bill watched, the cascading snow coated the freshly plowed street. In the distance a solitary figure appeared. Bill wondered who would be stupid enough to venture out in this mess. The parka clad figure turned when he reached the front of their house and struggled through the knee-high snow towards the door. Just as he neared the steps to the front porch, Bill recognized him. With difficulty, Bill broke the ice gripping the window and raised it. He leaned out and hollered.

"Hey, Clap! Are your crazy?"

James Clapper, Charles' best friend, had taken a job in Boston at Honeywell mainly to be close to his Seattle schoolmate. Clap, as Clapper was inevitably known, still hoped to fulfill his youthful dream of starting a computer business with Charles. Bill considered Clap and Charles immature for clinging to childhood fantasies, but he liked both young men; they shared a natural charm and infectious sense of humor.

Clap did not answer. He merely waved as he struggled onto the porch.

"Come on up," Bill called, then slammed the window shut. Clap's appearance destroyed the last of Bill's waning resolve. If he had to take a B on the English Lit exam, he would make it up in the second six weeks period.

"Christ what weather," Clap complained as soon as he opened the door.

Bill laughed.

"Where's trey?" Clap used the nickname favored by Charles Andrew Swift III's Seattle friends.

"Poker, where else."

As soon as he joined the poker game, Charles arbitrarily decided to change tactics. The others naturally assumed he would follow his usual cautious style, playing the odds and bluffing not at all. This made him a predictable if steady winner. Charles whimsically opted to bluff and bet heavily on his first five hands regardless of the cards he drew. He assumed this would unsettle the others and possibly upset the way they judged his hands later in the game. Five-card stud was the preferred game.

After the one dollar ante, the dealer dealt the first card face down and the second face up. Charles checked his hole card. A three of clubs. He showed a five on top, undoubtedly the weakest hand in the game.

"King bets," the dealer declared.

The boy on his left tossed a dollar bill into the pot. "You'll have to pay to see my kings back to back."

The others laughed mirthlessly and tossed their one-dollar calls into the pot.

"I'll see you and raise you three," Charles said as he tossed the pot limit into the center of the table.

"A pair of fives won't buy you anything," the dealer laughed and matched Charles' bid.

"It'll cost you," the player with the king showing called and raised Charles another three dollars.

The other players, assuming Charles had a pair of fives, met the raise and re-raise. All of them had higher cards showing, thus anticipated they could outdraw Charles' lowly pair. They wondered about Charles' tactics. It was unlike him to push a weak hand. Charles called. He knew he had the low hand and that the odds were five to one against pairing his unassuming three in the hole. The dealer dealt the next round. Charles drew an unhelpful eight, giving him a three in the hole and a five and eight showing. The boy with the king drew a jack, and the player on Charles' right matched his seven.

4

"Pair of sevens check to the kings," the boy said. Charles and the other players deferred to the king.

"It'll cost you the limit," the player said, smiling as he tossed three bills into the pot.

The thin smile told Charles that the bettor was bluffing, that he lacked the second king. Regardless, every hand showing had Charles beat including the pair of sevens and solitary king and its jack companion.

"See you and raise you," Charles tossed six Washingtons into the pot.

The other players groaned.

"A pair of fives won't do it," the boy with the sevens said. He raised the limit again. This time Charles called along with the other players who each held cards that if paired would beat the sevens.

The dealer dealt again. This time Charles drew a four, no help whatsoever, while the king paired and the two sevens drew a queen. The boy with the two kings stared at Charles' hand, laughed, and bet another three dollars. Three players folded, leaving Charles, the two kings and the two sevens in the hand. Charles made a show of studying his hole card.

"I'm feeling lucky," Charles said. "A pair of aces will beat kings." Charles bluffed, falsely indicating that he had an ace in the hole and expected to match it on the last deal.

An ace in the hole would explain Charles' betting even if he went against the odds and relied on being dealt a second ace in the last deal. Both opponents called.

"Read 'em and weep," the dealer declared as he dealt a final card to each player.

To Charles' delight, he was dealt a second five. This gave him a pair of fives showing, the lowest hand in the game, but gave him the opportunity to bluff that he had a third five in the hole, the winning hand.

"Oh shit," the pair of kings complained. He hesitated as he considered his bet. Finally, he surprised Charles by betting the limit. "I don't believe it," he declared.

The player with the pair of sevens hesitated. Charles waited patiently. The boy studied his hole card, obviously trying to will it into another seven. Finally, he threw his cards face down into the pot. "I've got Charles, but I can't beat kings." He capitulated.

Charles smiled. He had a pair of fives, a clear loser. "See you and raise you three," he said.

"Shit!" The boy with kings swore. "He's got three fives." He fingered his diminished stack of bills, counted out three, then hesitated. Charles waited patiently. The boy glared at Charles then turned his cards over and threw them into the pot.

Pleased with himself, Charles raked in the pot.

While Charles carefully stacked and aligned his growing stack of Washingtons, the others complained about lucky draws.

Using the same tactics, Charles bluffed his way through five straight hands, winning each time. The complaints turned bitter. Just as it was Charles' turn to deal, the door opened, and Clap and Bill entered. As soon as Charles saw who had joined them, he handed the deck to the boy on his left. "You deal, I'm out of here."

"Oh no," the others chorused.

"You've got to give us a chance to get our money back," the fat boy from Boise said.

"Tonight," Charles promised. The others looked at him skeptically and groaned in unison. They clearly wanted a chance to win their money back, but they really did not want Charles back. He took the fun out of the game.

Charles, Will and Clap left the room. Charles slammed the door on the loud complaints.

"Thanks for rescuing me," Charles said as he folded the thick wad of single dollar bills. "I was pushing my luck."

"Don't try to con us Charles," Bill said. "We know better. Luck has nothing to do with it."

"It did today. I bluffed for five straight hands. If I tried again, they would have had me." Charles turned to his long time friend James Clapper. "What brings you slumming among the elite?"

"They closed us down because of the snow. Martha wasn't so lucky. She's still working." Clap referred to his live-in girl friend.

"And you decided to drop in and treat your poor student friends to a brew," Charles said.

"You got it. Gonna join us Bill?"

"Poor lad has to hit the books," Charles said.

"I'm with you," Bill countered.

Bill and Charles retrieved their heavy parkas from their room then joined Clapper in the trudge through the still falling snow to their favorite hangout, a hole in the wall student tavern run by an outgoing Greek on the other side of the campus. They ignored the sidewalk that was blocked by drifting snow and followed in the snowplow's path where the snow had re-accumulated to two inches. Bill took the lead; Clapper and Charles followed. They traced a path along the deserted roadway. After rounding the square, they reached an area where some of the heaviest snow had been removed from the sidewalks. Bill led

6

them past the frosted windows of a corner drug store. Just as he began to pickup the pace, he heard Charles shout:

"Hey! Look at that."

Bill turned to find Charles and Clap standing side by side as they peered into the drug store window. Before Bill could ask what was so interesting, both Charles and Clap hurried inside. When Bill joined them, he found the two friends excitedly studying an issue of "Popular Electronics."

"They don't have any beer in here," Bill complained good-naturedly.

The other two ignored him. Bill, an English major, had no interest in electronics of any kind.

"Aw shit," Charles complained.

"The Taurus," Clapper said.

"What's the big deal?" Bill asked.

"It's a computer," Clap said.

"They beat us to it," Charles said.

"Let me see," Bill said, more interested in learning what had his friends so agitated than in the subject itself. He had heard them talking about someday building a computer small enough to set on a desk that people could use to do many tasks at the office and at home. Bill had been skeptical. A typewriter was good enough for him, but Charles always insisted that the day would come when people would use machines to think with just as they use typewriters to write with. Bill, always indifferent to electronics, usually let Charles win that argument.

Charles purchased a copy of the magazine, and they continued on their way. Bill again led as Charles and Clapper followed, chatting excitedly about the Taurus. Bill decided that it was going to be another of those days where Charles and Clapper talked electronics to the exclusion of any other subject and any other participant, at least as far as Bill was concerned.

After they reached their destination and ordered three draft beers, Charles opened the magazine on the table in front of him and again studied the lead article. Bill was able to read the larger print: "World's First Minicomputer Kit to Rival Commercial Models." The picture showed a box with lights and toggle switches in the front.

"It could be a computer," Charles said. "It's being built by some guys down in New Mexico."

"What does it do?" Bill asked. He asked this question every time he heard his two friends discussing computers. They had yet to provide a satisfactory answer beyond bland statements like "everything."

"I don't think they have any software," Charles spoke directly to Clapper.

"They're using the 8080," Clapper said.

"What's that?" Bill asked.

"A chip" Charles said.

Charles' answer told Bill exactly nothing. He usually kept his mouth shut when his two friends got into their electronic discussions. He had no idea what they were talking about and hated showing his ignorance.

"I can adapt my simulator to work with the 8080," Clapper said.

The thought appeared to excite him.

"What's a simulator?" Bill had to ask.

"Charles used it so he could program our traffic counting machine."

Bill had heard more of that subject than he needed. The two friends had joined with a third guy when they were still in high school to form a company that somehow computerized the product from the rubber hoses that the state and municipalities placed on the roads to count traffic. As best as Bill could tell, the company had been a failure.

"I could use BASIC to write software for this baby," Charles said. "I can do it in my sleep."

Bill sipped his beer and dropped out of the conversation. The question "what is BASIC" had been asked and answered several times without providing Bill with more understanding than he had had to start. It was some kind of computer language that programmers used to write instructions to the hardware. As Clapper and Charles chatted, it became clear that both were sincerely worried. They still dreamed of starting their own company and helping to create the new computer age. The Taurus, as described in the magazine, had caught them by surprise.

"We have to do something about it," Charles slapped the table for emphasis.

Charles started to rock back and forth from the waist up, a sign that he was thinking rapidly. He stopped suddenly, and his glasses slid down his nose. He pushed them back with his forefinger.

"OK. The Taurus is a breakthrough, but it doesn't do anything."

"They need an interpreter and software," Clap agreed.

"And we have it," Charles said. "Will your interpreter work?"

"Maybe," Clapper said.

"It'll work," Charles said confidently.

"I should be able to convert it to work with the 8080 chip." Clap seemed less confident than Charles.

"And I could write the BASIC," Charles enthused. "Let's write a letter."

Bill could almost see Charles' mind racing.

"We'll use our Traffic Counter stationary. That way it'll look like we're a real company."

"What will we say?" Clapper asked.

"We'll tell them we have a BASIC interpreter that will run on the Taurus. That's all we have to say. They'll know they could sell hundreds of machines if they can show that they have software that does things. Maybe thousands."

Bill was tempted to kibitz by asking again: "Does what?" but he did not. His friends were too earnest and too serious to take any sideline kidding.

"We won't sell it to them outright. They wouldn't offer us much that way," Charles said. He started rocking back and forth.

"Some kind of royalty system," Clap agreed.

"Right," Charles slammed the table, rattling their now empty glasses.

This attracted the attention of the Greek who leaned across the bar and warned: "Take it easy guys. You break you pay."

"Three more drafts," Bill interceded before Charles could give one of his sarcastic answers and get them thrown out. Again. The Greek liked their business, but he was a hot head who took no trash talk from the students.

"We'll set our price low enough that they can't refuse," Charles ignored the interruption. He took out the wad of one-dollar bills that he had won bluffing at poker and waved it about, pausing only to riffle the bills. "We'll ask for a royalty of one dollar for each computer with our software that they sell."

That caught Bill's attention. "How many will they sell?"

"There are thousands of hobbyists who will line up all night to get their own computers," Clap answered.

"You're the writer," Charles turned to Bill. "You can help us make the letter sound like a real company."

Bill smiled, now part of the action. "When?"

"Now." Charles drained his full glass of beer in one gulp. "Before someone beats us to it."

Charles stood, pulled three bills from his stash, and dropped them on the table. "The beer's on me."

"Don't get carried away," Clapper laughed. Charles seldom spent extravagantly. Even in high school, he had been the tightest with cash of the group.

The three friends returned to Bill and Charles' room. Bill sat in front of his Royal typewriter while Charles stood behind him in a position where he could see the page and dictated. After three tries they had a single paragraph on Traffic Counter stationary that satisfied Charles. They used the address for PAMS, the Taurus company in Albuquerque, and Charles insisted that it had to be mailed immediately. Bill declined to accompany the other two on their snowy trek to the post office and returned to his studies.

9

When the letter arrived at PAMS three days later, it immediately generated considerable interest. The orders for the Taurus had cascaded in following the "Popular Electronics" article. Already, the small company was overwhelmed, but the response had kindled their ambitions. The company owners, including the engineer who had devised the Taurus, were fully aware of its shortcomings. They knew that if they could obtain software to run on the Taurus they could increase their price by two to three hundred dollars, and they and their company would prosper geometrically. They had had a number of offers from hobbyists, but unfortunately none had a simulator and a BASIC program ready to go. The letter from Traffic Counter implied that it had both. One of the owners dialed the number on the stationary but got no answer. Charles and Clapper in their excited effort to appear businesslike had overlooked the fact that the telephone number on their old stationary was the home number of the parents of the third member of the juvenile company. The boy's mother had been out when the PAMS people attempted to call.

While Charles and Clapper waited anxiously for a response from New Mexico, Clapper worked at home on his simulator during his off-hours from his HP job. They had implied they had a compatible simulator, and he wanted to be able to fulfill that promise. On the other hand, Charles, confident he could write the BASIC code for the Taurus if the company responded positively, continued his indolent life style.

A week passed with no response. Clapper visited Charles to discuss their next step.

"We should call them," Charles declared.

"Right, call them," Clapper agreed. "The number is in the article."

Clapper retrieved the magazine from its prominent position atop Charles' cluttered desk and handed it to Charles who was lying on his unmade bed.

Charles held his hands up and declined the proffered magazine. "You do it."

"No, you," Clapper insisted.

"But you will meet him first, and you're older and more mature looking."

"And prettier," Bill Oldham laughed. The two aspiring businessmen were acting like children.

Charles glared irritably at his roommate. He assumed his roommate was putting him down. Charles knew all too well that he looked like a geeky kid.

"He'll want to talk about the BASIC," Clapper persisted. "Since you haven't written it yet, only you can answer any questions."

Charles was confident of his ability to write the code. He just wanted to avoid the tedious task until he was confident the Taurus Company was genuinely interested in it. It would be a waste of time to write it now and then learn that they already had their code. Charles knew, however, that Clap was right, but Charles did not want to be the front man. His youthful appearance could be overcome when he demonstrated his brainpower, but Charles assumed that a first impression would be important if they had to convince the PAMS people to accept their offer. He thought it imperative that the older employed Clapper represent them, not a young Harvard student.

"I'll compromise," Charles finally replied. "I'll talk to them on the phone, but I'll use your name. Then, if they want one of us to come down and demonstrate our product, you go."

"Do it," Clapper agreed. "Then, when you appear, I'll tell them you are my son."

"Funny," Charles frowned. He did not like jokes about his appearance. Nobody did.

Charles reached for the telephone that was on a table near the head of his bed, and Clapper tossed the magazine to him. Clapper sat down on Bill's bed. Bill, who sat at his desk, watched. As Clapper started to pivot, Bill spoke: "Take off the shoes first."

Clapper laughed and leaned over and unlaced his ski boots. "Some people are fussy. Some aren't." He turned his hand palm up in the direction of Charles who was lying with his boots on his bare mattress.

Charles dialed the PAMS number. To his surprise the company owner answered on the third ring.

"This is James Clapper of Traffic Counter," Charles said.

"Oh, Mr. Clapper," the owner immediately recognized the name. "We received your letter last week; we tried to call the number on your letterhead, but no one answered."

Charles put his hand over the telephone mouthpiece and spoke to Bill and Clap: "We screwed up. We used the Traffic Counter letterhead, and it had Jake's telephone number on it."

Bill laughed.

"I knew we shouldn't have used that stationary," Clap said.

"I'm sorry about that," Charles returned to the phone. "We're in the process of moving offices and have a new number." He gave the number of his and Bill's phone.

"That's all right," the PAMS owner said, oblivious to the fact that a comparison of telephone numbers would indicate the Traffic Counter Company had moved across the continent. "I'm glad you called. We received a lot of letters offering to write software, but you seem to be the only people with a finished program."

Charles grimaced. He had yet to write a single word of code for the Taurus. Before he could think of something to say, the owner continued. "We don't have enough memory cards yet, and we're overwhelmed with the response. We're getting over a hundred orders a day."

"That's good news," Charles said.

"We want to talk with you fellows," the owner said. "Why don't you give us a couple weeks then come on down."

"Sounds good," Charles said. "We'll call first."

As soon as he hung up the phone, Charles leaped and touched the ceiling and gave a loud war hoop. "We're in. They still need the software, and he wants us to come down and talk about it."

"What's his name?" Bill asked.

"I didn't ask," Charles said.

"Shit," Clapper said.

Charles ignored him and shouted: "We've got it, We've got it," as he danced about the room.

"We'll have to get our hands on a Taurus," Clapper said.

"No sweat. Don't need it," Charles said. "We can simulate it."

Clap recognized that Charles referred to using Harvard's computers. Bill did not comprehend what Charles was talking about. Although Charles was taking no computer courses—he believed he already knew more about computers than Harvard had to offer—he had talked his way into using the computers at Harvard's Aiken Computation Laboratory. The lab had a DEC PDP-10 that Charles liked to use for computer games. Once, a professor had challenged his presence, demanding to know what graduate classes he was taking, but Charles had impressed him with his knowledge and succeeded in getting tacit permission to use the computer when it was not needed by others. From then on, Charles had used the professor's name as his patron. However, what he planned was not strictly legal. He intended to use Harvard's machines to write the code for a private commercial endeavor that had nothing to do with the university.

Charles and Clap visited a local electronic parts store and found the 8080 manuals that they needed. Clap at that point continued work on his simulator, and Charles took advantage of the semester break to use the lab's PDP-10. He wrote code sometimes up to twenty hours a day. Charles was well versed in several forms of BASIC and decided to use BASIC Plus. Charles knew that his program would have to be limited. The Taurus had a minimum of memory. Once, Charles called Albuquerque to ask for information about the Taurus that was not readily available. The call seemed to impress the Taurus's owner because Charles, pretending to be Clapper, was the first to call and ask how to get characters in and out of the Taurus.

After he was satisfied with his simulator, Clapper continued his day job at HP and spent many of his evenings working with Charles in his room and at the lab. Before long, both young men were exhausted, but Charles in particular thrived on the fatigue. Stimulated by the challenge, he took catnaps in the crawlspace behind Harvard's computer and worked through the night.

Charles completed his BASIC program in three weeks, and he confidently opted to forego testing. They did not have a Taurus and buying one for test purposes meant they had to order a kit from New Mexico, then assemble it when it arrived. Charles decided they did not have the time for all that; they had to be the first to get their system for programming the Taurus to the factory. Clapper called the Taurus owner and told him they were ready.

"Then come on down," the Taurus owner said.

Charles and Clap did not have a final version of their work. There were bugs to be fixed, but Charles insisted that they had enough for a demonstration. He did not have to say how important it was to be the first. The night before Clap was to take their handiwork to New Mexico, Charles was too nervous to sleep and spent the entire night going over his code. He knew that even a small error would prevent the Taurus from starting and ruin their presentation. Finally, Charles made a paper tape of his program on Harvard's PDP-10 and gave it to Clap.

While he waited in his seat for the plane to take off, Clapper reviewed everything that they had done. Like Charles, he too was worried. This was their big break. If his simulator did not work, or if Charles' code had an error, someone else would seize the opportunity and their long held plans for starting their own company and becoming a part of the computer revolution that they both earnestly believed was coming would pass them by. Just as the rising plane broke through the clouds, Clapper realized neither he nor Charles had written a loader that they needed before the Taurus could receive the coded instructions from the simulator. Clapper knew what had to be done. He calculated he would need about fifty instructions that he could enter into the Taurus using its front switches. Clapper took some three by five cards from his pocket—the ones he always carried to make notes about technical matters he was afraid he would forget if he did not instantly record them—and began to write.

Chapter 2

"Dum! I can sell it if you can make it work," Stanley said.

Stanley A. Pitts, Stanley to his only true friend and Sap, S.A.P., to his many non-admirers, paced from the rear of the cluttered Pitts' garage to the open door and back while his friend Harold Dumbroski, Dum, concentrated on the tangle of electronic parts on the workbench in front of him.

"It works," Dum said. His tone made it clear that he really did not care what Stanley thought he could do.

Stanley paused to look over Dum's shoulder. Dum was carefully soldering chips. Dum refused to use the more popular method of wire wrapping that others at the Homebrew Computer Club employed. Solder took more time, but Dum liked its precise symmetry that facilitated troubleshooting.

"Every hobbyist will have to own one. You saw the way they reacted at the club," Stanley said.

Dum grunted. He had displayed his baby at the last meeting to a curious membership, but not a single person had asked for a copy of his schematics. Dum knew why. For economic reasons he had used a Motorola 6800 microprocessor while the Homebrew hobbyists preferred the Intel 8080. Dum understood that but what was a $24,000 a year technician at Hewlett-Packard to do, particularly when confronted with HP's offer of a discount.

Stanley ignored his friend's reaction. They understood each other despite their differences. Dum's father was an engineer while Stanley's father was a blue-collar worker who drifted from job to job, sometimes a laborer, sometimes a salesman, never very successful. The two friends were loners, neither was popular with classmates, but Dum, an easygoing introvert, did not inspire the outright dislike that Stanley's aggressive manner encountered. The other kids, from the first grade on, all considered Stanley odd. Most shunned him, but not Dum despite the four years difference in their ages. Dum, the older, immersed himself in the world of electronics from an early age when he discovered that he quickly mastered technical theories that the other students could not even come close to understanding. Stanley tinkered with electronics and liked to brag that he knew much more than he actually did, but Dum, starting with high school, allowed a friendship to develop with Stanley taking most of the initiative. Most of the time Stanley fantasized about how he was going to make both of them rich, using Dum's electronic inventions of course. This dream did not impress the other kids who had attended grade school and high school with Stanley. They were familiar with Sap's dreams and remembered how in the first grade he was going to get rich when he grew up by opening a candy store across the street from their school.

"When will you be finished?" Stanley asked.

"Never." Dum placed the soldering iron on the bench and stretched to ease his back. He had been working for three hours with a chattering Stanley pacing behind him. His ears hurt from the barrage of Stanley's grandiose plans.

Dum studied his friend and wondered how Stanley could even dream they could find the money to replicate Dum's computer let alone gather enough cash to build computers to sell. Nobody in his right mind would loan money to Stanley. Dum smiled at the thought. He pictured Stanley with his long hair, grubby clothes, and dirty bare feet in rubber sandals approaching a dignified, middle aged bank vice president in a suit. Dum had to admit that Stanley looked better today, despite the emaciated look engendered by his current fruit diet. Anything was better than Stanley during his Hare Krishna or commune days let alone how he looked with his greasy pigtail on his return from a search for inner peace in India. Dum at least understood how others reacted to his introverted approach to life, but Stanley did not have a clue. He thought the others all admired him and his passions.

Stanley dropped into the overstuffed chair with the protruding springs and padding that they had retrieved from the curb where the next door neighbor had placed it for garbage pickup. Dum pivoted on his work stool and waited for Stanley's oration, which he was sure would follow.

"We've got to come up with a name," Stanley said. "A real catchy one."

"What do you suggest?" Dum's attitude made it clear he could care less what they called it.

"Karma, what about Karma?"

Dum shrugged. He had no interest in perpetuating Stanley's India experience.

"OK." Stanley had no trouble reading Dum's reactions when he wanted to. "What do you suggest?"

"You decide. Something American."

"Karma's American."

"Not really."

"We need something that will appeal to the hobbyists. They're our market. You've got to keep your market in sight."

Dum said nothing. He wondered where Stanley thought he had acquired all this marketing expertise.

With only a slight hesitation, Stanley slapped the arm of the chair. "I've got it. Astron." A cloud of dust rose into the air.

Dum nodded a halfhearted assent, then turned back to his workbench.

"It means star, in Greek," Stanley explained. "It's a symbol of the universe. We live in the space age, so what would be better?"

"Sounds good," Harold tried to cut off his friend's fantasizing. The hobbyists at the club would care more about his circuit board than Stanley's fancy name for it.

"We're reaching for the future," Stanley ignored Dum's silent reaction. Stanley cared as much about Dum's opinions as he did the number of welds on the board. What he wanted to do was sell the damned thing. "The astronauts all use computers. The average clod will buy it because of the name."

"We're talking about a circuit board," Harold tried to bring Stanley back to earth.

"I know that," Stanley groaned. While Harold created the technology, Stanley scrounged for the needed parts in local electronic supply shops.

Harold carefully set his soldering iron in the rack he had designed from a coat hanger, then turned. "How many do you think we could sell? Really." Harold wanted to visualize how many boards he would have to make not how much money they would bring in.

"At least a hundred."

"The club doesn't have that many members. If we sell ten boards, we'll be lucky."

"Maybe, but those who don't buy will talk about it. Free advertising. There are tons of engineers around here, and they'll be interested."

"If you sell a hundred, it'll be a miracle."

Stanley ignored Harold's pessimism. He liked to think on a grander scale.

"We need seed money," Stanley said as he calculated the money he would need to buy parts that Dum could use to make a hundred boards.

"I've got five bucks," Harold laughed. He knew that Stanley did not have that much.

"I've got the van."

Harold realized finally that Stanley was serious. Stanley hated exercise, even walking; he depended on his old Volkswagon. Harold thought for a few moments. "I could sell my calculator." His Hewlett-Packard calculator was one of his most cherished possessions. He reasoned if he talked about selling his calculator, he might frighten Stanley off.

"You don't like Astron for a name?" Stanley returned to the subject because on reconsideration he did not like Astron either.

"If you do."

"I don't. You suggest something. You're the tech wizard.

"That's it."

"What is?" Stanley was irritated by Dum's blasé attitude. He knew the marketing name was important.

"Wizard."

"It's a damned computer not a person."

"Might be close someday."

"What do you mean?" Stanley anticipated the answer. Dum believed in artificial intelligence, that one-day machines might approximate reason. Obstinacy made Stanley try to force his laconic friend to say it.

"You're right." Harold had little interest in the subject. Stanley could call it anything he wanted. The technology was the important thing.

"We need something catchy," Stanley persisted. Then, inspiration struck him. "That's it."

Harold concentrated on his soldering. He knew his friend wanted him to ask a question but ignored him.

"The Wiz," Stanley said.

Harold nodded agreement, barely moving his head, body shorthand for a lack of enthusiasm.

"It's catchy. People can say it's short for The Wizard if they want."

They sat in silence for a short period. Harold concentrated on his soldering while Stanley contemplated the promotional prospects in the name. Already the idea had become his own even though it had been Harold who first had suggested Wizard.

"The Wiz. That's it," Stanley declared, acting as if he thought speaking louder would stimulate agreement.

Harold did not respond. He knew that if he did he would only provoke Stanley. In some ways, the nineteen-year-old Stanley was still a kid.

While he waited for Harold to kick in, Stanley had another inspiration. They had a product, now they needed a company and a business plan. He had read that yesterday in a magazine he had perused at the corner drug store while killing time.

"We need to form a company," Stanley said.

"OK," Harold agreed. The circuit board was his invention, but he was interested exclusively in the technology. If Stanley made a few bucks selling it, Harold would not object. He would call the workshop in Stanley's father's garage a company if it would shut Stanley down. "One condition."

"What's that?" Stanley could barely conceal his enthusiasm. He finally had Dum engaged in something other than his damned solder gun.

"I like Astron better."

"Why?"

"It sounds futuristic. Wizard is jerky."

"OK. Astron Computer." When he said the words, Stanley liked their ring. "Wizard is jerky. No style."

. Two days later, on April 1st, Stanley presented Harold with a draft of a partnership agreement that he had written himself. At first, Harold assumed Stanley was playing an April Fools joke, but after Stanley persisted Harold realized he was serious. He had named the company Astron Computers and called their single product, Harold's circuit board, The Astron. Stanley had not forgotten their conversation as Harold thought he would. The agreement made Harold solely responsible for engineering and production with Stanley taking administration, marketing and sales for himself. They split ownership fifty fifty.

Harold scanned the crude document that Stanley had typed himself. Harold wondered where the kid had gotten the form to copy as he laid it aside on his workbench.

"Well, sign it," Stanley demanded.

Harold noted that Stanley had already signed his own signature. He decided to humor his friend and affixed his signature on the line underneath Stanley's. It was no surprise that Stanley had put himself first. He always did, but this did not bother Harold. He did not consider such things important, and, after all, Stanley was only nineteen while he was twenty-four. Most of the time the age difference did not matter. He had only one friend, Stanley.

Stanley grabbed the paper, folded it carefully and placed it in a clean white envelope.

"What's next," Harold asked, regretting his question as soon as he asked it. He was sure it would provoke a windy oration.

It did. While Harold worked, Stanley described his version of the future. First, they had to sell Stanley's van and Dum's calculator.

At the next meeting of the Homebrew Club, which Stanley also attended, Harold proudly presented his upgraded circuit board. Harold placed his board on the table in front of him and patiently explained his schematics to the other members who always closely examined the product of fellow hobbyists' efforts. Each was deeply enmeshed in a study of the fascinating subject of computers and was always ready to learn and copy from the efforts of the others. Stanley joined Harold at his small table and, unfortunately, in Harold's opinion, tried to convince the others to buy the board. Harold could see that the others did not take to Stanley's intrusive demeanor. Although he was not a technician himself, not by education, training or self study, Stanley did not conceal the fact that he thought himself superior to the general club membership. His sarcastic style bordered on the caustic when in public, and his dress did not impress the hobbyists who were no fashion plates themselves. Even Harold sometimes privately wished that Stanley would give more frequent attention to his personal hygiene. For whatever reason, the presentation of Harold's circuit board proved to be a non-event. Despite the fact that they returned the treasured circuit board to the workshop in Stanley's parents' garage without a single order, Stanley remained enthusiastic about their project.

The next day, Stanley, while Harold was at work, collected the circuit board and set out for a local electronics store where he often scrounged for parts. The store, unimaginatively know as The Electronics Warehouse, had been started on a shoestring by an independent electrical engineer in his late thirties who had tired of working for others. His inventory consisted largely of electronic parts that local hobbyists like those at the Homebrew Club needed. The Electronics Warehouse sold a computer kit marketed by a small company in New Mexico. The kit enabled the dedicated hobbyist to build his own computer, called the Taurus. It did not compute much, just flashed a few lights, but it let the hobbyists build their own computer. Sales had been so brisk that the owner, Jake Foster, had opened two additional shops catering to local engineers in the valley and had gone so far as to take on the distributorship for the Taurus for northern California.

Stanley brazenly entered the Electronics Warehouse and found Jake leaning on the glass display case leafing casually through the morning paper. Jake looked up to greet the first customer of the day and was disappointed to discover it was only the grubby kid who was always scrounging for bargain parts. Jake, who had been in the army, frowned at the dirty long hair, unwashed clothes that had obviously been picked up at Good Will or a second hand store, and the dirty feet with jagged toenails in cheap rubber sandals.

"Good morning Mr. Foster," Stanley greeted the owner warmly.

Jake ignored him and returned his attention to the newspaper. He did not even know the kid's name.

Stanley dramatically placed Harold's circuit board on the counter next to Foster's newspaper, close, but not touching. When Jake did not look up, Stanley untied the string he had around the brown paper, carefully folded it back, and revealed the circuit board. He waited for the owner to react. When he did not, Stanley spoke a little too loudly in his attempt to emulate a television announcer.

"And here we have…" Stanley paused, waiting for the owner to look up.

Jake turned a page of his newspaper and pretended to read the comics.

"The electronic marvel of the age," Stanley continued.

"Beat it kid," Jake growled.

Stanley did not move. He silently waited.

Finally, Jake looked up. He found a smiling Stanley establishing direct eye contact. Jake returned the stare for a few seconds then looked down at the circuit board. The symmetry and precise soldering surprised him. He was accustomed to seeing sloppy workmanship and wiring but clearly whoever had done this work was a skilled craftsman.

"Did you do this?" Jake asked, his skepticism clear in his tone.

"I could have, but I didn't," Stanley countered. "Nice work, isn't it?"

Jake did not reply. He carefully examined the board. "Why are you showing it to me?"

"It's your big chance. Sell a few of these and you'll be underway."

The kid's arrogance made Jake laugh. With three stores, he already considered himself on his way to creating a chain.

"Where did you steal it?"

"Don't you wish. This is the first in a long line of products that will put Astron Computers at the top of the Dow Jones."

"What in the hell is Astron Computers?"

"My company," Stanley replied, unable to conceal his disdain.

"Your company? Kid, you didn't make this. Where did you find it?"

"Astron Computers. We're ready to go into production today."

"Really?" Jake laughed at the teenager's audacity. "What's your production quota for today?"

"How many do you want?" Stanley countered.

Jake almost said a hundred just to provoke the kid, but he hesitated. The hobbyists might go for the circuit board if the price were reasonable.

Jake rotated the circuit board and studied it carefully from different angles.

"I might take fifty," Jake said. "If they work and the price is right."

Jake's answer surprised Stanley, and his face showed it. He had been thinking in terms of ten or twenty boards at the most.

"And not just a circuit board," Jake decided to press his advantage. "My customers want computers not just circuit boards." Jake knew the hobbyists who composed his primary clientele would be delighted to have the boards and use them to assemble their own computers, but he doubted the callow youth in front of him knew that. If he could get his hands on fifty boards as well made as these, Jake knew he could sell them. He had watched the grubby Stanley and his friend Dumbroski at the Homebrew meetings. Dumbroski was obviously the engineer, and the kid was just a hustler.

"What do you mean?" Stanley asked. He had regained his composure.

"Computers is what I mean kid. Look at this. No monitor. No software. No keyboard. No box." Jake watched the kid's reaction as he pretended to study the board. Jake had already made up his mind. He could sell a hundred circuit boards if they were all as well put together as this one, and he knew enough to make sure of that before accepting them. Jake hoped they were not stolen merchandise.

Stanley thought quickly. He could find monitors and keyboards in other valley shops. Harold could write BASIC; he could handle the software. The boxes would be a problem, but Stanley confidently calculated he could find a metal fabricator to cobble them some decent boxes. The boxes would be important. Despite his lack of interest in his own appearance, Stanley knew from TV that image was everything.

"OK," Stanley said. "Fifty."

Jake laughed derisively. "Fifty what? At what price?"

Harold and Stanley had discussed the issue of price. They figured the cost of the finished boards at three hundred dollars each, thus they could make a tidy profit if they could get at least three fifty for each.

"I'll give you something between five and five fifty each," Jake said tentatively. He figured he could sell a hundred computers at seven fifty. "But it depends on the finished product."

Stanley calculated quickly. He had no idea what the peripherals would cost, but he was confident he could scavenge up some guaranteeing a grand profit. Fifty would probably be too many for Stanley to handle, but the owner of Electronics Warehouse did not need to know that. Never able to control his imagination, Stanley immediately began to speculate that the three Electronic Warehouse outlets might be able to handle more than two or three hundred boards. If he and Harold could sell three hundred boards, their company would be on its way.

"I'll put you down for fifty," Stanley smiled. "I'll check with our production people and bring in a contract." When Jake seemed to swallow that line, Stanley continued. "We can't let you have them for less than six hundred each. We have to make a small profit."

Jake laughed at the kid's bravado. "We got a deal at five fifty." Jake pretended to read the comics while Stanley re-wrapped his circuit board.

"Don't forget. We're talking finished computers here," Jake called after Stanley as he headed towards the door.

Stanley waved. "No problem."

Jake doubted that the kid could deliver, but if he did, Jake surmised he could make a tidy profit.

As he exited the door, Stanley began thinking about financing their enterprise. He bumped into a hobbyist from the club who he knew casually as the latter was entering the store. Stanley was too involved with his thoughts to speak.

Harold, concentrating on a difficult chip, was leaning over his workbench in the Pitts garage when Stanley returned. Rock blared from Harold's old radio.

"Turn that damned thing off," Stanley ordered as he entered. Stanley had trouble not shouting the news. Perversely, he wanted to make Harold pull it out of him. That way he could dramatize how good a salesman he had been. Now, Harold would have to accept Stanley's primacy.

"Just a minute," Harold said. He carefully completed the task he was working on then set the soldering iron back in its rack. "Now, what's up?" The radio continued to blare.

Stanley tossed the package containing the circuit board in the air and pretended to fumble it, catching it just before it hit the cement floor. Harold did not like seeing his precious circuit board treated in such a cavalier fashion but knew that Stanley was merely trying to provoke him, so he said nothing.

Stanley smiled. "We're going to have to rent us some bigger work space."

"With what?" Harold asked.

"With the order I just got," Stanley spoke softly, underplaying his news.

"With what?" Harold repeated. He was bored with Stanley's games.

"Jake agreed to take fifty."

"Circuit boards?" Harold found the news astonishing. No wonder Stanley was on a high.

"Computers."

"Computers?" Harold could not believe that Stanley had agreed to sell computers.

"Jake said he could sell computers but not circuit boards." Stanley grabbed Harold by the arms and danced him around the garage.

Harold acquiesced for a few seconds then pulled away.

"Don't worry. I'll get the financing," Stanley said. "I can handle it. We've got two thousand from my van and your calculator to start with."

"We'd need ten or fifteen thousand just to get started. And I have..." Harold reached into his pocket and pulled out a few coins. "I have exactly seventy-six cents."

"I'll get twenty-five thousand. We'll need at least that much for a hundred computers."

"I thought you said fifty." Sometimes Harold had difficulty in understanding his friend. Stanley frequently got carried away. Harold still could not tell whether it was from an excess of self-confidence, arrogance, or simple enthusiasm. Probably, a mixture of all three.

"Might as well build a hundred. I'll sell them to the hobbyists."

"It'll cost us at least twenty five thousand to make a hundred," Harold calculated. He could build the circuit boards, write the software and assemble the components. He knew that, but he did not know where they would get the money to buy the components.

The next morning Stanley donned a clean T-shirt and freshly laundered jeans. He thought about wearing his dirty tennis sneakers but opted instead for his rubber sandals. Shoes made his feet sweat. He pulled his long black hair back into a ponytail and fastened it with a rubber band. He studied his appearance in the mirror and was satisfied. He was confident that he would have no difficulty getting his loan. Jake's commitment gave him a lock.

Stanley was the only person waiting when a middle-aged man unlocked the door at the Peoples Bank of Los Altos.

"Morning son," the man nodded at Stanley.

"Good morning," Stanley replied. Stanley exuded confidence. "I'm looking for the loan manager." Stanley had thought about asking for the bank president but had settled on the loan manager as the best place to start.

"That's me," the man replied.

"I need a loan," Stanley said firmly.

The man smiled as he appraised the young applicant. Obviously, young, probably a teenager, clean clothes, but hardly the kind that would impress a loan officer, and dirty feet in rubber sandals. The long hair needed a shampoo badly.

"My name's Davis," he said and offered his hand.

Stanley grabbed it firmly and identified himself: "Stanley Pitts."

"Come over to my desk, Mr. Pitts," Davis said.

Stanley followed him across the wide lobby. Two cashiers who were counting their money getting ready for the day's business exchanged amused smiles when they saw the young man following the loan manager to his desk.

Davis sat down behind the desk and motioned for Stanley to take a seat opposite him. Davis folded his hands and rested them on the bare desk in front of him. "Now…" Davis began when Stanley interrupted him.

"I want thirty thousand," Stanley said. He had decided he needed backup operating capital. It would be easier to get it all at once than make two trips to the bank.

"I see," Davis said. "I assume you want to buy a car…"

"Not at all. I need a business loan."

"Really. What kind of business are you in Mr. Pitts?"

"The computer business."

The answer surprised Davis. "Would I recognize your company's name?"

"Astron Computers."

"I don't believe I've heard of them."

"We're a new company. When can you give me a check?" Stanley tried to hurry the man along.

"Do you have any collateral, something to guarantee the loan?" Davis decided to dismiss the boy quickly. He had not yet had his morning coffee.

"Oh, we have a prototype," Stanley did not explain that they had only one circuit board, "and a big order."

"I see. Do you have any property you can pledge to guarantee the loan?"

"No, but we will be signing a lease soon." They needed more work room, and Stanley planned to find a garage or barn or something cheap as soon as he got the loan.

"About this order?" Davis did not explain that a lease was not what he had in mind.

"Yes?"

"May I see your documentation?"

Stanley, who had nothing in writing from Jake, hesitated, not knowing what to say.

"Something on the customer's stationary confirming the order including number of computers ordered, price guaranteed, that sort of thing."

"Oh," Stanley said. "I'll be getting that this afternoon."

"You have only an oral commitment?"

"Yes. Is there something wrong with that?" The man was beginning to irritate Stanley. If he wanted to pass up an opportunity to make big money in the future, Stanley would take his business elsewhere.

"Oh yes. You see, Mr. Pitts, The bank has very clear guidelines for loans, particularly for loans of this magnitude."

"If you don't want my business, forget it," Stanley said rudely. He stood up and walked out of the bank. He was genuinely angry and did not even bother to smile at the cute cashier who had watched the exchange from across the room.

Stanley paused at the curb and considered his situation. There were other banks in town, but he calculated his reception would be much the same. He had reasoned that the Peoples Bank would be his best option. After all, his father had kept a checking account there for over ten years, and it held a mortgage on their home. Stanley suddenly realized he had been turned down before he had a chance to mention his father's account or the mortgage. Stanley vowed to remember Mr. Davis and make sure that Peoples Bank did not get any of his business in the future.

Stanley strolled down the street to the corner drug store. He entered, smiled at the young girl behind the lunch counter, and ordered coffee black. While he sipped it, he tried to figure out what to do next. He resolved not to waste any more time on banks. They were capitalist bloodsuckers anyway. He decided that the electronics stores where he scrounged for parts would be his best bet. They knew him, and he assumed they would jump at the chance to get in on a good deal. Besides, they, like Jake, were potential customers. The first store he visited he tried a different tack. Instead of asking for a loan, he offered to exchange shares in Astro Computers for the parts they needed. He calculated this would be a deal for the store. They would make a big sale and get something more valuable in return, shares in a start up company that was sure to grow. The storeowner turned him down flat. He had long harbored a dislike for the sharp-tongued young man who constantly tried to get new parts for a fraction of their cost.

Undismayed, Stanley visited four more stores. Some of the managers had the effrontery to laugh as he offered his proposition. That evening Stanley shared only a small portion of his failure with Harold. Although he said nothing to discourage Stanley, Harold was not disappointed. He had not expected Stanley to succeed. Harold could not imagine any competent bank manager willing to lend money to a couple of kids with nothing more than a dream. Harold was content to putter with his electronics, but Stanley remained determined.

The next day Stanley borrowed his father's car and visited every electronics store within a thirty-mile radius. Not one responded to his pitch. Undaunted, Stanley expanded his quest and for a week called at every business he could find that had even a limited potential for providing him credit to obtain parts. He did not get discouraged until his former employer Atari turned him down. Atari was willing to provide the parts he needed but demanded cash on delivery. Finally, an Atari employee who Stanley had known casually suggested that he try a Stanford professor who owned a small company Stanley had not heard of. The professor, apparently accustomed to dealing with students who looked like Stanley, patiently listened to Stanley's Astron Computers pitch. The professor had made a modest profit backing similar startup companies and was looking for opportunities in venture capitalism in a small way. After listening to Stanley for an hour, the professor decided to take a chance. He had no parts to sell but he owned a paper company with a credit line at an electronics distributorship. He agreed to buy parts for Astron Computers. With the professor's name and company as entre, Stanley approached the electronics dealership. Once again he repeated his pitch to a skeptical businessman who was not impressed with the kid with big ideas. The professor's backing and Jake's order for fifty computers inspired the man to agree to sell Astron computers $20,000 in parts on credit with no interest if the bill was paid in thirty days.

That evening when Harold appeared at the Pitts garage to work on the circuit boards Stanley greeted him at the open door.

"Christ, Dum, we're never going to get these boards done."

Harold ignored his friend and went directly to his workbench where he turned on the light and plugged in his soldering iron.

"How many are ready to go?" Stanley demanded.

"Ten."

"It took two weeks to make ten?"

Harold was tempted to note that he had to work eight hours a day at the plant before coming to the Pitts garage where he put in a minimum of five hours. His wife Helen who also worked had initially encouraged him to concentrate on his inventions, obviously interested in the money they might produce for them. She, however, had started to complain, only mildly. They had been married for six months. Harold assumed the intensity of her complaints would increase as time passed and Harold showed no profit from his labors. It didn't help that Helen did not like Stanley, but there was nothing Harold could do about that. No one but him liked Stanley.

"This is the last one," Harold said.

"The last one? What do you mean? We need a hundred."

Harold chuckled to himself. Stanley sometimes drifted out of touch with reality. Stanley did not have the responsibilities Harold had. Harold said nothing. Stanley had made no secret of the fact that he had opposed Harold's marriage. At the time, Harold assumed, and still did, that Stanley begrudged him the time he spent with Helen. Stanley had difficulty believing that any one could have interest in anything but Stanley's own plans and dreams.

"I can't make any more boards without parts," Harold said evenly. Harold was not a complainer. He enjoyed putting his boards together. Getting parts was Stanley's job.

"No problem," Stanley smiled.

Harold waited for Stanley to explain where he was going to get the money to pay for the things Harold needed, but Stanley said nothing. Harold, who was accustomed to Stanley's games, began work on his board. He ignored Stanley who paced behind him.

"I've got the stuff," Stanley shouted, no longer able to hide his excitement. "We're on the way."

Harold turned, still holding the hot soldering iron, and looked skeptically at Stanley.

"The stuff's in dad's car, and I'll get the rest tomorrow."

"We have to get a bigger place to work." Harold gestured at the packed garage with the glowing soldering iron.

Stanley paused before commenting. He looked around, apparently searching for useable space. "We don't have any money to rent a place," Stanley admitted.

Harold turned back to his bench.

"I've got a line of credit for the stuff. That's no problem. We just don't have any cash."

Harold continued to work. Those things were Stanley's problem and held no interest for him.

"What about your apartment?" Stanley asked.

Harold chuckled, but said nothing. He and Helen lived in a one-room efficiency. He could picture his workbench, all the stacked parts, and Stanley and Helen in the same room.

"I guess that won't work," Stanley said. He knew that Helen might object, but she would have to sacrifice if she wanted her husband to own half of a computer company. Stanley contemplated the problem, then made a decision. "OK. We can store stuff in my bedroom. You can work here."

To Harold's surprise, Stanley delivered on his promise to obtain the needed parts the next day. Harold helped Stanley distribute them between the garage and Stanley's bedroom. Before they were finished, both young men realized they would need even more room and help. Stanley had no difficulty persuading his mother to let him use the family spare bedroom that once had belonged to his recently married sister. Stanley stacked the few possessions she had left behind in cardboard boxes in the narrow hallway. Then, he and Harold stuffed the empty drawers, the closet and even piled the bed high with electronic parts. When they finished the two friends retreated to the garage.

"We're going to need help if we hope to finish these boards soon," Harold said.

"No problem."

Stanley had thirty days before his credit line expired and he had to pay for the parts. As it was, he had only what Harold needed for the first fifty boards. Stanley had promised Jake fifty computers, and they did not have the needed monitors, keyboards and software. Stanley assumed Harold could write the software—he could do anything electronically—but the hardware posed a problem.

"I mean now," Harold said.

I'll take care of it," Stanley said.

By the time Harold arrived the next day, Stanley had his sister employed as a helper. Harold patiently showed her how to do some of the simpler work on the boards. After a few days, she became quite adept at installing semiconductors. A slight problem arose when she began working in her old room and insisted on watching television at the same time. This produced a few missteps, but Harold easily corrected them.

Stanley rented a post office box and arranged for a telephone answering service. He tried to create the illusion of a prosperous company rather than a garage and bedroom startup. He also hired one of Harold's friends at Hewlett-Packard. The company had tried to transfer him to Montana, but he did not want to move. Stanley put him to work running the errands that had previously been Stanley's responsibility. Harold, who had expected help at the workbench, did not complain.

Stanley decided that a business needs a bookkeeper. He hired a junior college acquaintance. She agreed to work by the hour and to pick up Stanley's handwritten lists of expenditures and receipts once a week. Next, Stanley found

Harold an assistant. Stanley's companion on his low cost trip to India joined Harold in the garage where he helped assemble the boards under Harold's sharp supervision. Stanley's parents were soon infected with the young people's enthusiasm, and they began to help. His father lined the walls of the garage with plasterboard and installed additional lighting and electrical outlets. His mother took on the task of retrieving messages from the answering service, kept the coffee pot hot, helped organize the storing of the parts, and counseled Harold's anxious wife when she called to complain about her husband's extended absences.

As soon as the first fifty circuit boards were completed, Stanley and his helpers loaded the first twenty in Stanley's father's car, and Stanley drove to make his delivery to Jake at the Electronics Warehouse. Stanley needed the cash badly. His line of credit had expired, and he was overdue on the parts he had purchased. Stanley was apprehensive but tried not to show it. He knew that he was not living up to his commitment to Jake to provide computers complete with monitors, keyboards and software. If Jake refused to accept the boards, the dream would end.

Stanley parked by the hydrant in front of the store, retrieved two boards from the trunk and entered. He found Jake in his usually position at the glass case containing the more expensive components.

"Here they are. Right on schedule," Stanley announced.

"What's that?" Jake asked.

"Your order. I have more in the car."

Jake glanced at the circuit boards. He immediately recognized that the kid was trying to con him. He had ordered fifty computers not circuit boards. Jake returned his attention to the newspaper he had spread on the counter and resumed reading.

Stanley, not to be deterred, placed the circuit boards on Jake's newspapers. "Where should I put the rest?"

"Kid, I told you I would only take complete computers. I have all the parts I need." Jake waved his hand to indicate the loaded shelves that surrounded him. He then pointed at an assembled Taurus that set on top of a pile of kits.

"I know," Stanley admitted with a smile when he saw that his bluff was not going to work. "I'll be honest with you."

"That will be a change." Despite himself, Jake was beginning to admire the arrogant kid's chutzpah.

"Dum has the first fifty boards ready. Look at the great work."

Jake studied the top circuit board and had to admit that the workmanship was superior.

"Harold knows what he's doing," Jake said.

"He's working on the next fifty right now," Stanley said. Stanley knew that his words were not quite truthful. Harold was at work, but not at the garage. He was at his job at Hewlett-Packard. He planned to start on the next batch of circuit boards tonight if Helen did not create too big a fuss. She wanted to go to a movie or some such thing. Stanley decided that Dum would have to quit his job and devote himself full time to Astron Computers if Stanley pulled off this deal.

Jake laughed. He knew that Dumbrowsky still held a full time day job. "What about my computers?"

"Mr. Foster. I agree with you completely. The computers will be big sellers. It's just that we worked so hard to turn out these boards we didn't have time to assemble the peripherals."

Jake suspected the kid was not telling the truth, but he was accustomed to dealing with hustlers who promised everything and delivered little. Jake too was a hustler. He was negotiating to open three more stores with little more than his totally mortgaged three-store chain as leverage. Jake shared with his many customers and friends the conviction that the valley's technical industry was on the threshold of unimaginable growth. Jake intended to ride that opportunity. While Jake mulled the question whether he should take a chance and let the kid off his own hook, Stanley smiled confidently.

Finally, Jake decided. "OK. Bring in the boards. I don't know why I'm doing this. You're going to bankrupt me."

"You won't regret it," Stanley said. "We do outstanding work."

"Dumbroski does good work," Jake laughed "I didn't say he didn't. You're a work of art kid. I don't know if I can sell fifty boards."

"If you can't, Astron Computers will take them back."

Jake laughed. "Cut the shit, Pitts. You guys don't have a pot to piss in. If I give you any money that's the last I'll see of it. You might take the boards back, but I'll never see a cent of my money again." Jake thought for a few seconds. "Maybe I should just take them on consignment."

"That mean you won't pay for them now?" Stanley blurted.

Jake liked the kid's reaction. It meant he was desperate. "You got it. That's what consignment means."

It was Stanley's turn to think quickly. His letter of credit had run out. He had used every dollar to buy the parts for the first fifty boards and the first installment of the payback was due tomorrow. He needed twenty-seven thousand dollars for the store that had provided the parts on the professor's credit line, and he had yet to pay their new helpers a single cent. They all knew that he was delivering today and were looking forward to some cash for their hard work.

"Look Mr. Foster," Stanley said. "I'll be honest. We need the cash real bad. If you pay me for the boards, I give you my firm promise that the next fifty will be complete computers. Everyone will be an Astron 1." On the spur of the moment Stanley decided to assign a number to the name he and Dum had agreed on for the first computer. Calling it Astron 1 sounded like it was the first of a line.

"Astron 1, huh. Where did you come up with that fairy name."

"Astron is Greek for star."

"That's what you're shooting for huh? The stars?"

"I like that Mr. Foster. Shooting for the stars. We'll make that our company motto. We can name our other lines of computers after other stars. That will sell great."

"Other lines?"

"Mr. Foster," Stanley let himself get carried away. "This is just the beginning. Some day there will be an Astron computer in every home. Why the day will come when…"

"Spare me the speech kid. I know it by heart. Bring in the boards." Jake shook his head in wonderment at his own gullibility. The kid was something of a salesman despite his shabby appearance and arrogance.

"A check will be all right," Stanley said as he turned toward the door.

"My accountant is at lunch," Jake said. "Come back at three and I'll have a check for twenty-five thousand."

Stanley paused. "Twenty-five thousand? Mr. Foster we agreed on five hundred and fifty for each board. That's twenty-seven five."

"You're good at numbers too, kid. It's five hundred a board. This is a new deal. You promised computers and didn't deliver. Take it or leave it. I'm offering too much as it is."

Jake calculated that he might get six fifty for some of the boards in this and his other stores. The lower cost would leave him room to discount later if they did not sell. Some of the hobbyists would pay any price for good work, and these were well made. If the kid produced computers with keyboards, a small monitor and software of some kind, Jake could push the price up to seven fifty. The Taurus was just a kit that could not do anything, and the hobbyists had to assemble it. Jake was betting they would jump at the opportunity to get a ready to go computer.

Stanley thought about bargaining some more, then he decided not to press his luck. Foster had agreed to take the boards, buy fifty more computers, and to give him a check today. "A deal," Stanley said. "I've got eighteen more boards in the car and will bring the other thirty back at three when I pick up the check.

Chapter 3

James Clapper led the handful of passengers from tourist class as they rushed to escape from the confines of the Boeing 707 at the Albuquerque Airport. The bright sun hurt his eyes. As he hurried from the ramp toward the arrival lounge, Clap groped in his bookbag for his sunglasses. Nearing the entrance to the airport building, he gave up his search. Obviously, in his haste he had failed to pack them. He had also overlooked the difference in climate. He had no need for the heavy fur lined parka which he carried self consciously over his arm.

James pushed through the open door and to his surprise heard a deep male voice call out: "Clapper."

James had no difficulty identifying the man who had shouted his name. He was the only person standing inside the terminal near the door for debarking passengers.

"That's me," James smiled.

The tall, overweight man nodded. "I'm Ed."

"Mr. Johns?" Clap asked, assuming that Ed was Ed Johns, the inventor of the Taurus. He wondered why the company owner had taken the trouble to meet him at the airport.

"That's right," Johns said, offering his huge hand. "You got a bag?" Without waiting for an answer, Johns led the way towards a side door where baggage handlers were already unloading a wagon containing about ten suitcases.

Clapper retrieved his shabby suitcase, a hand-me-down from his father, and turned to find Johns already moving towards the front door of the terminal. With Clapper struggling along behind balancing his parka, overloaded bookbag, and suitcase, Johns led the way to the parking lot. Clapper, who was about five ten, was impressed by the bear of a man who preceded him. Johns stood at least six feet four and probably weighed two hundred and fifty pounds.

Johns paused in front of a battered, road weary pickup of indiscriminate age. Johns turned, took Clap's bag, and effortlessly tossed it into the back of his truck.

"You won't be needing that," Johns nodded at Clap's parka.

Clap nodded silent agreement. He was already sweating inside his corduroy slacks and wool sweater.

"Climb in," Johns said as he tucked the parka underneath the suitcase.

Johns made no attempt to converse as he sped down the highway. Johns drove fast and passed aggressively whenever they encountered a slower moving car. Clap, unimpressed thus far, wondered what he and Charles had gotten themselves into. Johns did not fit Clap's image of a successful computer manufacturer. Clap silently watched the passing landscape. After the forested and snow covered New England countryside, the New Mexico landscape was inhospitable. The highway, lined with gas stations, fast food restaurants and car dealers, looked like most others in the country. Clap saw nothing to envy.

"We'll stop at the shop, then I'll take you by a hotel," Johns said.

Since he had not asked for Clap's views, he said nothing. Clap worried that he and Charles had gotten excited over nothing.

Johns turned off Route 66, drove a few blocks, then pulled into a shabby strip mall. He stopped in front of a darkened storefront with painted windows. Overhead, a canted sign had a missing letter: "MASSA E."

Johns turned and smiled when he found Clapper staring at the odd sign. "Don't get all excited, son," Johns said. "That's not us. We're next door."

Clapper turned in the direction that Johns was looking. He saw another storefront. This one had a discolored sign over the door: "PAMS." Clapper immediately concluded he and Charles had wasted their time and money. He decided to tour the "shop" as Johns called it, then ask to be taken back to the airport.

Johns got out of the pickup, walked to the door under the "PAMS" sign, and waited.

Clapper dismounted from the passenger side then hesitated. He did not know what to do about his parka and suitcase.

"Don't worry. Your goods will be all right. We won't be long here," Johns said.

The words did not reassure Clap. The suitcase and parka were not what troubled him. He had not flown all the way from New England to God knows where just to take a quick tour of some yokel's shop. Clapper was anxious to demonstrate their software on the Taurus, and neither the building nor the owner impressed him.

Without waiting for a response from his visitor, Johns opened the door under the PAMS sign and disappeared. Clapper followed. Inside, he stopped in surprise. The shop was much larger than he anticipated. The wall on his right abutted the massage parlor. On the left, it appeared that PAMS had expanded to include what once had been two or three individual stores. The interior walls had been removed to create an impressive work area. The exterior walls were lined with floor to ceiling shelves stacked with parts. In the center front of the room, about twenty men and women were busy packing boxes with what Clap assumed to be components for the Taurus. Several of them looked up and smiled at Clap.

Fans swirled overhead. The back wall was lined with stacked boxes. In the center of the rear, the room opened on to a dock where three men were loading boxes into the back of an open van.

"This is it," Johns said. He had a deep voice but spoke softly for such a big man.

Clap had difficulty hearing him over the din of the shop.

"We've got R & D set up in back," Johns turned and started to the rear of the shop. "We're a little overwhelmed. Ever since that magazine article, we've averaged over two hundred orders a day."

They passed three women sitting behind tables littered with stacks of invoices and papers. All three were busy talking on telephones, and not one looked up as Johns and Clap passed. "This here's our business office," Johns said.

Near the back, Johns stopped in front of a workbench. "R & D," Johns said proudly, showing a little emotion for the first time since they had met. "This here is Bill," Johns nodded towards a slender middle aged man leaning over a nondescript, gray, sheet metal box that set in the center of the bench.

Clapper immediately recognized the box from its picture. He was looking at the Taurus, an alleged microcomputer with seven kilobytes of memory. Clapper tightened his grip on the bookbag that held his paper tape, notes, and Charles' code. Forgetting the tawdry condition of the company and its physical site, Clapper wished Charles were with him to share the experience. Lights on the front of the box blinked.

"OK," his laconic host said. "I'll take you over to your hotel now."

Clapper stared at Johns in disbelief. He had not come all this way to spend time in a shabby hotel room.

"I'm ready to run our software now," Clapper said with more assurance than he felt. He knew that Charles waited in his room as nervous and excited as he was. After all, they had not even tested his code.

"Sorry, tomorrow," Johns said and turned to leave.

Clap looked at the slender assistant for help.

"We're testing to see if we're ready for you," the man nodded towards the blinking lights.

Clapper suddenly realized the other two men were as nervous as he was. He had untested software, and they had a box that could not do anything but blink. They needed him. The thought gave Clapper more confidence.

"Tomorrow it is," he smiled before following the silent Johns back through the shop.

This time Clapper detected the employees surreptitiously studying him. He threw his shoulders back and tried to appear like a visiting executive. He had had his doubts when first viewing PAMS. It had not dawned upon him that Johns and his employees were also assessing him. Clap and Charles and Johns and his employees were all betting their futures on the Taurus. Clap now understood his host's demeanor. He had expected an engineer and instead he got a kid in corduroy slacks and sweater who obviously could not read weather reports. Clapper laughed. He at least appeared older than Charles who could pass for fourteen. Clap wondered what Johns' reaction would have been if Charles had gotten off the plane.

Clapper was relieved to see that his suitcase was still in the back of the pickup. Since they had the only vehicle in the parking lot, it appeared the massage parlor did not do much daytime business.

Johns drove silently while Clapper studied the forlorn countryside.

Johns glanced at him. "Not what you're used to, huh?"

"Drier," Clapper said, determined not to let Johns' gruff demeanor intimidate him.

"Better take off the sweater before your roast," Johns suggested.

These were the big man's first words that even came close to making Clapper feel welcome.

Clapper did as he was told.

"I was expecting some guy in a suit," Johns said.

"I'm a engineer. I work at HP."

"How old are you?"

"Twenty-four." Clapper wondered if he should have added a couple more years to his age.

Johns without warning turned suddenly and stopped in front of the entrance to the Hilton Hotel.

Clapper immediately began to calculate his cash. He had planned on getting a room at the local Y or a hostel someplace.

"I've got your reservations," Robert said. "Nicest hotel in Albuquerque."

"What about something closer to the airport," Clapper suggested.

"Nah. You stay here."

Clapper, not wanting to admit he could not afford such a grand place, retrieved his bag and parka from the back of the pickup. As he turned, a bellhop grabbed the suitcase.

"This way sir."

Carrying his sweater, parka and precious bookbag, Clapper meekly followed the bellhop and Johns into the hotel. The lobby intimidated him. He knew he could not afford it. He tried to remember if he had brought his checkbook with him. He wondered if they even would take a check.

"Room for Mr. Clapper," Johns looked down on the receptionist.

"Yes sir." The receptionist, an attractive young lady with a bright smile, slid a check-in card across the counter towards Clap and waited, obviously, for him to produce a credit card.

Clapper did not immediately react. He wondered if he should ask how much a night.

"How long will you be with us" the girl asked.

"Three days," Johns answered. He looked at Clapper. "Figure that will be enough. We test tomorrow, then you can go home the next day. Time to look around our part of the country while you're here."

Clapper filled in the card while he anxiously tried to devise a way out of his dilemma.

"Will you be using Visa or American Express?" The girl asked.

"Cash," Clapper answered.

"That will be one hundred and fifty dollars," the clerk smiled.

That was more than Clap had with him. He couldn't pay the hotel bill let alone eat for three days.

"Here, take this," Johns said, offering his personal Visa card. He had finally realized he had overbooked his visitor.

Clap spent a restless night. Worried about his untested software, he tossed and turned and watched the minutes chug past on the cheap radio clock that set on the table beside his bed. Finally, before dawn, he surrendered, rose from the tormenting bed and took a long hot shower before shaving. Then, he sat on the stiff hotel chair and tried to read a paperback. Unable to concentrate, he turned page after page without seeing a single word. His mind struggled with the potential pitfalls inherent in Charles' hurriedly written code that he carried on a paper tape in his bookbag. Finally, at five minutes to seven he tossed the paperback to the table and escaped from the room.

Clap was the first to enter the spacious breakfast room. He obeyed the sign that directed guests to wait to be seated. As Clap watched, a single server diligently straightened chairs. Clap scuffled his feet then coughed.

"Oh good morning sir," the young Hispanic male finally acknowledged him.

"Am I too early?" Clap asked.

"No sir. We get our share of early risers." The server glanced about the empty room.

As he led Clap to a chair near the window, two white coated males pushed carts into the room and began to set up a buffet.

"Will you have the buffet?" The server asked.

Clap sat down, considered the question, and decided he was not very hungry. "Could I see a menu please?"

"Certainly, sir."

The server retrieved a menu from a stack that set atop a nearby divider.

Clap opened it. The server patiently waited. Clap immediately noticed the prices. The cheapest item was two dollars and fifty cents for two eggs, toast and coffee. Clap knew he could get the same with a side order of sausage and home fries for a dollar and a quarter at any diner.

"Would you like more time?" The server asked, trying to hurry Clap along.

Clap was tempted to skip breakfast then remembered that he was dining on Ed Johns' credit card. "Give me the number three." Clap selected the five-dollar breakfast.

Clap had consumed his breakfast, read the morning newspaper thoroughly trying to learn as much about Albuquerque as he could, and was sipping his third cup of coffee when Ed Johns appeared at the restaurant door. By then, the room was half filled with patrons, mostly businessmen in dark suits. Clap waved and stood up to greet his host.

Johns nodded and sat down at Clap's table.

"Finished?" Johns asked.

"Yes. It was delicious," Clap said. "Will you have something?"

"Already have."

"Coffee?"

"No. Finish yours, and we'll get out to the plant."

"Did the tests turn out all right?"

"I was in at six this morning. All is ready if you are."

Clap decided not to address that comment honestly. He would have been delighted to join Johns earlier. Clap forced himself to take a sip of coffee, hoping to appear nonchalant. He raised both palms upwards, indicating, ready when you are.

"Would you like to order sir?" The server appeared and addressed Johns.

"The check," Johns grumbled.

The server made a show of leafing through the order sheets then ripped off Clap's page. The server placed it on the table. Johns grabbed it. "What's your room number?"

Clap thought for a second then remembered. "507"

Johns filled in the number, then signed the chit. Clap pulled a dollar bill from his wallet, dropped it reluctantly on the table, then hurried to follow the already departing Johns.

Again, they had a silent ride before parking in the deserted lot directly in front of PAMS.

"What's PAMS stand for?" Clap asked.

"Personal Air Micro Systems."

The answer puzzled Clap. He could see no connection between the Taurus and something with "Personal Air" in its name. The Taurus was a toy for computer hobbyists.

"Ground controls for model airplanes. We're not in that business anymore. We went belly up, but we kept the name. Already had the sign."

Clap followed Johns into the shop.

The place appeared to be as busy and disorganized as it had yesterday. He glanced at his watch. It was still only eight o'clock.

"We're having trouble keeping up with the orders," Johns said, having caught Clap's glance at his watch.

They passed the three person business office. All were talking on the telephones. "The time zones keep us hopping," Johns said.

"We're ready to go," the slender man who Johns had introduced as Bill stood in the R&D corner of the partitionless room. He smiled at Clap and pointed at the workbench.

The lights in the front panel of the rough gunmetal box were all blinking. Suddenly they stopped. Bill patted the box fondly. "She passed all the tests."

Clap hoped so. The magazine article had made it appear that the machine was already in production. Obviously, PAMS was acting as if it were. Boxes of components to be assembled by hobbyists were being hauled out the back door for shipment to customers. Clap was glad he and Charles had not prematurely purchased a Taurus.

"All right, let's see it," Johns turned to Clap.

Clap approached the workbench confidently. He opened his bookbag and took out his simulator and Charles' punch tape. He threaded the tape into the reader. He grabbed the notes he had scribbled on the plane then studied the switches on the Taurus. There was nothing there he could not handle. He fumbled a few times but flipped the switches.

"The loader program," he said. As soon as he spoke, he realized he should assume that Johns and Bill should know what he was doing. They had built the Taurus.

Neither responded. Clap glanced at his companions. They both were smiling. It suddenly occurred to him that they were as nervous as he was. Not one of the three of them expected the Taurus to actually respond.

Clap decided to quit posing. He crossed two fingers on each hand and raised them, asking luck to smile.

Clap pointed at the last Taurus switch.

"Let her rip," Johns ordered.

Clap flipped the switch and stepped back to watch catastrophe develop. He worried that he and Charles had been carried away. In their anxiety to seize opportunity and start on the path of realizing their dream of helping start a technological revolution, they had overreached. They weren't ready. Clap wished that Charles with his infectious enthusiasm and overweening self-confidence were there to assist him.

The tape flowed through the reader. That simple act seemed to impress both Ed and Bill, but Clap knew it meant nothing. Certainly, the reader would feed digital instructions to the Taurus. That did not mean the machine would understand them. It could look at the streaming ones and zeroes and simply shut down.

They waited for fifteen minutes while the tape rolled.

Ed and Bill watched silently. Disbelief was now etched on their faces. Both were obviously waiting for failure and disappointment. Clap assumed this was not the first time this little action had been tried. He now realized that he had to wait a day until PAMS was confident from its testing that their machine could not be accused of creating the inevitable failure. Clap grew more worried as the tape reached the end of its roll. He wondered how he could have expected the two inventors to have confidence in a kid who had appeared out of nowhere in his wool sweater. If he did not know how to dress for the climate, how could he possible be expected to write software for their temperamental machine.

Suddenly, the Teletype clacked. Johns and Bill crowded next to a surprised Clap to see what their machine had done.

"MEMORY SIZE."

An amazed Clap realized that Charles' program had communicated with the machine. He glanced at his companions who were exchanging surprised looks.

"What's it saying?" Bill asked.

"Something worked," Johns exclaimed.

Clap typed 7168 and pressed enter.

The Taurus responded immediately: "READY."

'PRINT 2 + 2," Clap instructed. He pressed enter.

"4," the Taurus printed on the tape.

"Hey, it works," Johns shouted.

Within seconds, everyone in the shop, some thirty persons in all, had gathered around the machine. Most stood with disbelief written on their faces. Others pounded each other on the back and shook hands.

Clap now stood confidently before the Taurus. He was more amazed than they were. Charles' code that had been written in haste over a three week period, mostly late at night in the deserted Harvard lab, had worked flawlessly. Clap wished that Charles were with him to share this marvelous moment.

All watched with unconcealed anticipation as Clap again approached the keyboard. He typed in a classic program, one that simulated a soft landing without running out of fuel. It was a well-known, simple text application. Charles' BASIC code seemed to be working much faster on the Taurus than BASIC did on the big machine at Harvard. Eventually, the Taurus froze, but this did not trouble the enthusiastic audience. They knew how sensitive these little marvels were at this stage of their development.

"Son, we're in business," Ed Johns declared with as much enthusiasm as Clap had seen him muster. Bill patted Clap on the back.

"Can I call my partner?" Clap asked, knowing how anxious Charles must be.

"Sure," Ed Johns said.

"You guys are amazing," Bill said. "I never thought I would see this baby do anything but blink." He patted the top of the now silent Taurus.

Johns nodded to one of the women who manned the telephones.

"This way sir," the woman, who was a good ten years older than Clap, said.

Workers touched Clap on the shoulder as he passed. Clap was not sure that he had ever felt as good in his life or ever would again.

In their Harvard room, Charles and Bill Oldham maintained their vigil. Also with them was another Harvard friend, Mike Hamilton. Mike, a business major, frequently joined Bill and Charles in poker and nocturnal revels. Mike, like Bill Oldham, knew nothing about computers and cared less. Mike was a math whiz and had the distinction of having outscored Charles on a math standard test by twenty points. This was not an accomplishment that greatly pleased Charles, but he submitted graciously to Mike's good-natured reminders. As one of Charles intimates, despite the fact that he lacked Charles' fascination with computers, Mike had joined Bill and Charles in the tense wait for news from Clapper. The fact that Clapper had not called the day he arrived in Albuquerque worried Charles and his friends.

"I don't know why he didn't call last night," Charles lamented again. "They should have tested it as soon as he arrived."

"Don't forget the time difference," Bill tried to distract the pacing Charles.

"What is it?" Charles asked.

Bill looked at Mike who shrugged.

"Two hours, I think," Bill said.

Charles stopped pacing long enough to look out the window. The roads were bare, sidewalks shoveled, and the sky gray. "Is it supposed to snow again?" Charles asked.

"Why not?" Mike laughed. "Why should today be any different?"

Before Bill could chip in, the phone rang. Charles grabbed it before the others had a chance to move.

"Yeah," Charles answered in his usual direct fashion.

"Tell me," he ordered.

Suddenly, Charles started hopping up and down. He gave a loud warhoop as he leaped to touch the ceiling. "It worked," he shouted at Bill and Mike.

Both cheered loudly. Someone in the room upstairs stamped on the ceiling.

Bill paced back and forth as far as the ten-foot cord allowed. He held the phone to his ear and listened. Both Mike and Bill could hear the excited Clapper on the other end talking loudly and excitedly.

"He's still at PAMS," Charles said. "It worked the first time and they want our software." Charles hopped into the air again.

He listened as Clapper described the test in detail. When he hung up, Mike suggested that they go out and celebrate.

"Right," Bill Oldham agreed, reaching for his parka.

"Can't," Charles said. "Have to get over to Aiken and work on the code."

"No way," Mike insisted. "First the Greeks and then you can do what you want."

"When is Clap coming back?" Bill asked.

"Tomorrow," Charles said.

Chapter 4

David Howard studied his reflection in the mirror and sighed. Dark hair, neatly trimmed, rimless glasses, classic nose, bent, firm chin. He stepped back and examined his freshly pressed black suit, his only one, white shirt and rep tie. The image was certainly not himself. Only a week before, he had considered a sport shirt, pullover sweater and jeans formal wear.

"Armonk," he said to the image and watched as it shook its head. "What in the hell am I doing in Armonk?"

David had expected his life to change when he graduated from Cornell, but not this much. Here he was, a small time jock out of Harrisburg, Pa. He had made Honorable Mention on the state's list of high school running backs. At six foot one hundred and eighty-five pounds, he had been too small to compete with the big boys. He had been good enough, however, to get an athletic scholarship at Cornell. Granted, it was only the potted Ivy League, and Cornell's reputation was for academics not sports, but it had been enough for David. Four years of small time glory and he had graduated with a modest B- average. He had known the best years of his life had to end sometime, but he had not anticipated Armonk.

An American History major, David had been an adequate student. He had played more than he had studied, but still he had been able to ride his surprising ability to crack tests to graduation. If he had to depend on his day to day classroom performance, he would have been lucky to graduate. He had learned that in high school.

A heavy hand rapped on his door. David glanced at his Timex. Eight-thirty.

"We're leaving," the unfamiliar voice warned.

David turned off the light and joined three of his new classmates in the brightly-lighted hallway. IBM's trainee dormitory impressed David as a cleaner, better-furnished college residence hall. Not like the shabby barracks where the jocks camped, but the nicer ivy covered brick dorms where the rich boys stayed.

David followed the others down the hallway and wondered if he had made a big mistake. When the IBM recruiter had offered David a position in the next trainee class, David had been flattered. The ten thousand dollars a year to start

41

certainly did not place him in the class of the engineering students or the athletes in the football factories, but for David it was more than he had expected. Neither his history major nor his athletic prowess had prepared him for life after college.

"Get a move on," Richard Gourd, the guy who owned the new Mustang that would carry them a mile across the IBM campus, stared at David who had fallen behind. "You might want to be late, but not me. I know better."

David shook himself from his musing and hurried to catch up. Like the other eighteen members of their trainee class, David had decided to keep an eye on Gourd. He should know what he was doing. He was a mustang. His father was one of many IBM vice presidents. The old guy probably had his son locked into a brilliant career.

Their first day as trainees had been a bitch, and David was not looking forward to this one. They had sat for ten hours and listened to a series of pompous executives, all dressed in the IBM costume of dark suits, spotless white shirts, and conservative solid color ties, expound on IBM protocol and traditions. They had devoted a full hour to explaining, teaching them the lyrics, leading the trainees in singing the company's anthem: "Ever Onward." David had felt like a young kid in music class.

The song session had been followed with lectures on dress codes, company culture, and the grander wisdom of the company founder, Tom Watson Senior. David hoped they would not be tested on the subject matter because he had tuned out the speakers after their first five minutes. The lecturers, company junior executives all, appeared to have been produced by the same cookie cutter. David found it difficult to imagine himself in a few years standing up there teaching an incoming class the company song, for Christ sake. Maybe, he could sell enough computers to make himself too valuable to spare for teaching duties.

"Know what's on the agenda for today?" A serious young man in the front seat asked Gourd.

David did not remember the guy's name.

"Company history," Gourd answered.

David's groan slipped out. The others stared at him. "I thought we heard all that stuff yesterday," David tried to recover.

None of the others commented. David assumed they all had to feel as he did but were afraid to speak honestly in front of Gourd.

As they walked into the building that housed their classroom, Gourd dropped back and looked seriously at David as he spoke in a low earnest voice: "A word of advice, Howard. You had better watch your attitude. These people are serious about their traditions. You won't last a week if you act like a lone duck."

David did not know what to respond. He did not know Gourd well enough to judge him, and most of all he did not know what a lone duck was. "Don't worry, I'm a company man," David said flippantly.

Gourd glared at David then hurried ahead to join the two others.

David squirmed in his seat for the next three days. He learned more about IBM's early days than he thought he would ever use. He decided the cookie cutter speakers were not indoctrinating them. They were brainwashing the trainees. By Friday, David realized that the other trainees were changing their dark suits every day. They sure had more extensive wardrobes than he did. His one black suit had grown wrinkled from constant wear despite the air-conditioning in the classroom. The students rushed outside during their short breaks to light their cigarettes and move around trying to loosen cramping back muscles. The heat took its toll on David's suit. He decided he would have to slip off Saturday morning and buy two more suits on his father's credit card. They would not be paid until the end of their first month. Then he could pay his father back, but until then, he would have to live off his largesse.

Somehow, David survived his first four weeks at IBM and drew his first paycheck. No matter how hard he tried, he knew he might never fit in. The company seemed to have rules and regulations covering every single possible contingency he might face professionally and personally. As a former running back who liked to dazzle the opposition with his moves and improvisations, he had to adapt to a new regimented system, and that was definitely not his style.

IBM projected the image that it was so big, so all encompassing, that it existed in its own world. The rest of the United States provided the environment for IBM's dazzle. IBM employees had no need for the outside world. The company health plan safeguarded their physical well being. The credit union managed their finances. The culture told them how to dress, talk, act, live. They resided in stratified IBM communities. Significant pay raises saw them trade houses and move to live with their new peers. They associated with IBMers. They did not need others.

This intradependence extended to all of their social activities. Their children went to IBM sponsored schools while the wives shared their, tennis, golf, gossiping with other IBM spouses. The males, of course, competed with each other, whatever their interests.

David was not fully attuned to this culture the day he learned of the existence of a company softball league. A neat three by five card on the bulletin board invited interested players to try out for a position on a team known as the Clippers. David decided he needed an outlet for his accumulating nervous energy. His body craved the physical exertion that eight years of organized high school and college sports had provided. David was confident of his ability not to embarrass himself playing softball. During his off-seasons throughout his school career, he had played both basketball and baseball to keep in shape. He never participated in league softball, but as a hardball player he had carried a 300 batting average and fielded third base adequately.

David availed himself of the telephone extension that IBM Administration had thoughtfully placed next to the bulletin boards.

"General Products Division," a pleasant female voice answered.

"May I speak to Ernie, please?" David said.

"May I say who is calling?"

"My name is Howard. I'm calling about Ernie's notice on the board."

"Oh," the female voice became a little less ingratiating. "For the softball team?"

"Yes."

"Mr. Hendricks," Ernie heard the woman call. "A guy named Howard wants to talk to you about softball."

"Hello Howard," Hendricks came on the line. His voice was firm and deep. "What can I do for you?"

"I'm sorry to bother you at the office," David said. "This extension was the only number on the card. The one on the bulletin board."

"You're new here," Hendricks said.

"Yes sir."

"Don't call me sir if you're a ball player. Any experience?"

David did not know how to answer the question. If he understated his ability, he might not be invited to play. If he overstated it, the other players would see it immediately.

"Where have you played?" Hendricks sensed his hesitation.

"I played football at Cornell," David answered. "I..." David began to explain he was a jock who kept in shape with baseball during the off season.

"You are Howard who?"

"David Howard."

"The running back?"

David was flattered that Hendricks recognized him. Not everybody followed Cornell's football. He had a modest reputation around the league and some notoriety in Ithaca.

"That Howard, huh. You kicked our butts. I graduated from Princeton."

David was in luck. He had had an outstanding game against Princeton last October. He had gained one hundred and fifty yards and scored three touchdowns.

"Yes sir."

"Ernie. Have you played any softball?"

"A little, but I was on the baseball team."

"Batting average?"

"Three hundred."

"Not bad. Position?"

44

"The infield." David was willing to try anything but pitching and catching.

"Third base?"

"Yes sir." That was his position. David had hit it lucky.

"Can you play tomorrow?"

The next day was Saturday. David had wondered how he would spend the weekend. He had yet to meet any interesting girls. "Yes Sir."

"See you at the field at nine. Don't worry about equipment. We can outfit you. Daddy IBM looks after his own."

This was the first time David had heard one of the IBMers speak in a cavalier manner about their employer.

"I've got my own glove and shoes." David figured his old Nikes would do fine.

"Glove all right. No shoes. We'll outfit you." Hendricks paused. "How new are you?"

"Four weeks."

"Our luck. If you are any good, we'll break you in right." Hendricks paused. "I think I should warn you. We play pretty serious ball here, and we're in a fight for first with the Penguins."

"I'll try not to embarrass you," David smiled. He had seen geezers playing slow pitch. They took it seriously.

"See you at nine," Hendricks hung up.

David had intended to ask where the field was located, but Hendricks did not give him a chance. David figured he could get directions from one of the many guards who patrolled the campus.

The next morning David, driving a two year old Volkswagon that he had bought on time with a loan from the Credit Union, arrived at the field. He was impressed to find a manicured field surrounded by solid bleachers, a grandstand and freshly painted green fences. He parked in the spacious lot partially filled with Cadillacs and Lincolns. He heard the sound of a sizeable crowd and the crack of bat against ball. He checked his watch. He was fifteen minutes early. The crowd oohed then applauded. David wondered what he had gotten himself into. This was much more than the friendly game of softball that he anticipated.

David wore jeans, an old Cornell sweatshirt and his Nikes. He carried his battered glove under his elbow. The gate to the back of the grandstand had a company guard.

"Player or spectator?" The husky man in the uniform of the IBM guard force asked.

"Player, maybe," David responded.

Penguins or Clippers?"

"Clippers."

The guard pointed to his right. "First door."

"After he passed the guard, the man called after him. "First game?"

"Yes."

The man smiled. "Thought I didn't recognize you. Good luck."

From his tone David inferred that he could also have said: "You'll need it."

David opened the door with the Clipper sign and entered. He was surprised to find himself in a well-equipped locker room, as good as anything they had had at Cornell. Lockers lined the walls. Trainer tables, weight machines, two whirlpool tubs and related equipment filled the room. An office on the right looked like a coach's office and on the left a door opened on to a glistening white tile shower room. Stainless steel fixtures indicated that no expense had been spared. Muscled men in various stages of undress sat on the benches in front of the open lockers.

"You must be Cornell," a voice from his right greeted him.

David turned to find a solidly muscled man wearing a white uniform with pin stripes and "CLIPPERS" in red letters across the front smiling at him from the doorway to the coach's room. While the man studied David's muscular six-foot frame, David appraised him. He estimated Hendricks to be somewhere between thirty-five and forty, but he did not appear to have an ounce of fat on him. He could have passed for a professional player.

"I'm Hendricks." He offered his hand.

The grip was firm, the palm almost as wide as David's. He had always prided himself on his big hands that had limited his number of fumbles when being pummeled by oversized linemen. Hendricks squeezed once then released David's hand, not making the gesture a competition.

"Hey guys, listen up," Hendricks spoke loudly enough to be heard by all. "This is Dave Howard. We're giving him a try."

Some of the players reacted. Some did not. A couple said "Hi," and two others nodded. Most continued to don their uniforms without reacting. Hendricks did not identify any of the players by name or position.

"A strange locker room," Dave thought. This was not turning out as he had anticipated. These guys were too serious about a simple game of softball.

"You can get your uniform over there," Hendricks pointed to a counter that led to a room adjacent to the coach's office. Hendricks led David to the counter and shouted: "Squirrel."

Immediately, a small man dressed in blue scrubs similar to those worn by hospital doctors rushed from what David assumed was a supply room.

"This is David. Outfit him," Hendricks ordered.

The man smiled and offered his hand to David. "Welcome. I'm Squirrel, your friendly trainer."

David took the hand and smiled back. His first word of welcome from someone other than Hendricks. Squirrel indeed looked like his name. He could also have been called chipmunk. He had large buckteeth in a narrow face.

Squirrel studied David then looked down at his feet. Without another word, he turned and entered the supply room. Within seconds, Squirrel reappeared carrying a stack of clothing with a pair of new spikes on top.

"Where did you get my sizes," David could not help asking.

"No problem. Six foot, a hundred ninety pounds, size twelve's."

David nodded assent. He was impressed with Squirrel.

He picked up the uniform and cleats. Everything was brand new.

Beside him, Hendricks glanced at his gold watch. "Let's go. We've got the field," he called to the team. He turned to David. "Get dressed and join us. We have the field for warm-up for thirty minutes. Then, game time."

Hendricks started for the door. He called back to Squirrel over his shoulder. "Show him a locker, then when he's dressed, let him pick out a glove if he wants one."

Squirrel ducked under the fold-up counter top. "Let's go, champ," Squirrel said.

"Call me David," David said. He was not about to let himself be pushed around by a runt named Squirrel.

"Right champ," Squirrel countered.

David dressed quickly. He knew he would need as much warm-up as he could get. He had played spring ball at Cornell, but it had been almost two months since he had worked out seriously. And he had to get accustomed to the bigger ball. These IBM hot shots might dress like pros, but David had no doubts about his abilities on a ball field. He had the body and the hand and eye coordination to excel at any sport, to standout among amateurs. He did not delude himself about his ability to compete with real professionals.

David decided to keep his old glove. The pocket was small for the larger softball, but he was comfortable with it. He would get a ball and work it in the pocket.

The ball field surprised David. The grass was tailored, the running paths raked smooth, and the bases appeared new. David doubted that there was a major league field better maintained. The Clippers were taking batting practice. Hendricks stood behind the cage at home plate watching a beefy player who looked more like a tackle than a baseball player pound the ball as the practice pitcher hurled it at the plate. The pitcher, with his windmill windup, had good speed. David knew that a good softball pitcher could really move the ball to home plate. While the beefy guy took his strokes, the majority of the Clippers stood to one side playing catch, warming up their arms and reflexes.

As soon as Hendricks saw David, he spoke to the batter then turned: "OK, David. Grab a bat and let's see what you have." He nodded toward the dugout where a ball boy guarded some thirty bats carefully aligned in a row.

"How do I know which are somebody's favorite?" David asked the kid.

"They're all yours. Take your choice. The others are in the rack." He nodded toward a bat rack inside the dugout where each player obviously kept his preferred wood. The bats were separated into compartments, each numbered.

David leaned over and picked up a couple of the bats from the middle. He was not sure what size he would need for softball. After a few seconds of testing, he selected a black forty-two. Not sure how he would react to the changed size and speed of the softball versus the hardball, he did not want a too heavy bat. Not at first. He would worry about power later. Now, he wanted to meet the ball squarely. He could worry about swinging for the fence later.

After he had made his choice, David turned to find Hendricks watching him. The batting cage was empty, and both the catcher and pitcher were waiting for him. Sports were David's thing, so he did not rush. He took his time and swung the bat several times, loosening his muscles and getting a feel for the bat. Finally, he walked up to the cage. Before entering, he took the bat in both hands, put it behind his head, and began twisting and turning. He was not about to pull a muscle first thing. Finally, he nodded at the catcher and stepped up to the plate. The warm-up pitcher faced David, whirled his arm in the peculiar windmill throw of the softball pitcher, and hurled a fast ball at the plate. It sailed past David, knee high, and he did not swing. He wanted to get a sense of the pitcher's skill.

Nobody said anything. The catcher tossed the ball back to the mound. Again, the pitcher delivered a fast ball. This time shoulder high. Again, David did not swing.

"Anytime now, college boy," the catcher grumbled.

His comment caught David by surprise. Obviously, Hendricks had briefed the team on his background. He now had something to prove.

The pitcher grooved one right down the center of the plate. David swung and the ball popped skyward to second base. He had swung underneath the ball, catching only the lower half. A popup.

The catcher said nothing. Hendricks watched without expression. The pitcher gave David a fast one, low and outside. This time David timed his swing accurately and caught the ball squarely. It whistled into the hole between third and the shortstop. David showed no emotion. He stayed in the box and waited for the next pitch.

After David had connected solidly with some fifteen pitches, sending the last five deep into the outfield, Hendricks spoke: "You'll do."

David stepped back. "I'm rusty."

"It comes back fast," Hendricks said.

David joined the other players who were aligned along the first plate line playing catch, just loosening up. He caught the first ball thrown him and used it to widen the pocket in his old glove. Occasionally, one of the players would throw David a grounder or toss it high over his head. David realized they were testing his fielding skills, but he handled them flawlessly. David could feel Hendrick's eyes on him, assessing him.

Two professionally attired umpires emerged from the field house. The one with the chest protector walked to home plate and shouted: "Play Ball."

The two teams retreated to their respective dugouts. From his position at the end of the bench, David surveyed the nearly filled stands. He estimated that they held roughly a thousand spectators, a large number for an amateur softball game. Hendricks delivered his starting lineup to the home plate umpire. When he returned, he stopped in front of David. "Starting our regular lineup. We'll keep our ringer as a surprise." Ringer in sports terms usually described a talented player from a higher level who was deployed against an unsuspecting opponent at a critical time.

It did not take David long to realize that the two teams were evenly matched. Both had quality players, clearly talented semi pros. A couple on each team was professional quality. Good enough to make David wonder what they were doing playing in a company softball league.

"Not bad, huh?" The husky player who had been in the batting cage when David arrived now sat next to him on the bench.

He offered his hand. "I'm Jacob."

"Dave," David took the hand. This was the first indication of acceptance by another player. "What's your position?" David asked, hoping it was not third base.

"Pinch hitter."

"I didn't realize we had specialists in this league," David blurted.

Jacob laughed. "Good hitter. No glove. No legs."

David did not know how to respond, so he turned back to the field in time to see the Penguins cleanup hitter smash a triple to the left field corner. A player who had walked came home. The Penguins led one zip.

"Shit," Jacob spat on the field in front of the dugout.

Half the people in the stands cheered loudly while the others groaned.

"This is a big game for us," Jacob explained. "We're tied for first place with those guys with only twenty games left."

David wondered why it was important. The Clippers succeeded in getting out of the inning without the man scoring from third. The score remained that way until the last of the sixth when with two out the Clippers right fielder drew a walk. Hendricks called time out and turned towards David. "Jacob get your bat." He was sending him in to pitch hit. "David, we know you can run so take first." He sent David in as a pinch runner.

David pulled his hat down low over his eyes and ran towards first base. "Go get 'em Cornell," the man said as David replaced him. "But watch out for that sucker," he nodded toward the pitcher.

The Clipper first base coach approached David and spoke in a low voice. "He'll come after you on his first two pitches."

David jumped up and down on the bag, stretching his muscles. The pitcher rubbed the ball vigorously then stepped on the rubber. He studied his catcher's signals and nodded, ignoring David. David moved away from the first base bag. He took a three-step lead, watching the pitcher cautiously. As quick as a cat, the pitcher stepped off the rubber and hurled the ball to first. David made it back to the bag safely, but barely. The first baseman slammed the ball against David's thigh.

"Watch it," David growled. He rubbed his thigh. Clearly, the man had tried to intimidate him.

Again the pitcher took the rubber, David moved three steps off first, the pitcher whirled and threw to first, and the first baseman slammed him in the thigh again.

David said nothing, recognizing it was part of the game. As a running back, he was used to bruises.

Finally, the pitcher gave up on David and went to the plate. David knew that if he took a four-step lead, the pitcher would try again, and David would probably be caught. With the count three and two on Jacob at the plate, David danced confidently. He would go on the next pitch. The pitcher ignored him and came in low, hard and inside to Jacob who swung and caught the ball squarely. David was off and running with the pitch. The shortstop went deep to his left, made a spectacular play in knocking down Jacob's smash, recovered the ball, and threw it to second base. David was ten feet from second when the shortstop's throw arrived. David was preparing to slide when he saw that the throw was wide to the second baseman's left, pulling him off the bag directly into David's path. David instinctively decided against the slide and barreled into the second baseman just as he might a linebacker guarding the first down marker. The only difference was that this time neither man wore pads. David's shoulder caught the stocky second baseman in the chest and drove him backwards. The man tagged David on the side of the head with the ball as he flew backward. David fell on the bag, momentarily stunned. He blinked his eyes and tried to ignore the ringing in his ears.

"You're out," the second base umpire shouted as he turned and pointed his thumb at the sky.

David, still sitting on the bag, could not believe it. The partisan Penguin fans roared. David turned and found the second baseman laying on his back, out cold. Unfortunately, he still clutched the ball in his right hand.

David dusted himself off and walked slowly back to the dugout.

"Nice try," Hendricks laughed.

On the field, they were still ministering to the second baseman.

"Get your glove and take third," Hendricks ordered, still laughing.

As David made his way to his position, he wondered why Hendricks was laughing. They were still behind one zip. The score stayed that way until the last of the ninth when David got his first at bat. The Clippers were in the midst of a small rally. They had men on first and third with no outs, thanks to a walk and a bunt single. David looked at Hendricks, expecting to have a pinch hitter. Hendricks said nothing. David selected his bat and walked to the plate. He began his warm up and looked to the third base coach for his signal. At that point, David realized he did not know the signals. He walked towards third and waved, calling the third base coach to join him. They met midway between home and third.

"I don't know the signals," David said.

"Hit away slugger," the coach growled and headed back to his coaching box.

"Let's get on with it," the sweating home plate umpire ordered.

David did not know the pitcher but had been able to watch him for nine innings. He assumed the first pitch would be high and inside. The pitcher was a hard head who would try to intimidate the new guy who had cold cocked his second baseman. David decided to take his chances. The pitcher stepped to rubber, wound up his windmill, and unleashed a hard one at David's head.

As soon as the pitcher had started his windup, David had opened his stance. He held the bat higher than normal. Fortunately for him, the pitcher's throw was lower than it should have been and came at David shoulder high. Relying on his acute hand and eye coordination, David caught the ball out in front of the plate. The ball whistled towards third base about twelve feet inside the line. The third baseman who had been playing close to the bag holding the tying run on base did not stand a chance. The ball curved as it flew and went into the left field corner. Two runners scored and David stopped at second base with a standup double. The game was over.

David turned to the red-faced second baseman that he had kayoed earlier. "I hope you're all right," David said sincerely.

"Fuck you, kid," the second baseman snarled as he pushed past David and walked off the field.

In the clubhouse, David was greeted by his jubilant teammates who introduced themselves and shook his hand. For the first time that afternoon, David was pleased with his decision to try softball. Only one thing puzzled him. Hendricks, for some reason, continued to be amused by something David had done.

Monday morning David reported to the Trainee Office to read the bulletin board as instructed. This was the week the trainees began their rotational familiarization tours to selected IBM Divisions. The rotationals would last four weeks until the next series of lectures began. Each trainee had submitted his preferences in priority order. David had selected the Sales Division as his first choice. From his conversations with other trainees, he deduced that Sales had been the choice of a majority of his peers. Realistically, David did not rate his chances of attaining his selection as high. Already, the class had sorted itself out, and most had a good idea of their position in the class pecking order. Others had made it clear that ex jocks ranked near the bottom.

David was not the first to approach the bulletin board. From the reaction of those students who had been there earlier, David assumed that most had achieved their goals. They were smiling. This surprised him because the trainees were all highly competitive. Assuming that everybody could not be sent to Sales, David did his best to prepare for disappointment as he waited his turn.

"Shit," the trainee in front of David swore and frowned as he pushed past in retreat.

David said nothing. He had ranked the bespectacled string bean shaped fellow as an intellectual superior. His class comments had always impressed David. Anxiously, David approached the board. He checked Sales. His name was not listed. He checked Production. His name was not listed. He had not attained either of his first two choices. Dispiritedly, he read the Personnel and Administration lists. There, he found the stringbean's name. David now understood the young man's reaction. No one wanted Administration. He did not bother to read Finance. He assumed that was reserved for the accounting majors. He began to worry if he had been overlooked altogether. Finally, he reached General Products Division. General Products was responsible for all the miscellaneous items that IBM produced, a catch all of products including typewriters, adding machines and the like. Nobody wanted that assignment. The trainees had learned early on from the snide comments of their lecturers that General Products was a graveyard. It was where the marginal employees who had not made the grade elsewhere were dumped. IBM did not fire its employees. It had a reputation to uphold as the foremost paternal conglomerate in the world of business, the one that cared for its employees. IBM did not have to force its deadwood out. It sent them to the General Products Division where they languished in the lower pay scales for a career or in desperation quit on their own.

David found his name listed with four others under the General Products Division heading. Immediately, David consoled himself. The assignment was only for four weeks. He would take it just like spring practice—something to be endured. He made a note of the room he was supposed to report to. It seemed familiar. As he made his way across the IBM campus, David realized why it was familiar. It was Ernie Hendricks' office. When David arrived, he found the four trainees who had been assigned to General Products already there ahead of him. He took a seat at the end of the line. He looked up and discovered an attractive brunette smiling at him.

"May I help you?" She asked. She sat behind a large, polished wood desk.

"Trainee, Howard," David said.

"Oh yes. I have your name here. We talked Friday, and I was at the game."

David remembered the voice that had answered the telephone when he had called to ask about the card on the company board.

"Nice game," the girl smiled again.

David's spirits rose. She seemed to be flirting with him. "I watched most of it," David said. He was not accustomed to sitting on the bench. He had been a starter in baseball and football since his freshman year in high school.

"I liked your base running."

David flushed. He was not proud of having clumsily flattened the Penguin second baseman. He should have slid.

"Mr. Hendricks," the girl called to the open door behind her. "Mr. Howard is here."

"About time."

David recognized Hendricks' voice. He saw the other trainees staring at him, two curiously, two with unconcealed jealousy.

"You may go in now," the secretary said.

David noticed the name plate on her desk. "Ms. Wilson."

"Thanks Ms. Wilson," David said as he passed her desk.

She pivoted in her chair and watched as David entered Hendrick's office. "My name's Margaret," she called after him.

David found Hendricks sitting behind a large desk. Except for a modest stack of papers resting on the desk's right corner, the highly polished wood surface was bare. Hendricks sat in a large, upholstered executive chair with his feet on the desk. Directly across from him was a large leather couch flanked by two leather-covered chairs. To his right, David could see the broad grass covered campus through the four large windows. On his first day at IBM, David had learned that offices were furnished according to rank. Clearly, Hendricks was much more senior than he had indicated Saturday.

"Welcome to General Products," Hendricks said. Noting the confusion on David's face, he elaborated. "On the ball field, I'm Ernie. Here, I am Division Chief. You can call me anything you want in private, but here sir will do."

"Yes sir," David said. He decided to call Hendricks sir. Otherwise, he might slip up at a bad time. At the very least, the other trainees would resent any familiarity with the boss if it came from the ball field.

"You're here because I asked for you," Hendricks said.

David hoped that was the reason he did not get Sales. He could do without such help.

"We need you for the team, so I want to make sure you don't get into any trouble right away. Don't worry about it. We all get in trouble. That's a way of life here."

The words struck David as heresy. Thus far, all their lecturers had extolled IBM as an employer. Not one had dared offer a negative word. David began to like Hendricks even better than he had on Saturday.

"I know you wanted Sales, but I thought it best that you avoided it for a while."

"Shit," David thought, trying to mask his true reaction by smiling.

Sales in IBM was where the action was. It was the breeding ground for future company presidents, but that was not all. IBM's primary moneymaker was the main frame computer. Once a customer hooked up, they belonged to IBM for life. IBM did not sell its computers. It leased them. With a lease, the customer looked to IBM for service, training, and software for life. The sales representative remained their main point of contact. He earned his steady commissions from the customers every year from sale to retirement. The salesmen did very well, and did not work hard at it either. They entertained clients at long funded lunches, played golf and tennis with them, and regularly traveled to visit them wherever their headquarters were located. David looked forward to seeing Europe and Asia.

"Why is that, sir? Avoid sales I mean." David hoped his blunt question would not provoke Hendricks.

Hendricks laughed as if David had said something uproariously funny. "Can you handle a foil?"

"Foil?" David had never fenced in his life. He was surprised that Hendricks who was an impressive softball pitcher could be that weird.

"I've seen them in the movies. That's all."

"The movies?"

"The three swordsmen. That stuff."

Hendricks laughed. He opened a desk drawer and took out a small stack of plastic sheets. Hendricks selected one, turned to the credenza behind his desk. It had a built in overhead projector, the kind that the coaches at Cornell had used to diagram plays. Hendricks put the transparency in place and flipped a switch. A diagram outlining the offices of General Products appeared on the far wall. Hendricks pushed another button and a screen descended from the ceiling. The diagram now perfectly fit the screen.

"Is that a foil?" David asked, not knowing what else to say.

"That's a foil."

"Ernie, it's time for your ten o'clock."

David turned to find Margaret standing in the doorway.

Hendricks flipped the switches again, the projector went out, and the screen receded into the ceiling.

"We have three other trainees waiting," Margaret said.

"Send them over to Jeff. He can give them their assignments, and I'll see them back here for a welcome handshake at two."

David rose and started for the door looking forward to joining the others and learning about his interim assignment. He had no idea what it could be. Something humdrum obviously. Donkeywork no one else wanted to do.

"Whoa," Hendricks said. "Stay here."

David stopped and turned back for instructions.

"Take these," Hendricks handed David three of the transparencies. "You'll also need this."

Hendricks handed David an object that looked like a silver ball point pen. "It's a pointer, pull on the end."

David did, and a silver rod telescoped out of the pen.

"It won't be difficult. Nothing like attacking second base," Hendricks laughed.

David chuckled, although he did not find the comment amusing. He knew he should have slid instead of acting like a fullback confronting a linebacker.

"We'll only use two of these slides. I will tell you to give me number one. You turn on the projector. In the conference room, the screen will lower automatically when you hit the projector switch. You put number one on the viewer. When I talk about something, you point at it with that," he nodded toward the pointer. "When I say number two, you change the transparencies and point."

"Yes sir," David said. "I should study these first," he waved the transparencies. He assumed he had his donkeywork but did not want to screw up. "Will we be briefing a customer?"

Hendricks started for the door. "The Management Committee."

David knew what the Management Committee was. He immediately started to worry.

"Good luck," Margaret called as they passed her desk.

As they walked down the hall to the elevators, David walked at Hendricks' side, one pace behind.

"I suggest you buy a pair of wing tips," Ernie said as they waited for the elevator.

David looked down. He had devoted an hour on Sunday to polishing his only pair of black shoes. He glanced at Hendrick's shoes. Wingtips. Two middle-aged men in black suits joined them. Wingtips. Suddenly, David grew self-conscious over his damned shoes. It was no wonder he did not feel like he fit in to the IBM culture. He not only did not dress right, he did not know it.

"Don't worry. You'll catch on," Hendricks said as they boarded the elevator.

Hendricks pushed the button for the top floor.

"Do you think this is a good idea?" David asked, referring to his attendance at the Management Committee meeting.

"What? Elevators?" Hendricks smiled.

David held up the transparencies.

"Something wrong with the foils?" Hendricks asked.

Realizing that Hendricks was making a joke, David surrendered with a shrug.

Hendricks led the way into a huge conference room. The thick green rugs, rich walnut paneling, and immense, highly polished table lined with chairs intimidated David. Most of the seats were filled with dark suited, white haired men. Hendricks pointed towards a seat next to an overhead projector on the wall opposite the huge window. David, aware that he was the youngest person in the room, hurried to the chair and sat down. Hendricks seated himself in a chair near the center of the table. The room was silent. Just as Hendricks was settling in his chair, the door to David's right opened, and an older man, probably in his early fifties, entered. He was accompanied by a mature woman, quite attractive with nicely coiffered white hair, who took a chair along the wall to David's right. She smiled at David, then opened her stenographer's pad and sat expectantly with a poised pencil. The man moved quickly to a spot at the head of the table where he had his back to the glistening windows. The sunlight flowed over his shoulders directly into the eyes of the other executives who waited before him. The phrase "below the throne" popped into David's mind. David recognized the Chairman of IBM from his picture in the lobby. James Archibald Crane was renown for being the first non-Watson family chairman of the company. Thomas J. Watson Jr. had personally anointed him.

David slumped in his chair, trying to make himself smaller. David looked around the table hoping to get a sense for how the others were reacting to the Chairman. He had read that body language could tell you a lot. Most were quietly waiting for the Chairman to take his seat. When David's gaze reached the head of the table, he found the man who sat on the Chairman's immediate left glaring at him, David. It was the Penguin second baseman who David had creamed on Saturday. David nodded. He did not know what else to do. He was beginning to understand why Hendricks had been so amused when David had charged into the man.

Crane, without amenities, opened the meeting. "All right, George, begin."

The man on the right continued to stare at David until after the man who sat directly opposite him began speaking. David assumed the man was the company's Chief Financial Officer. He spoke in a high, thin voice. At first, David had difficulty discerning what the man was talking about. As he droned on, David realized he was describing the company's current financial state, division by division. The other executives appeared to be listening closely. Some smiled when the CFO's statistics praised their Division. The others seemed to feign concern when the statistics reflected less than progress. Before long, David's attention began to lag. When the CFO mentioned the General Products Division, David noticed that several of the executives glanced derisively at Hendricks. David decided that Hendricks was not the most popular man at the table. He wondered if it was because Hendricks and the frowning second baseman were the two youngest executives at the table.

"Thank you, George," the Chairman said when the CFO finally finished his report.

"Thank God," David silently agreed.

"We will dispense with the usual order of business," the Chairman declared. "I have asked Ernie to give us a special report."

Two executives seated between Hendricks and Crane groaned. "Not again," one said in a low voice.

David had heard that IBM executives were encouraged to speak frankly in house. It was a carryover from the Watson days. Both Watsons, father and son, had run a tight ship, expected discipline from their employees as well as adherence to a well-defined code of behavior, but also they insisted on frank open discussions. The "contention system" required junior employees to share in decision making while at the same time forcing them to defend their decisions against all company comers, peers, subordinates, and superiors. The trainees had already heard the descriptors "open warfare" and "bloody corporate backstabbing" applied to company meetings at every level. The Chairman remained above the contention, not taking sides, not until he made the final decision that became inviolable law.

The executives at the lower end of the table moved nervously in their seats. All sat up a little straighter. Several nervously arranged the blank tablets and pencils that were already aligned in front of them. Only the Chairman had bare expanse of table.

The Chairman ignored the reactions. "The floor is yours, Ernie."

Hendricks remained seated but spoke in a calm, clear, confident voice.

"The subject is the same one you have heard before. The day of the microcomputer is upon us."

"That's not quite correct," David's second base victim said sourly. "The day of the microcomputer might be on for some of the marginal companies, but it is not on IBM."

David glanced quickly at the Chairman to see how he responded to the man's rude interruption. The Chairman did not react at all. He waited patiently for Hendricks to continue. Despite the Chairman's blank expression, David sensed that he seemed to look with favor on Hendricks.

"We all know that argument," Hendricks did not seem perturbed by the interruption. "John, whether you and the Sales Department are yet ready to accept it, the day of the minicomputer is dawning."

David then realized who his antagonist was. John Parsons was Chief of the Sales Division. That knowledge explained a lot. No wonder Hendricks had been amused when David had decided to flatten the second baseman. It was rumored that Parsons was the Chairman's anointed successor. One day he would run IBM. David in one split second decision had doomed his career. He might as well quit now. He certainly could not go to Sales as he had planned. At that moment, David decided to take second best, General Products Division. At least he seemed to amuse Hendricks. He would remember in the future, however, that company softball games were not just sport.

"We cannot hold back progress. We cannot live in the past. If we don't ride the wave of technical advancement, the wave will ride us until we sink and drown."

"Spare us the bullshit," John Parsons interrupted again. "Our business is main frames."

"Our business yesterday and today was and is mainframes, but we are facing the same kind of change or die decision that Tom Jr. faced when he decided to go with the 360."

David knew about the 360. It was the first of the true computers. Tom Jr. had bet the company on it. "Fortune" had called it the five million-dollar gamble. Integrated circuits replaced the vacuum tubes. It also meant they had to persuade all their customers to throw away their old, hardware and software and to replace it with the new if they hoped to take advantage of the revolution the 360 offered. This was almost heresy to IBM who had based its future on leasing machines to business, keeping customers happy by providing everything they needed, with each new advance being stacked on top of the old. Many had argued the 360 meant commercial suicide. But Tom Jr. had won. The 360 propelled IBM the top of the heap, making it one of the most successful companies in history. IBM raised its market share from twenty to seventy percent. By the seventies, when anyone thought computers, they thought IBM. And now, Hendricks had the temerity to propose a sea change when they were at the apex of the heap. David immediately began to worry about what this would mean to his own future. Carrying the football could be handled instinctively, but this was something different.

Parsons groaned. He smiled around the table. The majority of other executives seemed to agree with him. They appeared content, however, to let him carry the weight of the argument. They were prepared to ride on the heir apparent's bandwagon, but the Chairman seemed to be sponsoring Hendricks.

"There are already minicomputers on the market. A hobbyist product, to be sure," Hendricks spoke quickly to abort another objection. "But the day is coming when there will be a computer in every home and...."

"Our market is big business," Parsons again interrupted. "Every top one hundred company in the world uses our mainframes."

"Today," Hendricks agreed. "Tomorrow is another story. And if I may continue," he did not appear intimidated by Parsons, "the microprocessor, the microcomputer will replace the mainframe."

"Prove that," Parsons laughed derisively.

"I cannot, anymore than Edison could have proved that in seventy years the world would be wired."

"And the battery powered car will replace the gas combustion engine," Parsons said.

"It might. But I don't say that with the same certainty that I say the mainframes' days are numbered. We all know the symbiotic relationship between sales and the main frame. Once customers sign on for our mainframes, Sales has it made. The customer is locked into IBM for service, upgrades..."

"And that's the way it should be, what made us the company we are."

David expected the Chairman to intervene before the squabbling deteriorated further, but he sat silently, listening, noncommittal.

"And then the salesmen who every January face the New Year with eighty percent of their sales quota filled by past sales can look forward to a fine life of golf, lunches, dinners and foreign travel."

"Does anyone object to that life?" Parsons countered.

"I do, for one," Hendricks said. "While you are enjoying the good life, there are startup companies out there who will one day take your business away from you with superior products that they will sell for a fraction of the price of your mainframes."

"And where is the profit in selling cheaper?"

"In selling more," Hendricks replied.

"What does General Products recommend?" The Chairman interposed.

"General Products recommends nothing. I recommend we build a microcomputer."

"What will it do?" Parsons asked. "Replace your typewriter line?"

"The microcomputer will replace our typewriter line and our mainframes too."

"With 2 K of memory?"

"That will change."

"Do any of the rest of you have any comments?" The Chairman asked.

"Even if the microprocessor eventually is able to do everything you so eagerly predict," the CFO was the first to speak, "it will clearly put us in competition with ourselves. Our money comes from the mainframe leases. If we sell a cheap product that damages the mainframes, we'll be out of business in five years."

David saw that the CFO had clearly come down on the side of the Chief Sales. Their arguments made sense. He began to wonder if his benefactor was a starry eyed visionary.

Hendricks started to reply, but the Chairman silenced him with a flick of the fingers. The Chairman looked around the table. The others remained silent.

"All right, Ernie, I want to know how you would go about building a microcomputer. Give me a memo."

"But..." Parsons flushed red as he started to protest.

"I've made no decision," the Chair interrupted. "The meeting's over."

The Chairman rose and left the room with his secretary in tow. Surprised, but relieved, David realized he had not used the transparencies or the overhead projector. The others filed from the room. David rushed to catch up with Hendricks who held out his hand. David gave him the foils. "I didn't think we would need them, but I wanted to be ready."

"I guess you understand now why your day of becoming a hippo in sales will never arrive," Hendricks said as they waited for the elevator.

"Hippo?"

"IBM speak for high potential. Hi po. Hippo, get it."

David nodded.

"Guess you'll have to be my hippo."

David did not know whether to be pleased with those words or not.

Chapter 5

Stanley Pitts, wearing a clean white T-shirt, dirty jeans and rubber sandals on sockless feet, paused outside the Electronic Warehouse and took a deep breath. He was tempted to go right on past. Stanley was bone tired and discouraged; no one else seemed to care. Harold, his partner in Astron Computers, was totally indifferent to business matters. Harold worked late every night at the Pitts garage despite having put in eight long hours at Hewlett-Packard. This Stanley acknowledged, but Dum obstinately refused to discuss the future of their company or Stanley's contribution. Dum was simply obsessed with electronics. Stanley had implied that the Electronics Warehouse was willing to buy another fifty boards if they came complete with a box for the circuit board and a monitor. This was not exactly the truth; Jake had said he would consider buying another fifty boards if they came as a computer package including software, and the software was the problem. Harold was already talking about the improvements he was going to make to their next version of the computer, the Astron 2. Installing software on the Astron 1 did not interest him. Stanley had neither time nor energy to consider any Astron 2. He doubted that the Astron Computer Company would last long enough to see an Astron 2, at least not with Stanley Pitts at its helm. Stanley was thinking about going back with the Hare Krishnas or maybe becoming a monk in a Zen monastery.

They were down to their last two thousand dollars. Stanley had spent the last from the twenty-five thousand they had gotten from the Electronics Warehouse contract on parts, supplies, salaries for their part time help, advertising and a myriad of other expenses including rental cars that Stanley used when his father's car was not available. If it were not for his parents, the Astron Computer Company would have already failed. They provided the workspace in the garage and spare bedroom, helped as they could, but they had no money to subsidize expansion. It had been expensive trying to give Harold what he needed to package the next fifty circuit boards as a complete computer. Jake had accepted the simple metal box Stanley had designed to hold the circuit board, but now he was insisting on a device that would load BASIC code into the board.

Stanley had tried to persuade Harold to build something that would load BASIC from a cassette tape. Dum had admitted that he could, but he had refused to do it. He claimed he had no time. They had not argued. Dum had said no then turned a deaf ear to Stanley, refusing even to consider his arguments. This had irritated Stanley, and he had almost cried when he described the pressures he was facing. Harold had simply replied that he handled the tech side and Stanley was responsible for all business matters. The best that Stanley could get from Harold was the promise to see if he could find one of the engineers at HP to build the tape interface. The only problem with that was Stanley would have to come up with some money to pay the engineer.

"Don't stand out there mooning about," Jake appeared in the doorway of his store. "Get in here. I need to talk with you."

Stanley, who had just decided to visit his friend and spiritual counselor Hiro, felt trapped. Jake usually acted as if Stanley were an unwanted visitor. If he wanted to talk with Stanley, then, certainly, he had either complaints or demands, probably both. Stanley knew that under no circumstances could he take back any of the circuit boards.

"Good morning Mr. Foster," Stanley tried to sound confident. "I've got good news. Dum has the next fifty computers ready for delivery."

Jake laughed. He was angry with himself for letting the kid con him into paying twenty-five thousand for the first fifty boards, but he had to admire the kid's nerve. "Come into the back," he ordered.

Jake turned and headed to the rear of the store where he had a small office next to their crowded storeroom. Stanley forced himself to smile at Jake's clerk and followed the owner. The clerk pointed at the circuit board display and disdainfully pointed his thumb towards the floor. Stanley felt like mooning him. Jake sat behind his desk and glanced at a straight chair next to the door. Stanley dutifully sat down.

"When do you want them delivered?"

"You got to be kidding me," Jake responded.

"How are the boards moving?" Stanley asked, trying to get control of the conversation before it went into areas he did not want to go.

"They aren't."

"How many have you sold?" Stanley was afraid of the answer. He had visited the Computer Centers' two other stores and had been shabbily treated. The store managers had not seemed to like him. This to Stanley meant that they would not try to move the boards, preferring to make him look bad."

"Not a single one."

"Not one sale here," Stanley repeated the bad news. "What about the other stores?"

"Not a single sale anywhere." Jake let his irritation show. "Look kid, I told you. My customers want a complete computer, not circuit boards."

"What about the Taurus?" The tech magazines said it was selling big time.
"It sells."

"But it's just a kit," Stanley felt he had trapped Jake.

"Nice try, kid. Sure the hobbyists buy the kit, but now they want a computer that they don't have to put together. And they want it to be able to compute."

"The next fifty are just what you need. They'll be in a metal box that I designed myself." Stanley had copied the design of one of the boxes that he had worked on at Atari and had liked.

"Does it have a way to get BASIC on to your dumb board?"

"How does cassette sound?" Stanley mentally crossed his fingers hoping Harold could find an engineer to design the thing.

"If it works."

"You've got a deal."

"Bring me down a prototype. I want to see it work first."

"How about an advance?"

"You've got all the advance you're going to get."

"It costs money to build boxes."

"Tell me about it. I've had to pay a cabinetmaker to build boxes for those boards you stuck me with. Despite the extra expense, I'm going to have to lower the price fifty bucks. And, even then, I'll have no guarantee of unloading them."

"The price for the second fifty, boxes, monitors and all, will have to go up."

"No it doesn't. We have a deal. The price for the next fifty will be the same as the first. I told you I expected a working computer in a box and you stuck me with a bunch of boards. Now, give me what I asked for, or don't appear around here again."

A dejected Stanley spent the next few hours fruitlessly searching for additional sources of revenue to pay for what Jake demanded. Late in the day, he decided to visit his friend Hiro. Stanley did not know why Hiro had decided to call himself that. His real name was Schultz. Stanley had met him at the Hare Krishnas following his return from India, and then Hiro was still Schultz. A college dropout like Stanley, Schultz had migrated to a Zen temple in San Francisco before returning to Los Altos to open his own temple in an abandoned storefront in the center of town. Stanley was attracted to Zen. He liked the idea of seeking enlightenment through introspection and intuition. Stanley thought a lot about himself and liked to make decisions impulsively. He had tried a course in philosophy and had found it boring. Socratic reasoning had first appealed to him, but after a while he found all that questioning tedious. Spontaneous decision making by the unconscious appealed to him. Stanley decided that maybe he should boot the whole computer thing and become a Zen monk like

Hiro. He would take an Indian name, though, not a Japanese one. Hiro to him sounded too artificial.

The store front door was locked, and the shade drawn. Stanley knew that Hiro lived in the back of his temple, so he rattled the door. Getting no response, he wrapped loudly with his knuckles almost hard enough to break the dirty glass. Finally, just as Stanley was about to give up, Hiro folded back a corner of the torn shade and peered out. Seeing Stanley, he smiled. Hiro unlocked the door and waved Stanley in with a deep bow. Stanley responded by putting his palms together in from of his chest and bowing back, just as he had learned to greet the Buddhists in India. Stanley had disliked India with its poverty, hungry people, and arrogance but he always pretended the opposite.

"Hi Stanley," Hiro greeted him.

"Hiro," Stanley bowed a second time.

"Come on in, brother, I've got tea in the back."

Stanley followed Hiro through what had once been a mom and pop grocery. Only a cracked wood counter and a few broken shelves remained. Hiro had hung some cheap bells and crepe paper from the ceiling and placed a tin replica of a temple surrounded by candles on a three-legged table against the back wall. Hiro liked to pretend that was where he meditated.

In the back room, Hiro had a mattress on the floor, a table, and two wood chairs with rods missing from their backs. A dented pan steamed above a sterno burner. Hiro could not afford electricity or gas. In fact, he paid no rent and camped in the deserted building.

Neither spoke as Hiro retrieved a second cup and filled it with boiling water. He took the tea bag from his own cup and dropped it into Stanley's.

"How is your business?" Hiro asked.

"That's why I'm here," Stanley said. "I need to meditate."

"Drink your tea, and I'll leave you to your thoughts."

The two sat in silence and sipped their tea. Stanley who liked his tea strong realized than the bag had been used too many times. He decided that the next time he came he would bring a new box of teabags, maybe one of the stronger herbal teas made from raspberry leaves. Stanley was considering a return to his fruit diet. Stanley looked at Hiro over the steaming tea. Hiro, with his hair tied in a knot and saffron colored robe that he had made himself looked like a Buddhist monk, somewhat. "He's dirty enough," Stanley thought. Stanley almost laughed, then chastised himself for the unkind thought. He had come here to discuss his problems. Unlike Dum who seldom listened, Hiro let Stanley talk and nodded his head in agreement. A little dirt was not something that bothered Stanley who did not consider cleanliness next to godliness. Jesus Christ was no priss.

"Would you like a little grass?" Hiro asked.

Stanley like Hiro thought grass relaxed him and helped him meditate.

"That's exactly what I need," Stanley replied.

Hiro reached into his robe and pulled out a small sack that he wore on a string around his neck. He carefully opened it and took out a crumpled cigarette he had rolled himself. He lighted the grass with a wood match, took a deep breath and held it while handing the glowing butt to Stanley.

Stanley took a deep drag, held it until Hiro released his, and then Stanley followed suit. He took another drag, as much as his lungs would hold, then handed the butt back to Hiro. It was barely an inch long. Hiro held it carefully in his fingers and filled his own lungs.

Stanley felt the relaxing effects of the narcotic smoke. He exhaled. "I've got problems, Hiro," Stanley blurted.

"Want to discuss them or meditate?"

Stanley thought about the question. He decided it would be better to get Hiro's opinion. No one else listened when he wanted to discuss his problems. "Discuss."

"Spiritual or earthly?" Hiro asked.

"Both," Stanley said. He sipped the now lukewarm tea.

"Business trouble?" Hiro asked. Stanley had visited him a month ago filled with his plans for the computer he and Dum had invented. Hiro had let Stanley have that little bit of vanity. He knew that Harold was the inventor and that Stanley had only a smattering of technical knowledge.

"More important than that. I'm at a crossroads. I have to decide what to do with my life."

"Be a millionaire or go back to college," Hiro guessed.

"Even more. Should I become a businessman or a monk."

"A monk?" Stanley's comment surprised him.

"Yea. A Zen monk like you," Stanley said. He did not like his friend's reaction.

Hiro hesitated. He was on the verge of giving this life up and becoming Schultz the college student again. He was lonely and needed money. Besides, he had heard rumors that the owner was preparing to demolish his decrepit building.

"Meditation has its advantages," Hiro equivocated.

Stanley reached for the grass. The butt was down to half an inch. He took another drag then dropped it to the floor as the remaining paper ignited and burned his fingers.

"You need to get some better stuff," Stanley said when he finally exhaled.

Hiro did not argue. He waited, knowing from experience that Stanley would continue to talk about himself.

"Nobody appreciates how much I'm doing. Dum nods and works on his damn computer. Everyone else chews on me."

"I thought you had a big contract."

"The money's gone, the banks are fools, and we have to grow."

"What about your parents?"

"They're great. Mom helps all the time. Answers the phone for us."

"But no money?"

"No."

"It's always that way."

"Should I give it up. Become a monk?"

Hiro did not answer because he didn't know what to say. He couldn't admit that he was ready to become Schultz again.

Stanley leaned his chair against the wall. "I don't need it."

Not knowing what Stanley was talking about, Hiro remained silent.

Stanley interpreted his friend's silence as agreement. That's why he liked Hiro. He was self-contained and listened. Hiro was three years older than Stanley, and frankly, Stanley had not liked him in high school. He had been a wise ass. Zen had changed him though. Now, Stanley could talk to him.

"I'm carrying this business. I do all the work. I get the parts, visit the customers, supervise the employees. Dum, he just plays at his workbench."

Hiro was tempted to point out that Harold produced their single product while Stanley did the fetching.

"I'm no magician. I need help." As soon as Stanley said the words, he regretted them. He did not like to admit that he needed help. "Money, I mean. Foster wants a completed computer. All I get from Dum is circuit boards. He won't stop working long enough to build the other equipment we need to sell his boards. He wants to get some high priced dummy from HP to do the work, but where am I going to find the money to pay him. I don't need that. I'll be a monk."

Hiro thought about advising Stanley of the drawbacks of the monk business.

"Not like this," Stanley said thoughtlessly. "I'll go to a big temple where they know how to do it."

Hiro almost laughed. "Where will you find one."

"Japan maybe."

"I'm not sure that's a good idea."

"No, India," Stanley ignored Hiro's comment.

Hiro concluded that Stanley was bluffing. He had hated India and would never go back. Hiro slid from his chair to the floor and stretched out prone. He closed his eyes and folded his hands into the prayer position. "We should meditate on this Stanley."

"I'll need a name," Stanley was lost in his own thoughts. "What name should I use?"

"Swami something," Hiro suggested.

"I'm serious," Stanley insisted.

Stanley joined Hiro on the floor. They lay side by side.

"Let's meditate," Hiro suggested.

Stanley folded his hands on his chest, closed his eyes, but continued to talk. "Maybe then they'll learn they need me," Stanley said.

Stanley pretended to meditate for several minutes. Suddenly, he sat up and slid until his back rested against the wall. "Got another?"

Hiro immediately knew what Stanley wanted. He thought for a few moments about saying no. He really did not like Pitts any better than he had in high school, but now that he was a Zen priest, or at least pretended he was, he had to exercise control. He had been asked to leave the San Francisco monastery because of "excessive drug abuse." Hiro, Schultz, did not consider his occasional use excessive, but he had been expelled in any case. He had returned to Los Altos to open his own temple and drive the San Francisco puritans out of business, but he had not attracted many followers, just a few kids like Stanley who were seeking to hide from their outcast positions in their communities.

Hiro tried to ignore Stanley, but he persisted: "Come on Hiro. I'll bring a resupply tomorrow. I'm desperate."

Hiro agreed that Stanley was desperate, but not for the reasons he thought he was. Hiro relented and pulled his cloth sack from inside his robe. He deliberately selected the smallest roach he could find. He handed it to Stanley. "Enjoy."

Stanley took the roach, lighted it, then took a deep breath. "Mind if I light the candles?"

Hiro liked to save the candles and incense for his more promising followers, all three of them. Hiro nodded.

Stanley used a wood match to light the candles. Then, without asking, lifted the lid of the small bronze vase with the holes in the side that Hiro had bought second hand at a Sunday market and ignited the incense pillar.

Stanley lowered himself to the floor and sat cross-legged facing the candles and smoking incense. After a few minutes of silence, during which Stanley again feigned meditation, Stanley spoke: "I will become a monk."

"I'm not sure you're ready, Stanley," Hiro responded. He certainly did not want Stanley moving in with him.

"I could go back to India."

When Stanley said that, Hiro knew he was still bluffing. He said nothing.

"What do you think?" Stanley said as he finished the roach without offering to share it with Hiro.

"You must make your own decisions, Stanley."

"I treasure your advice."

Hiro almost laughed. Stanley did not treasure anyone's opinions but his own. "If you insist," Hiro paused for effect. He folded his hands in front of him in a prayer position and pretended to meditate.

Stanley tolerated the silence for thirty seconds then spoke: "Please. I'm dying here."

"Stanley, you should stay with the business. It will prosper, I'm sure."

"That's what I needed to hear," Stanley leaped to his feet. "I'll see you tomorrow."

Hiro bowed concurrence, but Stanley ignored him. He had already moving for the front door, leaving Hiro's precious candles and incense burning behind him.

Stanley exited Hiro's storefront temple and turned right. He felt completely relaxed. He knew it was due to the grass, not Schultz's posturing, but it had worked. Stanley was confident that he could turn things around. In his buoyant mood, Stanley hurried through the center of town. He wished he still had his van. As soon as the checks started coming in, he planned to buy a new company car. He could easily write it off as a business expense. Stanley avoided the block that housed the Electronics Warehouse. He did not need another dose of Jake's harping. As he passed Rosenbaum's department store, he noted a familiar pair of legs ahead of him. Stanley knew those legs well. He had studied them for an entire year when he had set one row over and one seat behind the owner in sophomore geometry. The legs were one of the reasons Stanley had flunked geometry, probably, thwarting a successful engineering career. If it were not for those legs, Stanley, reasoned, he might be building the computers and Dum or someone else would be hustling about doing the donkeywork.

"Hi Lois," Stanley called.

The legs belonged to Lois Reynolds. As he hurried to catch up, Stanley wondered if he would have any more success with Lois now that he was a businessman than he had in high school. Lois, a cheerleader, dated the star halfback. At the beginning of the year, she had been friendly enough. She used to chat with Stanley as they walked to their next classes that fortunately were located in adjoining rooms on the other side of the building. Then, Stanley had made his move, prematurely. He had called Lois at home and asked her to join him at a movie the following Saturday. Lois had politely declined. From that call onward, however, Lois had studiously avoided Stanley. When he had succeeded in cornering her between classes, Lois had responded coolly. As the year wore on, and Stanley persisted, Lois had become more frigid. Finally, the halfback, obviously at Lois' urging, had approached Stanley and warned him to quit harassing his girl. Stanley, not one to take risks that might invite violence, decided it was Lois' loss and deliberately snubbed her from that time on, until today.

Lois turned, saw Stanley, frowned, then waited for him to approach.

"Hi Stanley," she greeted him.

"Hi Lois. How're things?"

"I'm on summer vacation."

"College?" Stanley decided that Lois was still a snob. She had great legs but zero personality.

Lois turned and Stanley walked beside her. "Where you going?"

"Now?" Lois was on her way home. She had been scouting Rosenbaum's for her new fall wardrobe.

"No, what school?"

"University of Washington, where else? Certainly you know from the papers that Jeff is on the varsity this year."

Stanley got the message. Loud and clear. Lois was telling him it was still hands off. Stanley's luck had not changed. During high school, the girls had not liked him, none of them, not even the plain ones desperate for dates. It had always been one date and out for Stanley. Lately, he had not even tried. That was one of the reasons he had hung around with Dum. Harold at least seemed to appreciate him.

"I tried it for a couple of months, but it didn't work out," Stanley said.

Lois looked condescendingly at Stanley.

"I only went to LA Community," Stanley said, referring to Los Altos Community College. "The professors were all second rate." That wasn't quite true. Many of the professors were full time professionals who taught on the side. Most were quite expert in their fields. Stanley, who preferred talking to working, had decided to take a year off and meditate. That was when he had joined the commune. The truth was that Stanley would have flunked out if he had stayed in school. His highest grade had been a D.

"That's too bad," Lois said. She knew better. Pitts had not changed. She decided to escape before he came on to her. Fortunately, they were approaching the corner drug store. "It was good to see you Stanley," she said. "I'm supposed to meet Jeff here."

Her words were enough for Stanley. "Gee, I'd like to see him and discuss old times, but I've got a business meeting." Stanley waited for Lois to ask about his business.

Lois turned. "Good to see you Stanley."

"I'm in the computer business," Stanley called after her.

Lois waved her hand and disappeared through the door.

"Fuck her," Stanley muttered. "That's all the bitch needs."

Lois glanced out the storefront window to make sure Stanley kept going. She had lied. She had no date with Jeff. In fact, she did not even see him anymore. He had his eyes focused on bigger stars, but she did not need Stanley mooning after her again. He had one of the most unfortunate personalities. She wondered why the back of Stanley's T-shirt and pants were stained black. It looked like he had been rolling about on the floor.

Stanley returned home in time to join his parents for dinner. His sister had already departed for the small apartment she shared with her carpenter's apprentice husband. Stanley thought about going up to the spare bedroom to check on the boards his sister had been working on. She watched television while she worked and frequently got pins in backward. Stanley did not complain because she worked cheaply and waited patiently for her pay when Stanley was late. Stanley took his customary seat at the kitchen table. His father who worked the seven to three shift liked to eat early.

"What's for dinner?" Stanley asked.

"Meatloaf. Wash up," his mother replied.

Stanley did as he was told. It was a small price to pay for free meals.

"How's business?" Stanley's father asked as he joined the family in the kitchen. He had changed from his work clothes to his TV watching suit, jeans and T-shirt.

"Don't ask," Stanley replied.

"Keep at it. You're on to a good thing," his father said, dismissing the subject. He picked up the morning paper and ignored Stanley.

Stanley, who had two folded one-dollar bills in his pocket, thought about sharing his money problems but did not. There was nothing his parents could do. They had no money to share. Stanley knew he was imposing on them as it was. His father had to park his old car on the street, and two of the three upstairs bedrooms were filled with Stanley's computer clutter. Stanley ate hurriedly. The meat loaf was good, but Stanley was tired of it. Hamburg, chicken and hot dogs were the family staples. Stanley wished he could return to his vegetarian or fruit diets, but he did not have the money to feed himself.

When finished, Stanley retreated to the garage to wait for Harold's arrival.

"The back of your shirt and pants are filthy, Stanley," his mother called after him. "Change them if you go out. Put them in the basket and I'll wash them tomorrow."

"OK," Stanley conceded. He must have gotten them dirty at Hiro's temple.

Stanley continued to the garage. Harold always stopped off at home to say hello to his wife, wash up, and to eat a quick meal before retreating from his wife's complaints to the workbench in the Pitts garage. Dum was a hard worker. Stanley recognized that, but Stanley wished Dum would sometimes listen attentively to his recital of their business problems.

At seven on the dot, Harold arrived to find Stanley lounging in the overstuffed chair that he had pulled out into the driveway.

"Life of Riley," Harold greeted his friend.

"We gotta talk," Stanley said.

"So talk," Harold sat on his stool and turned to his workbench.

"Foster says he'll take the next fifty boards if we put the them in a box and make them look like a computer."

"You've got the boxes ordered." Harold plugged in his soldering iron and turned on his bench light.

"Yeah, but I don't have any money to pay for them."

Harold turned and smiled. He acted as if he knew something that Stanley did not. "Something will come up."

"I'm at a dead end, Dum. I going to become a monk."

"Don't give up Stanley." Harold turned back to his bench.

"And Foster says we have to design a system to feed software from a cassette tape to the board. Can you do it?"

"I can, but I won't. I don't have time."

"Then we're dead. You might as well stop working. We're out of business." Stanley stood up. "I'm out of here. I'm a monk."

"Stop the nonsense," Dum spoke as sharply as he ever did. "I can get one of the engineers at the plant to build a loader for us." Harold smiled, but Stanley could not see him as Harold was hunched over the board.

"And how are we going to pay him? We already..."

"From the five thousand," Harold said.

"What five thousand?" Harold had Stanley's full attention. He continued to lounge in the overstuffed chair, knees on one arm and head on the other, facing the workbench from his position outside.

"Our five thousand."

"You found some money!" Stanley shouted.

"Dad is loaning us five thousand," Harold said.

"Great," Stanley said. "Fuck the monks."

"He insists that we draw up a paper and sign it. Pay interest and make it legal like."

"No sweat."

"You know he doesn't trust you," Dum said, his back still facing Stanley.

"So what else is new?"

"He thinks you're ripping me off."

"Bull shit."

"He says I do all the work. The computer is my invention. You've done nothing."

"We're partners. I do the business, and you..."

"I know. I know. He just doesn't think you're carrying your fair share."

"Bull shit," Stanley said a little softer, not wanting his parents to hear. "If you want out, we can tear up our agreement. Fuck the company."

"I didn't say that. I'm just telling you what dad said. He insists that we meet. He wants to tell you what he thinks, then we sign the paper and get the $5,000."

"Where's he gonna get five thou?"

"Credit Union."

"A loan." Stanley thought a few minutes. "No wonder he's worried. OK. I'll sign the papers."

"And I'll get someone to build the cassette thing."

Stanley hated the thought of using their money to pay an engineer to do work that Dum could do, but he decided against starting that argument. He knew he would only lose because Dum would ignore him and do what he wanted. "OK," Stanley said, trying to sound as if he were authorizing the project.

"There's something else," Dum said. "I'm almost finished with the boards, and I am going to start working on Astron 2."

"But..."

"The next one will be much better."

Chapter 6

Clapper returned from Albuquerque carefully carrying an assembled Taurus.

"Its got 8K of memory," Clapper proudly informed Charles and Bill Oldham as they set it up on Charles' desk. "And we have an interface card that will support an audiocassette player or a teletype hookup."

"Let me see," Charles said.

Charles carefully examined the back of the machine. "Can we expand the memory?" Charles was thinking about the code for future applications.

"8K is the best we can do for now. This is the latest model."

"Plug it up," Charles directed, taking the chair directly facing the Taurus.

Clapper inserted the power card then reached past Charles to flip switches. Bill reclined on his bed across the room watching his two excited friends. He still had difficulty understanding why Clap and Charles were so excited about another electronic toy. He was frankly tired of all the computer talk. Charles and Clapper were convinced that the blinking box in front of them was the harbinger of revolution. Bill had his doubts. Adding one and one and getting two and blinking lights did not appear revolutionary. "What can it do?" Bill could not contain his skepticism.

"Anything we teach it to," Charles answered.

"The code still has bugs," Clapper said.

"I'm debugging it," Charles said. "Don't worry."

"Those guys were really impressed. You should have seen their faces when your code worked first time."

Charles flipped the switches and turned the Taurus off. "I'm going over to Aiken and work on it." Without another word, Charles donned his parka and left.

Neither Clapper nor Bill Oldham was perturbed. They were accustomed to Charles' abrupt mood swings.

"That's the last we'll see of him for a while," Bill laughed.

"This is really something," Clapper enthused, taking Charles' place in front of the Taurus.

For the next hour, Bill indulgently listened to Clap recount the story of his Albuquerque experiences including his embarrassment at the Hilton front desk.

For the next three weeks, Bill saw little of either Charles or Clapper. Charles sometimes appeared late at night after Bill had gone to sleep and departed before Bill awoke. When he asked Charles where he slept, wondering if he had acquired somehow a new girl friend, Charles had only laughed and admitted that he spent his nights at the Aiken Computer Center.

Twice Charles' mother called from Seattle wanting to speak with her son. On neither occasion was Charles present, and as best as Bill could tell, Charles in his preoccupation did not return her calls. Bill despised talking with Mrs. Swift, not because he did not like her, because he did, but because he hated being put into the position of covering for Charles. Charles had confided his delicate relationship with his mother. They were both highly independent, intelligent individuals. Mrs. Swift served on all kinds of boards and was accustomed to having her way. Charles loved his mother dearly, but he insisted on doing his own thing. He and his mother had a long history of conflicting wills. Charles joyfully recounted the time his mother had insisted on taking him to a psychologist for testing to determine if there was some way she could bend her stubborn and independent son to her will. After six months of consultations, the psychologist told Mrs. Swift that she would have to learn to live with Charles' idiosyncrasies, as he was not going to change. Since then, Charles and his mother had continued to test each other.

Finally, Mrs. Swift called a third time and demanded that Charles speak with her. Bill patiently explained that Charles was not present and seldom was.

"Charles spends all of his time at Aiken Hall," Bill said.

"Playing with computers," Mrs. Swift immediately understood. When Charles became enmeshed in a computer problem, he was totally inaccessible. "I thought he wasn't taking any computer courses."

"That's between you and Charles," Bill said. "But I don't think he is. He's working on code for his project with Clap."

"I knew it," Mrs. Swift exploded. "Be honest with me Bill. He's in New Mexico, isn't he?"

"No, honestly, he's at Aiken Hall."

"Bill, understand I'm not blaming you. But you tell that son of mine to call me immediately."

"I will, Mrs. Swift, but I can't promise he'll do it. I don't see much of him."

"He's your roommate," Mrs. Swift made no effort to conceal her disbelief.

"He sometimes naps behind the computers at Aiken."

"I wish he were sleeping with a girl instead."

"I'll make sure he gets your message," Bill promised.

Bill went directly to Aiken and found Charles typing code, staring at the monitor and muttering to himself. Bill relayed Mrs. Swift's demand that Charles call immediately. "Yeah, sure," Charles said without diverting his attention from the monitor. Bill surrendered and left.

Three weeks later Clapper loaded his meager belongings in his old Chrysler and departed for Albuquerque and his new job as Vice President and Chief of Software for PAMS. Clapper left his live-in girl friend behind but took with him Charles' latest version of code for the Taurus. Bill had some difficulty understanding the arrangement. Clapper was a PAMS employee, and Charles in Cambridge was not. Charles and Clapper agreed to form a company which would manufacture software, and this company would lease not sell the BASIC code for the Taurus to PAMS. The leasing was Charles' idea. Clapper would have sold the code outright, but Charles refused.

In March, Charles decided he had to see PAMS for himself. Without telling his parents, he flew to Albuquerque for a brief stay. Clapper met him at the airport, still driving his Chrysler.

"You're going to have to get rid of this wreck," Charles told his friend as they departed the airport.

"Things are going great," Clapper enthused. He did not share Charles' interest in cars. Charles liked speed and still yearned to get behind the wheel of the old red Mustang he had inherited from his father. Charles had been dismayed when his mother refused to let him drive it cross-country to Harvard where Charles was wheeless. "The orders are poring in."

Charles looked at his friend, wondering if he had become more a PAMS employee and less his partner. "We have to come up with a name for our company," Charles deliberately tested his friend.

"I'm really overwhelmed at PAMS. You'll have to continue to carry the load with the BASIC. By the way, Ed's delighted with your code."

"What about Digital Systems?" Charles asked, not to be deterred by talk about PAMS.

"We're a software company. We've got to have Software and Applications in the name."

"Digital Software and Applications Systems?" Charles laughed. "Real catchy."

"OK," Clap conceded. "What about Digital Software?" Clap did not expect Charles to agree.

Charles remained silent. "Not bad," Charles said. "Dig Soft, Digital Software."

"I like Digi Soft."

"Let's stick with Digital Software for now," Charles said. For him the discussion was over. "Christ, this is a dismal place." Accustomed to Seattle and the water and green around the lake where his father's house was located, New Mexico's arid desert appeared uninhabitable.

"Check the temperature," Clapper said defensively. He anticipated he and Charles might have to live in Albuquerque for some time.

"Not bad," Charles said. He had carried his parka from the plane, was wearing a sweater, and still perspired.

"Better shed the sweater," Clapper said. "It's sixty five already today."

"It was below zero when I left." Charles was not about to defend Cambridge's climate.

"Here's home," Clapper said as he pulled to a stop in front of Room 9.

Charles studied the canted sign: "The Sundowner." The motel had obviously seen better days. "Couldn't you find someplace cheaper?" Charles asked.

"Don't worry. The price is right."

Charles eyed the highway and the filling station next door. He had no wheels. He started to worry then dismissed the thought. He would not be staying long this time.

"Where's PAMS?" Charles asked.

"Couple miles down the road," Clapper answered.

"Anything I should know?" Charles made no attempt to get out of the car.

"Nope. They're all good guys."

"Let's check them out," Charles said.

"Let's dump your suitcase and that sweater. You won't be needing them."

Charles relented, grabbed his suitcase and parka when Clapper opened the trunk, then followed his friend into their temporary home. Room 9 was shabby. Clapper's belongings were strewn about the room, but it had twin beds and a small table that served as a desk. Charles slid his suitcase across the room toward the one made bed and tossed his parka and sweater after it. "Let's go."

At PAMS Charles found everything as Clapper had described it—the massage parlor with the painted windows, the parking lot with the buckling pavement, the storefront factory. Charles smiled when he saw Ed Johns' reaction to his appearance.

"Christ," Johns took the cigar from his mouth. "A kid." He offered Charles his hand; Charles took it. "Did he really write that code?" Johns turned to Clapper.

"Wait until you see what he's done with it," Clapper answered.

Without waiting to be asked, Charles strolled through the shop. He ignored the stares of the workers who watched him with open mouths.

"Better worry about child labor laws, Ed," one of the women called.

Charles winked, and the woman blushed.

"We're set up in R & D," Johns said, passing Charles and leading the way to the workbench in the back of the large room.

"My office is over there," Clapper said, pointing towards a desk a few feet from the workbench.

Charles noticed a hand lettered sign: "Vice President, Software." Then, the saw the open rear door that led to a loading dock stacked high with cardboard boxes.

"Shipping," Johns said.

Charles walked to the workbench and took charge. He opened his briefcase and took out a long roll of punched paper tape. Without asking for instructions, he fed the tape into the silent Taurus. In seconds, it began flashing.

The screen printed: "OK."

"I changed "READY" to "OK," Charles said. "Uses less memory."

Charles demonstrated the game of Tic Tac Toe that he had put on the tape. It worked flawlessly. Johns and a handful of PAMS employees watched. Charles completed one game. He won. "Want to try it?" He stepped back.

Bill Oldham was sitting at his desk when the telephone rang. With Charles in Albuquerque, the room belonged to him if one did not count the clutter on Charles' bed and the papers and silent computer on his desk. Bill hoped the caller was Jan, his current girl friend, who he had phoned earlier to invite him over "to study" together.

"Hello Honey," Bill said.

"Hi Bill," Charles' mother laughed. "I didn't suspect you cared."

"I'm sorry, Mrs. Swift." Bill was embarrassed at his impulsiveness.

"I'm flattered. Is my son there?"

"No. He's not. May I take a message."

"Can you tell me where he is?"

"Uhhh." Bill did not know what to reply. He assumed that if Charles wanted his mother to know that he was in New Mexico, he would have told her.

"Bill?"

"He's out of town," Bill mumbled.

"He's in Albuquerque, isn't he?"

"Yes ma'am."

"I told him not to go. He's in Harvard to study not play with computers."

Bill realized there was little he could say that would help Charles, so he said nothing.

"Excuse me, Bill, I know it is not your fault. That son of mine can be very irresponsible. Do you have a phone number?"

Bill hesitated. "I'm not sure where Charles is staying, but Clapper is at the Sundowner Motel."

"Then that's where Charles is. Thank you, Bill. Sometimes I wish..." Mrs. Swift caught herself then continued. "If Charles calls, tell him to phone me."

"Yes Mrs. Swift."

"Thank you again and goodbye," Mrs. Swift said and hung up before Bill could respond.

Bill contemplated phoning Charles and warning him but decided against it. Charles could deal with the formidable Mrs. Swift on his own.

Charles and Clap joined Ed Johns and his partner at the Hilton for a working dinner. Charles would have preferred a cheeseburger and fries then a session with Clap at the Sundowner to talk about their plans, but he acceded because Clap had been waiting for Charles to arrive and take over the negotiations with PAMS. Clap and Johns had not discussed the terms for the use of their software with the Taurus. The subject of price had not come up, and Clapper who had eagerly accepted Johns' offer of employment now found himself in the compromised position of selling a product to his employer.

Ed Johns and his partner Bill Yates were waiting in the Hilton bar when Charles and Clapper arrived.

"Join us?" Johns glanced at the two long necked bottles of beer standing on the table in front of him and Yates. As soon as he had spoken, Johns had wondered if he had made a mistake. He did not know how old Charles was. He looked to be about fifteen. The waitress would probably embarrass him by asking for proof of age before serving beer.

"I'll have a diet coke," Charles smiled as he solved the problem.

"I will too," Clap said. He would have preferred a beer but decided not to make an issue of it.

After the waitress had served the two diet cokes, the four men sat quietly. Charles and Clap waited for Johns to take the lead, while Johns and Yates were not quite sure what to say to the two young men they were supposedly negotiating with.

"Do you follow football?" Johns finally asked. Johns was a diehard Cowboy fan and assumed Charles and Clap had a favorite team.

"Not really," Charles smiled.

Johns, not a loquacious man, did not know what to say next. He was in no hurry to discuss contract with the boys.

"You boys got girl friends?" Johns finally broke the silence.

"Not at the moment," Clap answered. He had left an unhappy live-in behind when he moved to Albuquerque abruptly.

Charles did not respond. He did not consider it any of Johns' business.

"You boys gave any thought to what you want for your software?"

Clapper deferred to Charles.

Charles shoved his glasses back up his nose with a forefinger, then leaned forward. "That's why we're here, Mr. Johns. We'll give you a worldwide license to install our BASIC on each Taurus you ship. You must promise to make your best effort to promote and sell the software to your customers."

"We can't complain about that," Johns chuckled. "What's your bottom line?" Johns hoped that the kids would not try to set an outlandish price for their program. PAMS was getting many orders, now, and the software should generate more, but PAMS operated on the margin. They could not afford any unrealistic demands. He had set $5,000 as a fair price but had not discussed it with anyone.

"I'm glad you asked," Charles smiled. He rocked from side to side then continued. "How does $3,000 up front on signing sound?" Before a pleasantly surprised Johns could respond, Charles continued. "Additionally," he began.

Johns groaned. "There's always an additionally."

"Additionally," Charles smiled. "We will expect a royalty. We set the asking price low because we do not want to sell rights to our product. We will retain exclusivity to our BASIC but license you to package and ship it with the Taurus."

"And what is this royalty?"

"For every copy of the BASIC that ships, we will charge $50. That is for the 4K version. The price will rise incrementally with every version for 8K and higher."

"Hmm, that's interesting," Johns leaned back in his chair. "It gives me something to ponder. We were thinking along the lines of a flat fee."

"I'm sorry, Mr. Johns," Charles said. "We must retain exclusivity."

"I'll have to sleep on that," Johns said. "The exclusivity gives me a problem. I could go along with the royalty, but I would like the right to sublicense to other companies."

"With a time limit, and we get our royalty no matter who sublicenses it," Charles countered immediately.

"Ten years," Johns said.

"Five," Charles said.

Johns leaned back and wondered where this kid was coming from. PAMS could use that talent. "We're going to need you guys to work with us full time to help improve the product," Johns said.

Charles looked at Clap who nodded agreement. "One of us," Charles countered. "And only until the end of year."

"Then, we'll regroup in January. We might need help for a longer period."

"Only if the royalties amount to $2,500 a month."

Ed Johns finished his beer and waved for seconds for him and Yates. Charles and Clap shook their head no. Neither had touched their diet cokes. Johns extracted a cigar from his shirt pocket and made a production of lighting it.

"If royalties average more than $5,000 a month, I will need both of you."

Charles did not answer directly. Instead, he said: "We will want a security agreement included."

"A security agreement?"

"Yes. We have to protect our copyright. Software is just like books."

"And how would a security agreement work?"

"Anyone who buys the Taurus with our software has to sign an agreement promising that the software is for their personal use and that it will not be copied and given or sold to another user."

"Do you think people will buy that?" Johns asked.

"They will if they want our BASIC," Charles answered.

"I'm going to have to bring the lawyers in on all this," Johns said, hoping that reference to lawyers might scare the kid off.

"That's exactly what I think. We will need a firm, binding contract before you ship any computers with our software installed."

"We're already shipping," Johns protested.

This made Clap uncomfortable. As an employee of PAMS, he had assisted in getting the Tauruses ready for shipment. Now, he was caught in the position of having prematurely authorized shipment of his and Charles code.

"That's all right," Charles said. "Consider those authorized. We'll make the contract retroactive."

Yates, Johns and Clapper relaxed visibly at these words.

"What about filling orders between now and the signing of a contract?" Johns asked.

"You can ship all the computers you want. Please do not ship any with our BASIC until we have a contract."

Johns frowned. "Christ kid, you ought to be a businessman."

The meal passed with a minimum of small talk. Neither Yates nor Johns were gregarious men, and both Charles and Clap were too excited to talk about the weather. Johns used his credit card to pay the bill. "We'll write it off as a business expense," he laughed despite the shock of the three hundred-dollar check.

After they had parted from the PAMS owners, Charles and Clap retreated to Clap's Chrysler for the return to the Sundowner.

"Where did you get all of those ideas?" Clap asked.

"I read up on things." Charles dismissed the query.

"I would have just sold them the code."

"No way. The real money will come from the royalties for a long, long time. We don't sell any of our software outright to companies. We license it."

"Sounds good to me," Clap deferred. He and Charles had long ago established their pecking order. Charles expected to be the boss in any endeavor he joined, cards, business, games, whatever. "How much do you think we'll make this year?"

"Enough to replace this heap," Charles laughed as he patted the Chrysler's dirty dashboard.

When they reached the Sundowner, they found a torn sheet of paper attached with tape to their door: "Check with the desk."

"Have you paid the rent?" Charles asked.

"In advance," Clap answered.

"And you told them I would be staying for a few days?"

"Yes."

"We better see what they want."

Clap and Charles entered the dingy room that served as the motel's reception area. They found the motel's elderly owner sitting in his rocking chair watching an ancient television set.

"Hi boys," he greeted them without taking his eyes off the set. "You like Perry Mason?"

"Sure," Clap said. Charles shrugged. He seldom watched television.

They waited but the old man said nothing.

"You left a message saying we should check with you," Clapper said.

"Yeah." He pointed to the telephone that set on the counter where guests checked in. "Some woman's been calling every fifteen minutes. One of you named Charles?"

"Oh shit," Charles exclaimed.

"I reckon your mother wants you sonny. You from Seattle?"

"Thanks," Charles said and grabbed the stack of message before starting for the door.

"Sonny," the old man called, still staring at the set. Perry Mason was now in the process of solving the mystery.

"Best call now," he pointed toward the phone. "I would appreciate if I don't have to take another message."

"You have to call from here," Clap said.

"Collect be best," the old man said.

Charles reluctantly negotiated the collect call with the operator. He knew from experience that his mother would not give up until she reached him. The worst thing was he knew exactly what she would say.

"Hello," Mrs. Swift answered the phone on the first ring.

The operator interceded, asking if Mrs. Swift would accept the collect call. She did.

"Hi mom," Charles put as much cheer into his voice as he could muster.

"Charles!" Mrs. Swift spoke sternly. "What are you doing in New Mexico?" Before Charles could answer, she continued. "No. Don't answer that. I know what you are doing there. Why are you not in Cambridge? You promised you would go to class and not play with those damned computers."

Charles hesitated, waiting for her continue. When she did not, Charles responded: "Mom, I am going to class."

Clapper grimaced at the bald lie. Charles seldom went to class even when he was in Cambridge. Charles ignored him.

"Not in New Mexico, I hope. What are you doing there?"

Charles smiled. She seemed to accept his comment about classes. He was halfway there. "I'm just here for three days, helping Clap with a problem."

"Doing what?"

"A little software thing. Nothing important. Don't worry."

"All right, Charles, but your father and I want you to return immediately to Harvard and stay there as you promised."

"Yes mother." Charles waited for her to continue, but she seemed to have run out of steam. "How did you learn I was in New Mexico?" He knew but perversely wanted her to say it.

"From that nice roommate of yours. I wish you would follow his example."

"Yes mother."

"Do you have a girl friend?" His mother changed her tack.

"Certainly. Several of them."

"Good. But don't get serious. Your father and I want you to graduate first."

"Yes mother."

"Don't try to con me Charles. I'm serious. You can play with your computers during summer break. Until then, go to class and try to improve your grades."

"I'm doing fine."

"How many B's?"

"A couple."

"Then raise them. I don't want anything less than A's out of you. I know you can do it if you would only try."

"Yes mother. How's dad?"

"He's fine. Working hard as usual. But don't think you can use him to get around me. He wants you to return immediately to school."

"Yes mother. Give dad my best."

"Do what I say and call me as soon as you get back."

"Yes mother."

Mrs. Swift, believing she had attained her objective, hung up.

"Everything OK?" Clap asked.

"No problem," Charles smiled.

The next morning Charles and Clapper were approached by a straggly dressed young man just as they were preparing to enter the Chrysler and drive to PAMS.

"Hi guys," he called as he approached the driver's side of the battered Chrysler. "You wouldn't be going to PAMS would you?"

Clap looked at Charles and shrugged. He knew who the guy was. He had appeared at the PAMS plant yesterday. Ed Johns considered him a pest. He had long dirty hair tied in a pigtail with a rubber band and wore grubby jeans with bleach stains and a dirty T-shirt. Ed said he had ordered "one of everything" and had flown all the way to Albuquerque from Seattle when the overwhelmed PAMS had not been able to fill his order immediately. The harassed Johns had promised him his Taurus kit including one of everything for the next day, but the young man had irritated them all by checking out every phase of PAMS' operation.

"I'm staying here too," he indicated the Sundowner behind him. "A real dump, but cheap," he smiled. "I'm a PAMS customer and need a ride to the plant."

"Hop in," Clapper decided he could not be rude to a customer.

The young man climbed in back. "I'm Harry Miles," he said as he offered his hand over the seatback to Charles. Charles shook it.

"Charles Swift."

"And I'm James Clapper."

"You're chief of software," Miles said. "And you're the kid who wrote the BASIC."

Miles' aggressive approach irritated Charles, but, like Clap, he did not feel he could cold-shoulder a customer. If Johns signed their agreement, Mile's order meant fifty bucks.

"I sure would like to get a copy of that code," Miles said.

"Where you from?" Charles evaded the comment.

"Seattle. I'm a hobbyist. I belong to a computer club. We have fifty members. If I get my hands on your code, I would be the first one with a Taurus that can actually do something."

Charles said nothing.

"I guarantee that every one of our membership would buy a Taurus if they saw it work."

Charles silently calculated: "Fifty times fifty would mean twenty-five hundred dollars."

"We don't have an agreement with PAMS yet," Charles said. "It's currently under negotiation."

"That's OK," Miles was undeterred. "Just give me a paper tape with the code."

"We can't do that," Clap entered the conversation.

"PAMS is shipping Tauruses with BASIC," Miles countered. "I saw that yesterday."

"We negotiated an agreement last night. It's in the hands of the lawyers, but no computers will ship with BASIC included until it is signed." Charles said.

"That's different," Miles leaned back.

Charles suspected from Miles' attitude that he had not surrendered.

They arrived at PAMS and Charles and Clap retreated to Clap's corner desk while Miles headed straight for Ed Johns' small office.

"That guy is a pain in the ass," Clapper observed.

A short while later, Ed Johns joined Clap and Charles.

"Morning boys," he said. "I talked with my lawyer. He'll have the contract ready next week."

"Good," Charles said. "I hope you don't ship any Tauruses with BASIC until it's signed."

"We've got one problem," Johns frowned.

"Miles," Charles said.

"Yes. He's ordered one of everything and insists on the BASIC too."

"We can't do that, Ed," Charles said.

Johns looked at Clapper, obviously expecting help from his employee.

Clap looked in turn at Charles who did not react.

"Charles is the boss. He wrote the code," Clapper said.

"That puts me in a tough position," Johns said. "I promised Miles he could have the program. He might get us more business."

"When did you promise?" Charles asked.

"Five minutes ago."

The answer irritated Charles. Obviously, the pushy hobbyist had tried an end run after their turndown. Charles thought about the potential profit to be made from the Seattle hobbyists and decided to compromise. "OK, Mr. Johns," Charles said. "But give him a copy of the first BASIC."

"The one with all the bugs," Johns chuckled.

"Tell him that," Charles said. "Tell him it's full of bugs and that's why we were reluctant to give him a copy. Tell him in a couple of weeks we'll have tested the bugs out."

"I'll make sure he understands that," Johns said. "I'll say he's getting an experimental copy."

"OK, Mr. Johns," Charles said. "But we should not ship any more BASIC until we have an agreement."

Johns nodded assent but did not appear happy.

"That'll give me time to work on the bugs," Charles said.

Two days later Charles returned to Cambridge. The PAMS lawyer was still working on the contract. Ed Johns assured Charles that everything was fine, and Clapper promised to keep after it. Charles called his mother then returned to the computer center at Aiken Hall to work on his BASIC for the Taurus. Charles was determined to make the code as clean and bug free as possible because he intended to sell it to companies other than PAMS.

For two weeks, Charles' life ran smoothly, then disaster struck. Charles was sitting at a workstation connected to the labs' big PD-10 working on his BASIC when Doctor Jenkins, the Assistant Professor in charge of the laboratory, approached.

"Swift, I've been looking for you," Jenkins forced Charles to break his concentration. "What are you working on?"

Charles hesitated before answering. He normally worked on the PD-10 later in the evening when he was confident Jenkins was not around and the machine was not being used by others. Charles did not have university sanction to use its big computer. Charles knew that Jenkins would assume that if Charles was not in a class, not monitoring a class, he was working on a private project. Charles early in the semester had hustled one of Jenkins' graduate assistants to give him a password granting access ostensibly because Charles was familiarizing himself with the PD-10 in preparation for future computer courses he planned to take. This was not true of course because Charles considered himself advanced far beyond anything the university had to offer in its computer courses.

"Just familiarizing myself with the PD-10," Charles said. He decided to stick with his initial ploy. He certainly could not explain that he was using the PD-10 to simulate the Taurus so he could write code for it. Using Harvard's computer for a commercial purpose was certainly not a legitimate use.

"That's interesting," Jenkins smiled thinly. "Let me see your monitor."

Jenkins leaned over Charles' shoulder and studied the lines of code that filled the screen. "Looks like you're writing something."

"Just experimenting," Charles said. He knew he was in trouble. He had had similar troubles as far back as grade school. Each time his enthusiasm had led to being banned from the use of computers.

"Why is it, Swift," Jenkins stepped back," that your password has been used to log on far more than any other?"

"I practice a lot."

"If I am not mistaken," Jenkins continued, "you, an unauthorized user, spend more than ten hours a day on the PD-10."

Charles thought about cracking that he needed repetition to learn but thought better of it.

"Usually late at night. Don't you sleep?"

"I don't need much sleep, sir. I nap a lot."

"I find that strange. But I'm also intrigued. You seem to be using the United States Government's ARPLANET net to store programs at Carnegie-Mellon."

"I simply wanted to see how it worked," Charles tried. He had used the government's net to store his BASIC program in the Carnegie computer where it would not come to the attention of busybodies like Jenkins.

"Unauthorized use of ARPLANET it a serious matter. Now tell me the truth. What are you working on?"

Jenkins had Charles in a corner. He decided he would have to tell a partial truth because Jenkins could study his files and determine for himself that Charles had been using Harvard's computer for a commercial venture.

"I'm trying to write software that could be used with the Taurus," Charles admitted.

"Any success?"

"I've gotten it to work on the Taurus," Charles grudgingly admitted.

"Have you been in contact with PAMS?"

"Yes sir."

"Then you are using Harvard's computers in a commercial enterprise."

"Yes sir." Charles hoped his youthful appearance would work for him. "Is there something wrong with that?"

"You know there is, Swift. Don't try to play the innocent with me." Jenkins had encountered Charles before, and he did not like the youth's arrogant attitude.

Charles said nothing.

"Shut that down and don't come back to Aiken again without my specific authorization."

"But...sir..."

"Don't but sir me, Swift. I know you. I have no choice but to report this encroachment to the administrative board for investigation."

Charles logged off the computer and retreated to his room. There, he explained to Bill Oldham what had happened.

"What are you going to do?" Bill asked.

"Work here," Charles said. "I can use the Taurus and the teletype to log on to a public time share service."

"That'll be expensive."

"I know, but I'll write it off as a business expense."

"What about Jenkins' investigation?"

"I'll dance around that."

"You're not going back to Aiken Hall."

"No. That's out."

For the rest of the semester Charles, to Bill's dismay, worked on his BASIC in their room. Charles' nocturnal work habits proved particularly exasperating because Bill liked to sleep at night. Finally, they reached a compromise. When Bill went to bed, Charles had to shut down his teletype. Charles was called before the administrative board that had the authority to expel him from Harvard. This was a real worry. Charles wrote a long report justifying his use of the computer, relying heavily on the graduate student's approval. He tried to downplay his use of Harvard's computers for private commercial gain, but this was difficult. In the end, the board compromised. Charles had to agree to put his current version of BASIC in the public domain and to refrain from further use of Harvard's computers for commercial purposes. Charles recognized his tenure in Aiken Hall was at an end and was not bothered by the requirement that he make his current version of BASIC available to the public. He had already begun work on a much-improved version.

Chapter 7

"Dum," Stanley sipped from his frosted can of V8 juice and waited for his friend and partner to acknowledge his presence.

"Dum," Stanley repeated, signaling that he wanted to do more than talk to Dum's back.

"Yes, Stanley," Dum turned. Dum needed uninterrupted time to work on his new computer. Normally, Stanley was content to talk to Dum's back, wanting an audience not a conversation.

"We're on the threshold. I sense it."

Dum nodded noncommittally. He had no interest in hearing Stanley's newest version of the future. Stanley's dreams seldom worked out.

"Foster has ten stores now," Stanley said.

That caught Dum's attention. He had heard that Foster was extending his chain of Electronics Warehouse outlets deep into the valley, but ten stores seemed like a lot. "He's overreaching."

"Maybe, maybe not," Stanley countered, pleased that he now had Dum's attention. "He was delighted with our last fifty computers. They are selling at $666 each. We now have ten stores selling computers that earn us $300 per crack."

"Not bad," Harold mused. "We're solvent."

"Not yet," Stanley cautioned. "We don't have the money in hand. Foster refused to give us an advance."

"We're rich when he does."

"Hardly. We've got to spend money to make money."

Harold turned back to his bench. Stanley had almost trapped him again. It was another of Stanley's daydreams. He sometimes wished that Stanley had more than one friend. He was a burden. Harold's father kept saying that Stanley was getting a free ride, just taking advantage of Harold's work.

"How is Astron 2 coming?" Stanley now stood behind Harold and peered over his shoulder. "Can you get it to do color?"

"Certainly," Harold said. He knew that the hobbyists at the club did not think he could do it. They had laughed when he confided his plans, but Harold was confident.

"Then we'll go," Stanley said. Stanley had already made up his mind. Now he needed to persuade Harold to join him.

Harold thought about not responding to Stanley's bait, but he could not help himself. "Go where?"

"Atlantic City."

"Where?"

"Atlantic City, New Jersey."

"What for?"

"They're having a computer fair. We have to expand beyond Foster and his crummy chain. We need to advertise."

"Why New Jersey?"

"Because all the computer brass will be there. To see what's new."

"And to steal ideas," Harold said. Going to a computer fair appealed to him. He would be able to see how his work measured up with the pros. "Where are we going to get the money?"

"We can swing it with what we're getting from Foster."

"What about what we owe dad?"

"Tell him not to worry. We'll make enough contacts to double his money."

"Stanley, you're too much. You've never been to a computer fair in your life. How do you know we'll make any contacts."

"I've got faith in Astron 2," Stanley patted the circuit board that Harold was working on. "Your color will blow their minds."

Two weeks later, to Harold's complete surprise, they were in Atlantic City. They had brought three of their first circuit boards and their only Astron 2. Stanley fretted that the metal case that he had built for the Astron 2 was too plain. Harold worried that his electronics would not work.

They arrived the day before the convention and had taken the bus from the airport. Lacking reservations, Stanley had asked for directions to the cheapest lodging within walking distance to the Grand Hotel where the fair was being held. This led to their staying in a back street boarding house about two miles from the hotel. In order to keep their expenses to a minimum, something that Harold's father had insisted upon, they had not arrived until late afternoon on the day before the fair was to open. By the time they found their lodging, it had been too late to check in at the fair. Thus, at eight o'clock on the morning of the fair opening, Harold found himself trudging behind Stanley as he hurried to the hotel.

Stanley carried three circuit boards while Harold struggled with a large cardboard box that contained the computer, a monitor, and a keyboard. Their secret was that when Harold typed the startup password using the simple BASIC code he had written, the Astron 2 responded, after a time, with the words "ASTRON 2" and "READY" in bright red letters. The computer could not do anything else. That was all the code that Harold had had the time to devise, but Stanley was convinced it was enough.

After a block, Harold stopped and lowered his box carefully to the ground. "Stanley, You'll have to help me with this."

Stanley studied the big box, then made a decision. "Wait. I'll find us a cab."

Traffic was heavy on the street, but offered few cabs. Those that passed were filled, most going in the wrong direction. Harold waited. This was all Stanley's idea, so he was content to let him solve their problem. Either that or help carry the big box. Harold assumed that one way or another Stanley would find them a ride. Lifting was not among his priorities.

After a half an hour passed, Stanley began to alternate his worried gaze between the passing traffic and the moving hands on his wristwatch.

"It's getting late," Stanley said.

Harold responded by turning one hand palm up towards the box that now set on the curb near where Stanley waved futilely at passing traffic. Suddenly, an empty taxi appeared going in the opposite direction. Stanley shouted and waved" "Taxi, taxi."

The driver saw them in time. He pulled to the curb across the street and waited for a break in the traffic. When it came, he made a U-turn cutting in front of a large truck that honked its loud horn in protest.

"Where to boys?" The driver asked.

"The Grand," Stanley said.

The driver popped the lid on the trunk and waited while Harold deposited his box.

Harold and Stanley climbed into the back, but the driver waited. He turned. "That'll be five dollars boys."

"But it's only a mile," Stanley protested.

"Five dollars. Minimum fee for passengers with luggage."

"That's not luggage," Stanley said.

"Pay up or get out," the driver smiled.

"OK," Stanley surrendered.

Still the driver did not move.

"In advance, boys."

"I've never heard of such a thing," Stanley protested.

"There's a first time for everything. Now you have."

Stanley dug five dollars from his wallet and handed it to the driver.

He still waited. "No tip?"

"When we get there," Stanley smiled.

"No way. In advance, or get out."

Stanley took a dollar bill from his wallet and handed it to the driver. He looked at it as if it were beneath him to take such a paltry sum. Stanley started to draw it back but the driver grabbed it. "The Grand it is. You boys guests?"

Stanley did not answer. Harold did. "Exhibitors."

"Exhibitors? Pretty fancy. What are you exhibiting?"

"None of your business," Stanley snapped before Harold could answer.

The driver laughed. "Don't worry boys. I'll give you a receipt and you can write it off your taxes." The thought that the two young men, Harold with his scraggily beard and Stanley with his T-shirt, jeans and bare feet, were businessmen apparently amused the driver. He was still smiling when they pulled up to the Grand front entrance.

"Would you prefer the service entrance boys?" the driver asked. He winked at the doorman who was studying the passengers.

"Just wait until we get our box out of the trunk," Stanley ordered.

"Need help sir?" The doorman asked.

"We can handle it," Stanley said. He held the passenger door open while he waited for Dum to retrieve his box from the trunk.

"I'll bet you can," the driver laughed.

A man and woman exited the hotel. "Taxi?" The man called.

The doorman elbowed Stanley out of the way and held the door for the couple to enter the taxi. The man slipped a dollar bill into the doorman's hand. "See you boys," the driver called as he pulled away.

The doorman merely watched as Stanley and Harold entered the hotel. They followed the signs that pointed to the convention room where the "Computer Fair" was being held. Stanley checked his watch. They had five minutes before the exhibition was scheduled to open. They found a line of about twenty people waiting outside a closed door. Stanley with Harold following with his burden marched up to the closed door and opened it. Inside, he found a table facing the door. A fat, middle aged man in a shirt and tie sat behind the table. A cardboard sign with the words "Atlantic City Computer Fair" was taped to the front of the table. Behind the table the huge room was lined with gaily decorated booths. Most appeared professionally done.

"I'm sorry boys," the man greeted them. "But we're not open yet."

"We're exhibitors," Stanley said haughtily.

The man carefully appraised Stanley and Harold.

"Company name?" The man addressed Harold.

"Astron Computers," Stanley answered coldly.

"Are you registered?" He glanced at a clipboard on the table. "I'm not sure..."

"Right there on top," Stanley pointed at the first entry on the list."

"So it is. From Palo Alto. You boys came a long way."

"Where is our booth?" Stanley demanded.

"We have only one left," the man made no effort to hide his sarcastic tone.

"Well. Where is it?" Stanley demanded.

"Last space on the right," the man said. As he spoke, he placed a large check next to the name Astron Computers. "Better get set up boys. The customers are arriving."

The man rose from his chair, walked around the table, and opened the door to the corridor. The waiting viewers streamed around Stanley, Harold and his box.

"Let's go," Stanley said. He waited for Harold to pick up his burden then led the way to their booth. They passed over twenty exhibitors, some of whom represented well-known firms like H-P, Atari, Radio Shack and Commodore. All had their products nicely displayed. Most were manned by attractive young ladies with bright smiles and a ready manner.

"I think we made a mistake, Stanley," Harold called as they made their way along the line. Already, interested hobbyists and viewers gathered around the displays offered by the major companies.

Stanley did not respond. They found their booth, last on the right, only it was not a booth. Just an empty space and a folding table. A sign, "Astron Computers," was taped to the table.

"Let's get set up," Stanley said.

"Why bother?" Harold asked.

"Have faith," Stanley said. "Wait until they see your color."

While Harold unpacked Astron 2, Stanley arranged the Astron 1 circuit boards and commandeered a single straight chair that he found behind the IBM booth that was located near the entrance to the hall. He placed the chair on one side of the table and sat down. Stanley watched as Harold anxiously fiddled with his computer. After several adjustments, he had the lights on the computer's front panel flashing. That was hardly an achievement, but Stanley thought they looked nice.

"What shall I do with this?" Harold asked, turning from his computer to Stanley. Harold held the cardboard box in his hand.

"Put it under the table," Stanley said.

Harold did as he was told. "Now what?"

Stanley ignored him. He was too busy ogling the girl in the mini skirt who hawked Commodore's wares at the next booth.

"How come they all have fancy booths?" Harold asked, "while we get nothing but a table."

"Don't worry," Stanley said. "I'll take care of it."

"You and your big plans," Harold said. "We don't have a single customer. Who are we kidding?"

"Watch the shop while I take a reading," Stanley said. He smiled, trying to maintain a confident front. He was as worried as Harold was. He had never been to a tech fair before, and he had not expected it to turn out like this.

Stanley went from booth to booth. He estimated that there were some hundred people in the hall with others straggling in. They congregated near the entrance where the name companies had their booths. He tried to listen to the spiels in order to pick up some pointers but spent most of his time staring at the girls' legs. After two rounds of the hall, Stanley decided that he liked the Commodore girl best. Stanley glanced at Harold who sat alone behind his table glaring at Stanley. Stanley smiled, signaled with his hand palm down, indicating Harold should relax. Stanley pushed his way to the front row. Most of the spectators appeared to be engineers. Almost to a man, they had plastic folders in their shirt pockets containing an assortment of pens. Commodore's display featured business machines, adding machines, typewriters and the like. They had a couple of clunky looking computers the size of washing machines.

When the girl finished her pitch, she retreated to the side where she stood smiling brightly, offering pamphlets to any interested attendee, and watching while a company engineer answered questions.

"Hi Ann," Stanley smiled at the girl, picking up on the nametag she wore high on one breast.

"Hi," she smiled back.

"That was an excellent presentation," Stanley said. In truth, he had not listened to a word of it. "Are you an engineer?"

"Oh no," the girl laughed. "I'm public relations."

"You do this all the time?"

"Only at trade fairs, sir." She turned and smiled at the man in the dark suit standing next to Stanley. "Would you like some of our literature?"

"I would," Stanley responded first. He reached out and took the pamphlet from her hand. "I'll confess something," Stanley tried to recapture her attention.

"I'll take one Miss," the suit said. Stanley glared at him and noted the IBM badge clipped to his suit pocket.

"I'm an exhibitor too," Stanley blurted.

The man looked at Stanley's ponytail, T-shirt and sandals. "I'll bet you are," he dismissed Stanley and turned away.

Stanley stared at the man's back as he departed. "Asshole," Stanley mumbled. He turned back towards the Commodore girl and found she had retreated to the other side of the exhibit. Stanley thought about following her but decided against it. He could make another try later. Stanley joined Harold in his lonely vigil with their display.

"Struck out," Harold laughed.

"Don't worry," Stanley said. "I got the lay of the land. I'll drum us up some business so get ready."

Harold leaned back in his chair, placed an elbow on the table, and rested his fist against his check. "I'm ready. Don't get trampled."

Stanley ignored Harold's sarcastic tone. Stanley decided he liked Dum better when he was leaning over his workbench and only half listening.

Stanley purposefully walked past the Commodore booth and selected the small crowd in front of the Digital display. They were demonstrating a small black box. Stanley listened as the girl brightly discussed RAM, slots, and storage. "Do you have color?" Stanley interrupted.

"Color? Sir?" The girl appeared confused.

Stanley pointed at the monitor that was displaying a black and white document.

"No, Sir, but Digital is working on it."

"You better check out the Astron Computer display," Stanley said, turning away from the presentation.

The girl shrugged then resumed her pitch.

Stanley worked his way from one presentation to another, each time asking the same question and getting a similar response. When he reached IBM with its freezer-size computer, Stanley glanced back the length of the hall to see if he had generated any interest for Harold. To his surprise, he saw a group of some ten men standing near Harold's table. Stanley hurried back. Harold's monitor displayed "Astron Computer" in red letters with bright blue stars in the background. Harold's turgid description of the Astron's technology that produced the color seemed to be striking a responsive chord in the viewers. Stanley decided it would be best if he let Dum explain although Stanley knew he could do it better. The viewers would probably recognize him as the shill who had sent them in this direction.

For the next two days, Stanley and Harold used the same technique. The presenters at the fancy booths learned to ignore Stanley's question, but he persevered changing his approach to side comments to other attendees. Stanley would softly say: "They have color at Astron Computer," and each time some of the hobbyists would excitedly approach Harold and his display.

By the end of the tech fair, both Harold and Stanley had wearied of the process, particularly Harold who kept protesting that it had been all a waste of time and money. They had provoked a modest amount of interest, but they had not solicited a single order. On the long flight home, Harold's complaints provoked a disagreement.

"Dad's right about you," Harold started the argument.

"In what way?"

"He says you're taking advantage of me. I do all the work and you take half the company. He says I should own it all."

"That's not fair," Stanley protested. "If you think selling your computers is easy, you try it."

"This trip was all your idea. A big bust."

"Wait and see," Stanley said confidently.

"We had as much success selling computers to that crowd as you did with that girl at the Commodore display."

"She doesn't know what she missed," Stanley said.

"I'll bet she does," Harold laughed.

"If you and your father think you can run the company, it's all yours," Stanley bluffed. He hated the way the girl had ignored him, and he wanted to change the subject.

"I've been thinking about that," Harold said after a pause. "The Astron 2 is all mine. Maybe I'll sell it myself."

Harold's comment jarred Stanley. He had always been able to control his good-hearted friend. The Astron 2's color had been their one success at the fair. Stanley knew they were on to something good. "How can you say that? I thought you were my friend," Stanley blurted.

Harold did not know what to say. Stanley was a big pain in the ass sometimes, maybe most of the time, but he was Harold's only friend. Harold decided to let the matter drop. He had no interest in business, and he was realistic enough to know his father did not have the time or inclination to help— he simply did not like Stanley.

One week after their return from Atlantic City, Stanley was sitting in the garage wondering if he should give the whole idea up when his mother opened the kitchen door and called: "Stanley, somebody called Commodore is on the phone."

Stanley jumped up. He snapped his fingers triumphantly. He knew the Atlantic City trip would pay off. He raced for the kitchen where he grabbed the telephone from his mother and sat at the kitchen table. "Pencil, paper," he whispered to his mother.

"Astron Computers, Stanley Pitts speaking," Stanley tried to sound as formal as he could.

"Mr. Pitts," a deep male voice greeted him. "My name is Myers. I am in charge of new product acquisition for Commodore Business Machines. I assume you've heard of us."

"Yes, Mr. Myers. Astron Computers had a booth next to yours in Atlantic City."

Myers hesitated. Stanley wondered if he had made a mistake referring to their single table as a booth. Somebody in the background was speaking to Myers, but Stanley could not hear well enough to know what was being said. "Yes." Myers continued. "We're interested in talking with you about your Astron 2."

Stanley took the pencil and pad from his mother and wrote: "Commodore. Myers."

"May I inquire just what is your interest?" Stanley said stiffly. At the same time, he began to tap the pencil on the side of the table.

"Certainly. If you have a free moment sometime next week, I would like to have one of my representatives visit you to discuss that very subject."

Stanley grimaced. He could imagine what a big deal Commodore businessman would think of their factory, his dad's garage. "If you wish," Stanley countered, "I could come to your office. I have some time on Monday."

"Oh that won't be necessary. I believe you already have met Ms Page."

"Ms. Page?" Stanley wondered if Ms. Page was the Ann with great legs who had been in the Commodore booth at Atlantic City."

"Yes. Ms. Page. You may recall her from the fair."

"Yeah, great," Stanley enthused. He figured he could hustle Ms. Page.

"Would Friday a week suit you?"

"Let me check my calendar," Stanley rattled the notepad. "Friday morning is clear for me."

"Could you give me your address?" Myers said.

"I'll pick Ms. Page up at her hotel," Stanley countered.

Myers apparently covered the receiver as he conferred with Ms. Page. "That will be fine," Myers said. "Ms. Page will be at the Palo Alto Holiday Inn."

"It's a date," Stanley said. "I'll pick her up at noon." Stanley figured he could take Ms. Page to lunch, give her a song and dance about Astron Computers, and then see what might develop. Stanley began to fantasize about a visit to Ms. Page's hotel room.

The Commodore appointment stirred Stanley into action. Without conferring with Harold, he decided to use the expression of interest by Commodore as a lever for selling the company. He visited his old employer Atari, and they turned him down cold. He visited Hewlett-Packard, taking care not to let Harold see him there, and asked Harold's employers if they were interested without mentioning Harold. They saw no future in microcomputers. Unable to sell the company, Stanley decided their only alternative was to expand. For that, they needed additional capital, more and better workspace and advertising.

Assuming that finding rental space would be easier than locating capital, Stanley borrowed his father's car and began his own search. Within an hour, he located a concrete block building in a rundown section of town about a mile from Stanley's home. A car repair shop occupied the first floor, and it had a sign in the front window announcing: "Space for rent."

Stanley did not like the appearance of the building. It was obviously in a rundown condition, and the garage had a number of vehicles parked around it that looked like they were being scavenged for parts. Assuming the rent would be modest, Stanley decided to have a look. Stanley parked at the curb and approached the office where he found a grossly overweight woman in a baggy one size fits all dress.

"Yeah son, what can I do for you?" She greeted Stanley. Her desk was piled high with paper, most of which was marked with grease stained fingerprints. Someone in the shop behind her was banging loudly on metal.

"You have space for rent," Stanley pointed at the sign.

"Upstairs."

"Can I see it?"

"We don't rent to no hippies," the woman chuckled.

"Don't let my hair mislead you," Stanley said. "I'm President of the Astron Computer Company."

"And I'm the Virgin Mary," The fat lady laughed, setting her many chins in motion.

"We need to expand," Stanley persisted.

"I don't," the fat lady patted her ample belly. "I've already done enough of that."

"May I see the space?" Stanley persisted.

"We charge four hundred a month, one month in advance," the woman became serious.

Stanley waited for her to continue.

"The stairs are around back. The door's unlocked."

"Can I go through there?" Stanley pointed to the open door behind her.

"Best go around the outside. The shop's dangerous. We don't allow customers back there."

"I'm not a customer."

"Go around the outside, son."

"Don't worry about the mess. We'll clean it up," the woman called after him.

Stanley circled the building and found a back yard filled with the skeletons of old cars. Rusty iron steps led to the second floor. Despite their appearance, they seemed secure. Stanley pushed open the door and entered. He was surprised by what he found. The entire upper floor was one room, no partitions. Boxes filled with old automobile parts were stacked along one wall. Another had a long, greasy workbench with racks that once held parts and tools. Old newspapers and other debris littered the floor. Stanley decided that if the owner cleaned the place up, he and Dum with a little help could make the place look like a decent shop. They could move everything from the garage along with the parts and stuff from Stanley's bedroom and his sister's old room. They could move Harold's tools and get cardboard boxes to fill space. Harold could get some his engineer pals from HP to act like employees. With a little luck, they could persuade Commodore that they were a small but thriving business.

Stanley returned to the office.

"What do you think son? Good enough for your computer company?"

"I'll give you two fifty a month, take it or leave it," Stanley said.

"Three fifty," the woman countered. "Cash in advance with one month rent as security deposit."

Stanley did not have seven hundred dollars. Stanley had assumed he could finesse the rent problem for a month. He only needed a place to show Ms. Hot Pants from Commodore. "I'm going to have to think about it. I'm not sure its what we need anyway.

"Take your time, son. You know where to find me."

Ten days later Stanley woke in a confident mood. Since he had failed in his effort to acquire additional working capital, thus could not expand, he had convinced himself the thing to do was to sell Astron Computers to Commodore. He acknowledged this would require some fancy footwork on his part because all they really had was the workshop in his father's garage, a few circuit boards and Harold's new computer, a work in progress. Harold refused to even estimate when it would be done and beyond that Harold had begun to act as if the new computer was exclusively his property not the company's. This did not worry Stanley. He knew that Harold's father was behind it all. Stanley was confident that he could persuade Harold to go along with any plan he devised. All he had to do was sell the company to Commodore without getting too specific about its assets and then persuade Harold to agree to turn over the Astron 2 to them.

Stanley decided to dress up for his luncheon with Ms. Page. He gathered his hair behind his head and fastened it into a ponytail with a new rubber band. He put on his only white shirt and buttoned it, leaving the collar open. He opted for a pair of clean jeans, worn but not ripped, and traded his sandals for his scuffed but serviceable Nikes. He considered socks, but decided against them.

Stanley was a fashionable forty-five minutes late when he debarked from the city bus in front of the Holiday Inn. He spotted Ms. Hot Pants as soon as he cleared the front door. She looked as good as Stanley remembered, only this time she was dressed a little more formally. She wore a medium length skirt that did a good job of concealing her shapely, muscular legs, a soft white blouse that concealed everything but hinted at more, and a jacket. Stanley assumed the jacket and blouse were her version of a man's suit. Stanley pretended not to see her and strolled towards the reception desk.

"Mr. Pitts," Ms. Hot Pants called before Stanley could speak to the receptionist.

Stanley turned and feigned surprise at finding the Commodore representative waiting. "Ms. Page. Welcome to our fair city."

"You're forty-five minutes late," his guest frowned.

"I apologize. Traffic was just terrible," Stanley smiled. He was confident of his ability to handle women.

"Did you have trouble finding a parking place?"

"Oh no. It was such a nice day," Stanley improvised. "I decided to walk. I guess I got distracted enjoying our scenery. When I realized time was fleeting, I hopped on the bus. That's where I made my mistake. We got caught behind an accident several blocks down."

"I understand." Ms. Page's tone and stern expression signaled that she did not believe Stanley.

"I believe we're scheduled for lunch," Stanley decided to change the subject. "Do you have a preference?"

"I'll leave that up to you," she said. "Someplace quiet where we can talk business would be best."

"I know the very place," Stanley said. He had already decided to take his guest to his favorite restaurant, a small semi-vegetarian hole in the wall that had three booths and a small counter with four stools. Stanley had frequented the place often with friends during his Hare Krishna days. "Shall we take a taxi or would you like to walk?"

"You decide."

"We walk. It'll give you a chance to see some of our fair city."

"I'm quite familiar with the city," Ms. Page said.

Stanley reached to take her arm to guide her out the door. Ms. Page pulled away. "You lead. I'll follow," she said.

Stanley shrugged. "See you," Stanley waved to the receptionist who had been closely listening to the exchange.

"A few blocks to our right," Stanley said as he led his visitor through the motel parking lot to the sidewalk. Although it was the noon hour, the streets were virtually empty.

"Your traffic jam seems to have dissipated," Ms. Page said.

Stanley smiled. Despite her lawyer suit and sour disposition, Ms. Page was an attractive woman. He hoped that at least one of his old high school classmates would see them together to bear witness to Stanley's good fortune.

"Can you tell me something about Astron Computers while we walk?" Ms. Page asked.

Stanley picked up the pace, forcing Ms. Page to hurry along on her high heels. Stanley was glad he had worn his Nikes. "Why don't we save the business talk until we get to the restaurant. That way we can both concentrate."

"I can walk and talk at the same time," Ms. Page smiled.

Stanley ignored her comment. He decided that showing up fashionably late had not been a great idea.

"I have a five o'clock plane to catch," Ms. Page said.

"No problem," Stanley smiled. He could describe Astron Computers in three sentences, but he had prepared a lengthy harangue about the potential of the Astron 2 and his plans for the future. He could have Ms. Page back to the motel by three.

Stanley chatted aimlessly about the city, and Ms. Page chugged silently beside him for five blocks. When Stanley stopped, waiting to cross yet another street, Ms. Page grabbed his arm. "What's going on, Mr. Pitts," she challenged. "Does this restaurant really exist or are we on our way to San Francisco?"

"We're almost there," Stanley said, knowing they had at least five more blocks to cover. He figured if he could tire her out she would be more susceptible to his pitch.

They silently covered the next five blocks. Just as they approached the restaurant, Ms. Page stopped again. Stanley raised his hands in surrender. "We're there."

Ms. Page turned and studied the dirty glass storefront. She glanced at the single sign taped to the window: "Semi-Vegetarian."

"Holy shit," she exclaimed. "You've got to be kidding me."

"It's great. Take my word for it," Stanley said seriously. "You like vegetarian?"

"I like vegetables on my cheeseburgers."

"They've got a veggie burger that's out of this world."

"That I believe," Ms. Page laughed for the first time since meeting Stanley.

Stanley held the door and bowed, gesturing for his guest to enter.

The booths and counter were empty. A fat man with a dirty apron knotted around his waist smiled at them from his position behind the counter. "Hi Stanley. Long time no see."

"Hi Walter."

"Lady," the fat man looked at Ms. Page. "Do you know what you're doing keeping company with this guy?"

"No, I don't," Ms. Page answered honestly.

"I see you're no longer in the Khrishnas," he turned back to Stanley.

"A passing fancy of my youth," Stanley said. He had not counted on Walter still working at the restaurant.

"Do you have money Stanley? You stiffed me last time," Walter said.

Stanley had forgotten that he had come up a dollar short on his last visit to the restaurant over two years previously. Stanley ignored the comment and led his guest to the back booth.

"Could we see the menu?" Stanley asked.

"Menus? In this dump? Give me a break," Walter laughed.

"What do you recommend?" Stanley asked coldly.

"Why don't you try a veggie burger," Walter said.

"What would you like to have?" Stanley asked Ms. Page.

"I'll try the Semi." Ann tried to make a joke of the sign "Semi-Vegetarian." When neither Stanley nor the fat man reacted, she continued. "Could I have a Nicoise salad?"

"One salad. Niswoze," Stanley mangled the pronunciation.

"What's that?" Walter asked.

Stanley looked at his guest.

"Lettuce and tomato with tuna fish, black olives, anchovies, oil and vinegar and..."

"One tuna salad coming up," Walter interrupted. "What do you want Stanley?"

"Give me a veggie burger and a small side salad."

"What kind of dressing?"

"Blue cheese?"

"French," Walter smiled.

"OK French," Stanley conceded.

After Walter disappeared behind a beaded curtain in the back of the restaurant, Ms. Page spoke first: "All right Mr. Pitts, cut the bullshit and tell me about Astron Computers."

"Call me Stanley."

"OK. Stanley, talk."

"I've been thinking price," Stanley tried to look serious. He took a ballpoint pen from his shirt pocket and wrote a figure on a paper napkin. After writing, he made a show of putting the pen back in his shirt pocket and pushed the napkin towards Ms. Page. "Take a look at this, Ann. I can call you Ann can't I, and tell me what you think."

Ms. Page stared at Stanley, ignoring the paper.

"I know your name is Ann. I heard them call you that in Atlantic City."

Ms. Page did not say a word.

"Tell me something, honestly, I've been wondering ever since we met. Are you related to the grocery store."

Ms. Page chuckled. "Don't try that old line on me. I've heard it all my life."

"Ann Page. A & P. I think that's neat. What's it like to be rich?"

"I'm not related to the grocery store, and I'm a working girl. I'm not rich."

She picked up the napkin and read. Stanley had written "$100,000." This time she laughed sincerely. "What are you talking about Stanley?"

"That's the bottom line, Ann," Stanley said.

"You've got more gall than Carter has pills," Ann said.

"That's a cliche if I've ever heard one."

"You don't even know what a cliche is," Ann countered. She shoved the napkin back towards Stanley. "Now tell me about Astron Computers. Who built that microcomputer with the color that you had in Atlantic City?"

"Ahh, now we are getting to the heart of the matter," Stanley said.

Ann waited for an answer.

"Harold and I did," Stanley lied.

Ann did not believe him, but she let it past. "I want to see your factory as soon as we can get out of here," Ann said.

"Certainly," Stanley agreed, wondering if he could get by with showing her the large room over the garage that he had visited. He decided against it. If the debris did not give him away, the fat lady would.

The fat man emerged from the kitchen. He placed a large wood bowl in front of Ann. "Tuna Salad," he declared.

Ann saw a tomato and lettuce salad with a slice of onion, a round blob of tuna fish, all covered with what she assumed to be oil and vinegar. The tuna looked like the man had opened a small pop top tin of tuna and dumped it on the lettuce.

Walter set a plate holding what resembled meat in a hamburger roll and a small side salad striped with French dressing in front of Stanley. "Don't try to get out without paying Stanley. You still owe me a buck from last time." Walter said. "I'll be watching you." He smiled at Stanley's guest. "Miss, keep an eye on your wallet around this guy."

Stanley ignored him. He picked up his veggie burger and took a large bite. "That guy's a real character."

Ann did not respond. She studied her "tuna salad" with obvious apprehension.

"Don't worry, it's great," Stanley said before continuing with his pitch.. "And we'll expect two jobs to go with the company. One for my partner Dum, and one for me, thirty-six thou per year each."

Ann, who had taken a tentative bite of her salad, almost choked. "Stanley, I don't know what you're talking about. We're not buying anything."

"I thought that was the purpose of this meeting. Commodore wants to buy Astron Computers."

Ann took another forkful of salad, this time with a piece of tuna included. It was not bad. She hoped the lettuce had been thoroughly washed. "We just want to get a feel for your company at this juncture."

Stanley smiled. He knew what he would like to get a feel of but did not say it.

After Ann had paid the bill—Stanley did not even make a pretense of fumbling for the check—they left the restaurant.

As soon as they were on the curb, Ann stopped and turned to Stanley: "OK, Stanley, let's see your plant."

Stanley hesitated. "We can't do it today. My car is in the shop."

Ann laughed and turned towards the hotel. "That's OK. I have a rental car. If you had been honest with me, we wouldn't have had to make this twenty-one mile hike."

"It's not twenty-one miles," Stanley protested.

"Were you ever a Boy Scout?" Ann asked.

"What for?" Stanley said.

"How many employees does Astron Computers have?"

"Enough."

"Stanley, if you want to sell this company of yours, you have to answer my questions."

"Six," Stanley said grudgingly.

"All full time?"

"Almost."

"Stanley, I'm out of here." Ann picked up the pace, forcing Stanley to walk two steps behind her.

"Four."

"Four what?"

"Four full time. We have two part time moonlighting engineers from HP."

"Name them."

"Who?"

"The full time."

"Me, Stanley Pitts, my partner Harold Dumbroski, an electronics engineer and two others."

"Name them."

"My mother and my sister."

"They both work full time?"

"Almost."

"Now we're being honest. Where is your shop?"

"In my father's garage."

"Now that wasn't hard, was it?" Ann decided she had learned enough. There was no way she was going to recommend that Commodore pay $100,000 for a home garage based company with four employees."

That evening a dejected Stanley joined Harold in the garage.

"Want a diet coke?" Stanley asked.

Harold paused long enough to shake his head then returned his attention to the Astron 2. Stanley thought about sharing his troubles with Dum then decided against it. He had not told Dum about his meeting with Ms. Hot Pants from Commodore because they had still not resolved their disagreement over the Astron 2. Dum's father continued to insist that it was Dum's invention and belonged exclusively to him.

"Dum, there's one thing we learned at Atlantic City," Stanley said.

"Yeah. Don't go again."

"No, I'm serious."

Stanley waited for Harold to give him his full attention, but Harold continued to diddle with his computer. Finally, Stanley attacked. "We need someone to advertise us."

Dum did not respond, so Stanley continued. "Every successful firm has a top notch outfit to sell it."

"No shit, Stanley," Dum turned and looked at Stanley. He realized that Stanley was on another kick and would pester him until he went along with it. "Who are we going to hire to advertise this," Harold looked around the cluttered garage.

"Thomas Edison started small," Stanley said.

"And he ended up small. He's dead. He can't help us."

"We need help, Dum. I can't do everything myself."

"I know that," Dum turned back to his workbench. Harold felt like saying that he was the one who did all the work while Stanley hustled about insulting people.

This time Stanley waited.

"And besides, what will you use to pay this great advertising firm?"

Stanley smiled. Dum had asked the question that Stanley wanted. He sipped his diet coke and waited, now playing Harold's silence game.

"Well?" Harold turned.

"We don't have to pay them."

Harold snorted.

"No really. Advertising firms don't take retainers. The good ones are not crooks like lawyers."

"They're not interested in making money?"

"Certainly they are, stupid. They'll take a percentage if the firm has potential."

"So we give them the company, and they make all the money."

"I'm only talking a couple percent." Stanley was not sure what he was talking about, but he was determined to sound like he did.

"I'm not stupid," Harold said, turning back to the workbench. "Let me concentrate on this."

"Then you agree. We need an advertising firm."

"You handle the business side."

Stanley finished the coke. "I'm going down to the library and do some research."

Stanley who preferred to intuitively create the big picture to donkeywork like research found exactly the firm he wanted in less than an hour. In fact Stanley did not find the name on his own, he had simply asked the lady with the silver hair at the library research desk if she could tell him where to find the names of the best advertising agencies in the valley. She referred him to the business section of the San Francisco Examiner, particularly the column that profiled startup companies in the valley. The library kept the Examiner on microfilm. Stanley struggled with the viewer until another patron loaded the roll of microfilm for him and showed him how to spin the wheel to turn the pages. After about fifteen minutes of scanning, Stanley began to lose his patience. It was exactly at that point that he found the article about the Rathborn Agency. The article praised the new company that specialized in representing valley technical firms. When the article was written, the Agency had just signed its first big client, Hewlett-Packard. Stanley immediately decided that he did not need to look any further. If the Rathborn Agency was good enough for H-P, it was good enough for Stanley.

The next morning at exactly ten o'clock, Stanley telephoned the Rathborn Agency.

"Rathborn Agency," a pleasant voice responded.

"May I speak to Mr. Rathborn please." Stanley had decided to start at the top.

"I'm sorry sir. There's no Mr. Rathborn. Can someone else help you?"

Stanley paused. He decided to brazen it out. "There must be a Mr. Rathborn. This is the Rathborn Agency, isn't it."

"No sir, yes sir," the pleasant voice responded.

"What?"

"No sir, there is no Mr. Rathborn. Yes sir, we are the Rathborn Agency."

"If there's no Mr. Rathborn, why do you call yourselves the Rathborn Agency?"

"You will have to ask one of our executives about that," the woman responded.

"I will," Stanley said. "Please connect me."

"To whom?"

"The top executive."

"That would be Mr. O'Hara."

"Then let me speak to O'Hara."

"I'm sorry sir, but Mr. O'Hara is in a staff meeting."

"Then let me speak to someone else."

"Who sir?"

"Another executive."

"I'm sorry sir, they are all in Mr. O'Hara's office. May I take your name and number and have one of them call you back."

"I am the President and Chief Executive Officer of Astron Computers," an exasperated Stanley declared.

"How do you spell that sir?"

"P R E S..."

"No sir. I mean the name of your company."

"A S T R O N."

"And your number please."

Stanley reluctantly left his mother's home telephone. "Have O'Hara call me as soon as he is free if he's interested in my business," Stanley blustered.

The Rathborn Agency never called Stanley back.

Chapter 8

Bill Oldham sat alone on the old bench swing on his parents' front porch on Horner Street in Clarksburg, West Virginia. He had been home only a week, and he was bored beyond belief. After the first enthusiastic greetings on his return for the summer, Bill had immediately fallen into the same routine he had followed since high school. This time, the situation was even worse. The problem was not the days in his father's shoe store. He accepted that. He needed the money to supplement his scholarship. Harvard was expensive. He had known that when he chose the rich boys school. All of his friends had gone to Morgantown to the University, and Bill had privately enjoyed their unconcealed envy when he had selected Harvard. If he had it to do over, he would choose Morgantown. Harvard was a grind, and Bill had few friends with the exception of Charles and Clapper. The rich boys tended to look down on the scholarship students. Bill now knew what it was like to be considered lower class.

Bill was having trouble finding things to do to occupy his free time. He had called his top three best girl friends from high school; all three had turned him down. One was married, one engaged, and the other pinned to a fraternity boy from WVU. Not normally discouraged, Bill had phoned even deeper into his list. Number four had moved to Atlanta, Number 5 had a boy friend, and number six pretended not to remember him. Bill had gone out once with his old high school group, but he had felt like an outsider. Drinking 3.5 beer at a noisy roadhouse was no longer his thing. Thus, he sat alone on the front porch doing exactly what he used to do in grade school, pondering what to do.

The telephone inside rang, but Bill ignored it. He no longer had a line of Clarksburg friends anxious to share his company.

"Oh, hi, Charles," Bill heard his mother say.

Bill wondered what his roomie wanted. They had parted two weeks earlier. Charles had finished his exams sooner than Bill and had flown immediately to Seattle to confront his mother. Bill had had to delay three days longer to finish his own finals then had stayed to see the final grades posted. He had gotten four A's and two B's, just enough to keep his scholarship. Charles had aced his exams, straight A's. Charles had planned to stay a week in Seattle, then drive his red Mustang to Albuquerque. Liberating the Mustang had been Charles' main objective in returning to Seattle. His mother had refused to let him drive it across country to Harvard, and Charles had truly missed his wheels. Bill would have liked to have had the luxury of a roommate with transportation, but he understood Charles' mother's concerns. It was not the long trip that bothered her. Charles was a good driver, but it was the speed that concerned her. Charles had amused Bill with his tales of his constant battle of wits with Seattle law enforcement officers determined to stop speeders, especially wild teenagers like Charles.

Bill heard his mother ask about Charles' family. Charles had once spent a Thanksgiving break with Bill in Clarksburg, and Bill's mother had been charmed by him. Bill waited patiently for his mother to finish her chat. After a few minutes Bill began to fidget. He entered the house and held out his hand. "Mom, it's long distance," he protested.

"He's here," his mother finally admitted. "Give my regards to your parents."

She smiled and handed Bill the phone.

"Charles. What's up?" Bill asked.

"I made it. I'm in Albuquerque, and I've got wheels."

"And your mother agreed?"

"She's eating out of my hand."

"What are you going to do?" Bill was not sure whether he and Charles would share a room during the coming semester or not. When they had parted, Charles hoped to escape Harvard and stay with his business in New Mexico. Bill, doubted that Charles would be able to swing his parents, particularly his mother, to support that plan.

"Oh, I had to compromise. Mom agreed to my doing the summer in Albuquerque with the Mustang, but I had to promise to come back for the fall semester. They really think I should graduate before getting involved in business."

"So do I," Bill said.

"I knew you would say that, but you have to seize the opportunity when it presents itself. I have a chance to get in at the start of this computer business."

"It'll wait for you."

"No it won't. The future is now. You've got to come out here and see this," Charles said. "It's exciting."

"I know you think so." Bill had heard Clapper and Charles discuss the coming computer revolution at length. Bill did not understand computers, was not very good at math, and was more interested in writing than technology. He dreamed of becoming another Thomas Wolfe.

"How are you set up for the summer?" Charles asked.

Bill groaned. When Charles put his mind to something, he did not give up easily. He was accustomed to getting his own way. If Charles' mother could not control him, Bill doubted that anyone could.

"I'm selling shoes," Bill said.

"Great work if you can get it," Charles laughed.

Bill envied Charles. He was doing what he wanted to do and having fun at it while Bill was tied to the shoe store. He could think of nothing more boring than running shoes, that is moving the boxes to fill the holes left by the sold shoes. And he hated trying to persuade young girls to buy the ballerinas and flats that style dictated, particularly when he knew that the shoes women bought gave them no support and virtually guaranteed that all would have bad feet when they grew older.

"I hate it Charles," Bill confessed. "I'm bored stiff."

Bill's mother who was in the kitchen groaned. "Poor baby," she called loud enough for Charles to hear. Bill wished he had some privacy.

"I am about to solve your problem," Charles said. "Come out here. My company will pay for your ticket."

"Fat chance," Bill said. "I need the money. I can't play around this summer."

"You tell him," Bill's mother agreed.

"How much are you making a week?"

Bill laughed. "My boss is a skinflint. Forty a week plus board and room."

"How does this sound?" Charles asked. "Airfare, free lodging, you can buy your own food, and $2,000 guaranteed."

"Two thousand. A week?" Bill joked.

"Nope. For the summer. I'm serious. I guarantee that you will return to Harvard with $2,000 in your pocket. You might have to ride cross country with me in the Mustang."

It sounded too good to be true. "Charles. I can't joke about something like this. I really need the money."

"Don't worry. I guarantee it."

"What about your mother?"

"She has nothing to do with this money. Your deal is with "Digital Software.""

"What's that?"

"Our company. Me and Clapper."

"I don't know anything about computers," Bill began to consider the idea. He knew his father could get along without him at the store. His job was just make work, helping out, as he had since grade school.

"You don't have to. I'm writing the code. Clap is helping while he works full time for PAMS, and we have four employees now, three college kids who write code in their spare time and an assistant who handles all the booking and errands."

"And me?" Bill did not want charity.

"We need manuals. Every Taurus that's shipped with BASIC needs a manual. And we need a Security Agreement. I don't have time to write them myself."

"And besides you wouldn't write anything that a normal human would understand," Bill said.

"Well," Charles seemed offended. "These are mostly for hobbyists anyway. But we need help. Will you do it?"

"What else will I have to do?"

"Anything we need." Charles hesitated. "You can handle public relations."

"What does that mean?"

"PAMS has a bus that travels around demonstrating the Taurus. When your writing is caught up, you can ride along and demonstrate our Digital BASIC."

"I don't know how."

"We'll show you. Don't worry. What about it?"

"Let me talk to my parents," Bill said. His mother was now standing in the kitchen doorway.

"That means yes," Charles laughed. "Great. I'll see you Monday."

Albuquerque was a hot one hundred degrees when Bill stepped out of the cool comfort of the DC 8 and crossed the burning pavement. His parents had driven him from Clarksburg to Pittsburgh where he had caught his flight, and the weather had been hot, at least it had seemed so for summer in the mountains, but it was nothing like this. Bill was sweating when he reached the air-conditioned sanctuary of Albuquerque's unimpressive airport, and he was carrying nothing but a book bag filled with his summer's reading.

"Ho," Charles greeted him at the debarking portal. "Mr. Oldham, welcome to Albuquerque, home of the world's leading software maker."

"And who is that?" Bill laughed.

"Me," Charles grinned.

Bill studied his friend. He did not look like a business mogul. He wore a white shirt, open at the collar, sleeves rolled above the elbows, jeans, and Nikes without socks. With his baby face, horn-rimmed glasses and long Beetles style hair, Charles looked to be sixteen at the most.

"Charles, you're still a baby," Bill laughed.

"What do you expect in two weeks." Charles frowned. He did not like references to his youthful appearance. He knew what he looked like and so did the girls who refused to take him seriously. As soon as he saw their reaction to his appearance, Charles tried to overwhelm them with his intellect, and this always rebounded against him. They then tagged him arrogant, and nobody likes an arrogant kid.

As soon as Bill's bag appeared, Charles grabbed Bill's book bag and left Bill to struggle with his heavy suitcase. Charles led the way to the parking lot.

After a short walk—the parking lot had few cars—Charles paused in front of a red Mustang convertible.

"So this is the famous Mustang?" Bill commented. He knew that the car was a 65, over eleven years old, a hand me down from Charles' parents who had purchased it new.

"The very same," Charles said proudly. He patted the trunk lid before opening it. Bill struggled to load his suitcase in the diminutive trunk.

"Bring everything you own?" Charles asked.

"I wasn't sure what a business executive in Albuquerque needed. I brought both of my suits."

"You won't need them," Charles laughed.

The Mustang burned rubber as Charles backed out of the parking space. He gunned the engine and raced towards the highway. Bill, a cautious driver, reached for his seatbelt. Charles laughed. "This baby is loaded with power. Eight cylinders."

Charles slowed but did not stop as they approached the main road. He glanced left then gunned the engine, sliding through the stop sign, pulling out in front of a speeding semi. The driver honked his protest, and Charles tromped on the gas petal. In seconds they were up to seventy miles an hour. Bill spotted a speed limit sign. Forty-five.

"Christ, Charles," Bill said as he grasped the door for support. "You know what the speed limit is?"

"I should," Charles said as overtook and passed a yellow Cadillac convertible.

The driver, a peroxide blond with double chins, blew her horn.

"Check the glove compartment," Charles said.

Bill pressed the button and a stack of paper fell out. On top were three tickets issued by the New Mexico State Police. All were for speeding.

"No wonder your mother refused to let you bring your wheels to school," Bill said. "You know you could lose your license."

"Nah, they're just for speed traps."

They stopped at a red light, and the woman in the Cadillac pulled along side. She honked her horn, smiled, and then gave Charles the finger. As soon as the light changed, she jumped the green and crossed the intersection ahead of

Charles. Deliberately, she cut back across in front of him, raised her finger a second time, honked again, then sped away from them.

"Shit," Charles complained as he accelerated in an effort to catch the speeding Cadillac. "There's only one thing worse than a female driver. It's a fat female driver."

"Where we going?" Bill asked. "If we manage to get there. Slow down Charles."

Surprisingly, Charles did as he was told. "The office."

"Where are we staying?"

"Same place."

Suddenly, a siren sounded behind them. Charles pulled to the side and waved the police car with the flashing red lights past. "I spotted that sucker," Charles laughed.

The cop glared at Charles as he passed, waved one hand in a fluttering motion that ordered slow down, then flashed past them. Charles held the car at exactly forty-five miles per hour. A few miles down the road, they passed the stopped Cadillac and police car. Charles waved his hand at the Cadillac as he passed, raising his forefinger and shaking it in the traditional no no sign.

"Thank God for cops and your mother," Bill said, vowing not to ride with Charles again.

"PAMS is just down the road," Charles said as he pulled to a stop in front of a square building with a flat roof. Totally lacking in charm, it was little more than a brick box with a single door. It had a number but no name.

"Four apartments," Charles said. "We're upper right."

Charles unlocked the trunk, and Bill retrieved his suitcase and book bag. Charles tilted the backs of the two front seats forward to protect them from the burning sun, and he left the top down.

"Not afraid of rain?" Bill joked.

Charles laughed. "Don't expect too much. We're a little crowded," Charles said as he led the way up the narrow stairs.

At the top, he turned right and opened the door. "Hey gang, this is Bill," he called.

Bill followed Charles through the doorway, then stopped abruptly. He found himself standing in the L shaped living room/dining room of a small apartment. One shabby couch stood along the interior wall flanked by an overstuffed chair with the same dubious fabric. Dirty mats were stacked along the outer wall under the double windows that were barely masked by ratty curtains with some kind of floral design. To the left, a drop leaf table was wide open, and two young men and a girl sat in straight chairs along three sides. Another straight chair was jammed against the wall with an ancient looking typewriter in front of it.

"That's your office," Charles said, pointing at the typewriter.

"And my room?"

Charles pointed at the mats.

Bill dropped his suitcase and bookbag. "Where shall I put these?"

"There will do."

The two young men and the plain girl laughed. Bill moved his suitcase and book bag to the wall next to the chair.

Bill was not discomfited but he faked a frown. The apartment resembled a typical student abode. He decided it would be an interesting summer. He wished, however, that the girl were a little less plain.

"Where's your office boss?" Bill asked the smiling Charles.

Charles, who was standing next to the table, pointed behind him with his thumb.

"And your room?"

Charles repeated the gesture. "And here's the kitchen." He turned and looked into a small alcove to his right. Bill pretended to study the small room, which contained a sink, an ancient refrigerator, a stove with two burners and a small cupboard with an equally diminutive counter. The sink was filled with dirty dishes. A metal garbage container was piled high with diet coke cans.

"It's the maid's day off," Charles said. "We have a company share policy. All the diet coke you can drink, and clean up your own mess."

"I can see it's rigorously enforced," Bill said.

"I'm Paul Harbinson," a tall, skinny kid with a beard rose from his place at the table and offered his hand.

Bill recognized the name. Paul was a computer genius that Charles knew from high school. Paul had just graduated. Charles who admired brains above all else had given Paul high praise.

"Glad to finally meet you," Bill said. "Charles does nothing but talk about you."

"Yeah sure," the boy said, sitting back down. "He runs a sweat shop here. He's never heard about child labor laws."

"I'm Dave Marks," the second young man offered his hand.

Dave wore thick horn-rimmed glasses and was on the pudgy side.

"You haven't heard of me. I'm the new guy. I've only been here seven days while the others all have a week on me."

Bill laughed. "Are you from Seattle too?"

"Nope. I'm a home boy."

"Found him at Desert U," Charles said. "He's a computer sci major."

"University of New Mexico?" Bill asked.

The pudgy boy nodded. "We don't have a single leaf of ivy."

Bill recognized that he referred to Harvard. He had learned in Clarksburg that many were jealous or resented those who had been lucky enough to attend Harvard.

"I'm just a poor mountain boy," Bill tried to break the ice.

"And I'm Joy," the girl waved across the table. To Bill's surprise, she no longer appeared so plain. She had taken off her glasses and pushed her long blond hair back.

"Hi Joy," Bill said. He tried to think of a smart remark to impress her with but failed. So, he just smiled.

"I do the heavy lifting and anything else that needs done with two exceptions. I don't cook and don't clean so don't treat me like your mother."

"I'd never do that," Bill tried to flirt, but Joy ignored his weak effort.

A silence descended on the room, and the three at the table resumed their work. Joy was studying scribbled notes, Paul was writing code, and Dave had a computer screen in front of him.

"We've got a time sharing plan with the local school district. They've got a PDP-10, and Clapper worked out a real deal, at least for the summer."

Bill nodded as if he knew what Charles was talking about. "Where's Clap?"

"He spends his day at PAMS, then nights he works here."

"Nights? You mean evenings?"

"Nights," Charles smiled. "Come see my office."

Charles opened the door behind him. The apartment's single bedroom did not contain a bed. Again, there was a pile of mats against one wall, and three tables, two had Tauruses on them, and one a teletype with keyboard. Bill understood that Charles wrote code on the paper tape then fed it into the computer. That's all he knew or cared to know about computers.

Bill turned to Charles. "Charles, I think we both made a big mistake."

"I didn't," Charles said.

"I know zilch about computers and have no interest in learning anything."

"You will before we're through with you. One of these days you writers will be lost without your computers."

"Computers?" Bill stressed the plural.

"Yes. Little ones you will carry with you, and bigger ones that you'll have on your desktop at home and the office."

"To do what with?"

"That's what everyone asks."

"Well?"

"You'll sit at home and do research and write on them, and you will carry the little ones to take notes with. And some day they will even communicate with each other."

Bill was skeptical but he saw no advantage in debating the subject. He had made his choice and would have to live computers for two and a half months. Before Bill could change the subject, Charles checked his watch.

"It's two o'clock. I have to get to work."

With those words, Charles disappeared into his office. Bill, weary from his trip, collapsed into the one overstuffed chair. He expected Paul or Dave or Joy to ask him some polite questions, but no one did. After about fifteen minutes, Joy smiled in his direction.

"There's diet coke in the fridge."

"Great idea," Bill said. Some caffeine sounded exactly what he needed.

When Bill had the fridge door open, Joy called: "You can bring me one too."

"And me," Paul and Dave chorused.

Bill gathered four cans in his two hands and deposited them on the table.

"She's always pulling that," Paul said.

Joy laughed. "They think that serving their drinks is women's work. Not around here."

Bill sat in the chair that Charles had indicated would be his office. He popped the can and leaned back expecting to chat. Paul and Dave silently continued to write code. Bill said nothing, aware from his experience with Charles, that code writing requires concentration. Finally, he picked up a stack of paper that was covered with Charles' swiggles. Joy smiled with a shrug of her shoulders that Bill assumed was an attempt to convey commiseration. After reading a few pages, Bill understood why Charles needed a writer. A layman, he had difficulty understanding what Charles was writing about. The technical jargon could just as easily pass as hieroglyphics.

Bill returned to the first page and began to list words that he did not understand in the margin. He decided that after he had compiled a lengthy list of indecipherables, he would confront Charles then try to express the same ideas in simple clear prose.

After he had worked for a while, his back began to bother him, and he checked his watch. Five o'clock. "When is quitting time?" He asked.

The other three laughed. "We have no set hours. Charles has all of us on salary, not by the hour."

Bill recognized what that meant. He too was working for a flat rate. Charles had made no mention of hours. He immediately realized that he had made a mistake. The money had sounded great, but now Charles was his employer not his friend. This simple change took away Bill's independence.

Bill, irritated with himself, decided to confront Charles with the thought. He stood up just as the door to the apartment burst open.

"Hi slaves," Clapper shouted.

"Clap," Bill greeted him over the groans of the others.

The two shook hands, and Clapper collapsed on the couch. Bill took the chair.

"What a day," Clapper complained.

"How much time do you put in at PAMS?" Bill asked.

"Nine hours minimum, then I start over here."

"When do you sleep?"

"Don't ask. When did you get in?"

"About one."

Charles emerged from his office. "Time for a break, slaves," he announced.

The three at the table immediately stopped working. Charles joined Clapper on the couch. Joy went into the kitchen and opened the fridge. "I'll take one," Charles, Clapper, Paul and Dave called in unison.

Joy took two cans of diet cola from the fridge, handed one to Bill, then dropped to the floor where she sat with legs crossed and back against the wall beside Bill's chair. Bill could not help admiring her long legs. Joy's skirt had risen to her knees, but she acted as if she had not noticed.

"Did Ed sign the agreement?" Charles asked Clapper.

Clap shook his head negatively.

"That's it," Charles declared. His face looked as menacing as a baby face could.

To Bill it resembled a pout.

"That puts me in the middle," Clapper complained. "I am a PAMS employee."

"Don't be stupid," Charles said. "That's what Johns wants you to think. You're a partner in Digital Software first."

Clap did not respond.

"Just tell him that if he doesn't sign that contract we gave him, I'm out of here. Tomorrow," Charles said.

Bill noted that Paul, Dave, and Joy were staring at Charles. "Where does that leave us?" Paul asked. Bill wondered the same thing. He had just arrived and it appeared he was out of a job. He knew what his father would say if he suddenly reappeared in Clarksburg. That would be embarrassing to say the least.

"Charles' father is acting as our lawyer. He drafted the contract," Clap explained to Bill.

"I'm not kidding," Charles said with emphasis. "You guys don't have to worry. I'll take care of you. PAMS won't be our only customer. Johns is bluffing."

"He's planning on announcing your BASIC 2 tomorrow," Clapper said. "As Director of Software for PAMS I'm supposed to brief a reporter."

"That's why I say he's bluffing."

Bill was surprised by Charles' demeanor. He had not seen Charles acting in his businessman mode.

"I think he's stalling just to get you to become a PAMS employee," Clap said.

"That won't happen. If I were his employee, he would be able to claim that our work is his."

Bill acknowledged that Charles had a point that had not occurred to him.

"And we wouldn't be able to license it to anyone else. Please, just do what I say, Clap," Charles said.

"OK boss," Clapper acquiesced. "If he doesn't sign, we're all out of here."

"That's OK for you guys to say, but I live here," Joy laughed. She was having fun working with the others, despite the long hours, but she knew she could easily find a less demanding job. She hoped that would not happen. Charles made them feel as if they were all participating in something big.

"Don't forget me," Dave chimed in. "I live here too."

"What's for dinner?" Paul asked. He did not consider Charles and Clapper's problems with PAMS any of his concern.

"Pizza or burgers?" Clapper asked.

"Pizza," Paul and Dave said.

"Want to come along?" Clap asked Bill. "I'll show you where to get the best pizza in the southwest."

"That's something I'll need to know," Bill joked.

"You bet your ass it's something you'll need to know," Paul said.

As he walked toward the door behind Clap, Bill looked back expecting to see his friend and roommate following. Instead, he found Charles retreating to his office. "One sausage and one pepperoni," Charles called over his shoulder.

"And one cheese," Joy added.

"And three cases of diet coke," Paul said.

"I'm not going to ride with Charles anymore," Bill said as Clapper turned his old Chrysler onto the highway.

"I don't," Clapper laughed.

"Why is Charles so uptight about the contract?" Bill asked.

"He's right," Clapper said. "We're taking a lot for granted with Ed. I don't think he's deliberately stalling. Ed just likes to take his time about things. Anyway, that's just Charles."

What do you mean?" Bill did not understand.

"Charles likes to be boss. He redid our partnership."

"Digital Software?"

"Yes. We started off fifty, fifty, but Charles decided that was unfair. Now its sixty-five, thirty-five"

Bill waited for Clap to explain.

"As usual, Charles was right. I work full time for PAMS while he does most of the work at the company. Also he contributed more capital to start up than I did."

"How much more?"

"I put up $606 and Charles $910."

"That's not worth complaining about."

"And Charles wrote most of the BASIC."

"It's something you two created together. It ought to be fifty fifty."

"Charles likes to be boss. I remember once in high school when he caught me doing something I shouldn't have. Charles quit. We needed him, and Charles agreed to come back only if he could be boss."

"But you're older," Bill said.

"And Charles is smarter. Smarter and tougher than anybody else."

"It still doesn't seem fair."

"Sure it is," Clapper said. "With Charles in charge we all are going to make out."

"I don't think I would like working for him," Bill said.

"You'll find out this summer."

The next day Ed Johns signed the contract. Digital Software got pretty much what Charles had demanded. PAMS paid Digital Software $3,000 on signing and agreed to a royalty for each copy of the software that was shipped. PAMS promised to pay Digital Software $30 for the original Basic, $35 for Basic 2, and $60 for Basic 3. In turn, PAMS got a license to ship BASIC with each Taurus and the authority to sub-license. Clap had to work full time until January 1 supporting the program. PAMS agreed to require security agreements for the software from each purchaser.

The wording of the security agreement fell into Bill's lap. He had no sooner finished that than he was informed that he was expected to edit "Computer Notes" a newsletter started by PAMS which informed customers about their Tauruses and future company plans. Before long, Bill knew more about computers than he ever wanted to learn. He drew the line at writing code or opening the computer cases. By August, he had mastered much of the jargon, a feat he was determined to forget as soon as he departed Albuquerque to return to Harvard. Charles surprised him at the last minute by revealing that Bill would have to return to Harvard without him.

"I'll give you a rain check on your cross country drive," Charles promised.

"Not necessary," Bill replied with emphasis. He had worried all summer about surviving such an ordeal with Charles at the wheel of the Mustang.

Bill's relief was such that he did not complain about Charles thoughtlessly forcing him to find another roommate. Joy gave Bill a check on Digital Software for two thousand dollars. When they parted, Charles warmly thanked Bill for his help and suggested he not make plans for next summer. Charles' final words dealt with his mother.

"When my mother calls, you can tell her I decided to skip this coming semester."

"Charles, you should tell her yourself," Bill admonished.

"I will. When she calls," Charles smiled.

Bill planned a stop in Clarksburg on the way to Cambridge and from there he could phone Mike Hamilton to see if he could share Bill's room in Charles' place. Bill doubted that Charles would ever grace Harvard's merry halls again.

Chapter 9

Stanley decided to go with the young and hungry. For him that meant the Moss Agency. His research—another quick trip to the library and two calls, the Chamber of Commerce and the Better Business Bureau—convinced him. The Moss Agency in two years had become the talk of Silicon Valley. Richard S. Moss, the Moss of the Moss Agency, had moved from San Francisco to Cupertino to launch his own firm. As his clients of choice, Moss selected the startup high technology firms. Stanley decided that Astron Computers fit the Moss specialty list to a T. Stanley realized that Astron Computer's current home address, his father's garage, and diminished bank account, $1,000, were not impressive, but it had two things going for it, the potential of Dum's new computer and Stanley's business acumen. Others did not recognize the latter fact yet, but Stanley had the ability to sell anything, at least he thought he did. That was his attitude when he phoned the Moss Agency.

"Moss Agency," a perky female voice answered.

"Good morning Ms.," Stanley spoke authoritatively. "This is Stanley A. Pitts of Astron Computers. Please put him on."

"Who did you wish to speak to sir?"

"Is he in?"

"Who sir?"

"Young lady," Stanley said, "I wish to speak with Mr. Richard S. Moss."

The phone clicked as Stanley was placed on hold. After a two-minute pause, the secretary came back on line. "I'm sorry sir. Mr. Moss is not available. I'll put Mr. Witt on the line."

"Wait," Stanley started to protest but failed.

"This is Thomas Witt. May I be of service, sir?"

"Oh shit," Stanley mumbled. He wanted to speak with the top dog not some flunky. "I asked to speak with Moss," Stanley blustered.

"I am Mr. Moss' assistant," Witt said. "How may the Moss Agency help you?"

"My name is Stanley A. Pitts. I am CEO of Astron Computers." Stanley paused to give Pitts time to comment. He did not. "We need an agency to represent us, and I wanted to ask a few questions to see if you measure up."

"Just a minute sir," Witt said.

Stanley heard Witt speaking to someone, and then he heard pages ruffling. He assumed that they were checking on him. He waited.

"May I ask where your company is located, sir," Witt said.

"Here in the valley."

Again silence. Then, Witt spoke again: "I don't seem to find Astron Computers listed."

"Of course not," Stanley said. "We're a startup. If you would just listen, I'll tell you what I need."

"I'm not sure that is how it works, sir. If we decide to take you on as a client, we will make our own study of your company and then provide our recommendations."

"Well, that's what we need," Stanley capitulated. "Why don't we get together and discuss the matter."

"That would be very good sir. Why don't we meet at your office."

Stanley did not want that. "Let's have lunch instead."

"Very good," Witt said. He suggested a date two weeks in the future, Stanley demurred, but finally capitulated when Witt claimed that was the only day he had free this month. "I will pick you up at your office," Witt insisted.

Unable to quickly think of a reason why not, Stanley agreed and gave his parents' address. "We are temporarily working out of a structure on my parents' property," Stanley blurted.

"Oh?"

"Yes. I told you we are a startup company."

"And you work out of your parents' garage?" Witt guessed. His job at Moss was screening clients, especially call-ins.

"Yes," Stanley said.

"Mr. Pitts I'm not sure we're the agency for you."

"How can you say that?" Stanley could feel him slipping away.

"Witt and Pitts is a lousy name for a team," Witt laughed.

"Yes," Stanley did not laugh. "I'll see you a week from Friday, then," Stanley said, then hung up before Witt could cancel the appointment.

Worried that Witt would not show up, Stanley called the Moss Agency the next day. The perky secretary alleged that Mr. Witt was in a meeting. Stanley left his name and his mother's phone number. Witt did not call back. Stanley called again that afternoon. The secretary, somewhat embarrassed, claimed that Mr. Witt was out of the office. The next morning Stanley called again.

"Mr. Witt directed that I take your message, sir," the secretary said.

"Tell him I think it would sound better if he added an 's' to his name," Stanley cracked.

"Sir?"

"Pitts and Witts would sound better." Pleased with himself, Stanley hung up.

Assuming that he had the upper hand, Stanley made no more demeaning phone calls to the Moss Agency. He assumed Witt would show up as planned. As it turned out, it proved to be a close decision for Thomas Witt. He had worked for the Moss Agency for a year and had met his share of crackpots. He was tempted to dismiss Astron Computers out of hand, but he could not. Richard S. Moss, the boss, was betting the shop on the technology revolution, and this was based on the belief that the future rested in the hands of the valley's hobbyists and engineers. He frequently pointed out that Hewlett-Packard started in a garage. This of course placed Witt in a difficult position. He figured ninety-nine percent of those working out of their garages were nonstarters, but he could not pass up the opportunity to get the agency in on the ground floor with another HP. It was for this reason that Richard S. Moss had set his agency's fee policy. He knew the one thing these startup companies lacked was cash, so he levied no fees for his services. Instead, he took a percentage of the profits of the new company. This meant that when they hit it, as some had, they hit it big. Unfortunately for Thomas Witt, this increased the pressure on him. He could not easily dismiss out of hand any new company, no matter how small, with a product that had potential. He spent a lot of his time visiting garages in the valley.

On the morning he was supposed to visit Astron Computers, Thomas told his secretary: "Call Astron Computers," he laughed, "and remind them that Witts is ready to meet Pitts."

"Sir?" the secretary said.

Thomas smiled. The agency secretary had a good telephone voice, a great body and no sense of humor. "Please confirm my appointment with Mr. Pitts," he ordered.

Stanley was in the kitchen brewing a cup of green tea, his current favorite, when the doorbell rang.

"I'll get it," his mother said.

Stanley checked his watch. Exactly, twelve noon. Witt was on time.

Stanley scooped in two teaspoons of sugar as he heard his mother greet the visitor.

"Oh yes, Mr. Witt. If you will have a seat, Mr. Pitts will be with you shortly." Stanley wished his mother did not take her role as secretary so seriously. Witt was their first business visitor at home, and his mother was playing it to the hilt.

Stanley sipped his tea at the kitchen table as he waited two minutes to give the impression that he was a busy man. His mother joined him in the kitchen, and he winked. "Wait a minute, then bring my tea in," Stanley ordered.

"Yes sir," his mother smiled. She liked playing business.

Stanley entered the living room to find Witt staring out the front window.

"Right on time, Mr. Witts," Stanley said, deliberately adding the s.

"Good morning Mr. Pitt," Thomas dropped the s.

Both chuckled. Witt's response hid his disappointment. Stanley was dressed in his wrinkled gray T-shirt, jeans and sandals. Stanley ignored the dapper businessman's scrutiny. Stanley did not approve of his uniform: dark blue suit, white shirt, red tie, black shoes. Thomas Witt decided he had found another loser and immediately began to figure out some excuse to avoid lunch.

Stanley gestured toward his father's recliner that faced the television and sat himself on the worn couch. "Did you have trouble finding us?" Stanley tried to make small talk.

"That's my job," Thomas Witt replied.

"And that is?"

"I screen valley companies for potential clients," Thomas answered honestly. He calculated he could take a quick look at the kid's shop, listen to a few half-baked ideas, then escape.

"Mr. Moss accepts your recommendations?" Stanley was not impressed with Witt who appeared to be only a couple of years older than Stanley. He acted like a pompous college graduate.

"Here is your tea, sir," Stanley's mother entered the room balancing two cups and saucers on the family's one small silver tray.

"Thanks mom," Stanley said.

She served Witt first, then Stanley.

"Thank you Mrs. Pitts," Thomas said.

Stanley's mother smiled as she left the room

Stanley, tired of the bullshit, took his cup and left the saucer on the coffee table. "Bring your tea. I'll show you the shop."

Stanley, without waiting for a response, headed for the garage. Thomas carefully balancing cup and saucer followed behind.

"This is it, for now. I'm looking for a larger place," Stanley said. "We've got a big contract with Electronics Warehouse, and we're about ready to announce the Astron 2."

"The Astron 2?"

"That's it," Stanley pointed toward Dum's workbench. Stanley had carefully showcased the same one of a kind model that they had taken to the trade fair. "This is only a working model. I'm designing a sleek plastic case for it."

Thomas was impressed. He had expected to find a circuit board and a jumble of electronics parts. He knew little about electronics, but he understood packaging. "Does it work?"

Stanley flicked a switch on the front of the metal box. Silence ensued for a few seconds, and then lights began to flash. The monitor displayed "ASTRON COMPUTERS" and "READY" in bright red letters on a gray screen.

"Color?" Thomas knew that color was something that everyone in the valley was working on.

"This is state of the art," Stanley said. "I'll let you in on a secret." He paused and waited for Thomas' reaction.

"Yes?"

Stanley pretended to look around the shop. "I need a copy of our secrecy agreement."

"Is that necessary?" Thomas immediately assumed that Stanley was bluffing.

Stanley halted his phony search and turned to look Witt directly in the eyes. "I guess not," Stanley said. "Do I have your word?"

"Our conversations with clients are always confidential."

Stanley liked his visitor's use of the word "client." That gave him the upper hand. "Then this remains confidential."

"I said it did."

"We also have sound," Stanley lowered his voice to a whisper.

"Let me hear it."

"Sorry," Stanley said. "Dum is still working on the speakers."

"Plural?"

"Stereo," Stanley confirmed, still whispering.

"There are just two of you?" Thomas asked.

"Six."

"Engineers?"

"Most." This was almost accurate. Dum and two of his part time helpers from HP were engineers.

"Including you?"

"I'm CEO. I do design, administration, personnel, management, sales, the whole thing."

"Tell me about your sales."

"We have a standing contract with a major electronics outlet. I' m working on others." The last was only partially true. Stanley was looking for more customers but had yet to find them. "I am negotiating with Commodore." Stanley did not say that Commodore had turned them down.

"What does Commodore want?"

"Until we have an agreement, I am not at liberty to say," Stanley smiled.

"What do you see for the future?"

"The sky's the limit." Stanley fell back on the old cliché.

"Can you be more specific?"

"This little baby here is going to make big money," Stanley patted the Astron 2 which was still flashing "READY" in bright red letters. Stanley did not say that Harold had not decided to make his invention a part of their partnership.

"I've seen enough here," Thomas Witt said. "Let's grab some lunch and we'll talk some more."

When they parted at two, Thomas was surprised at himself. He believed only about half of what Stanley said, but if even half were true, it was possible that the two kids in the garage were on to something. They looked like the kind of wager Richard S Moss liked to make. First, however, he wanted to check out Astron Computers further.

Thomas Witt visited Jake Foster at Electronics Warehouse. Jake bluntly provided a negative appraisal of Stanley and his Astron Computers. Jake complained that the unsold circuit boards on his shelves were sad reminders of his experiences with Astron Computers. He opined that Stanley A. Pitts, an arrogant young man, was not to be relied upon. Foster admitted that the Astron 1 boards that had been enclosed in wood cases had begun to sell; the workmanship on the circuit boards was particularly appreciated. He had no opinion on Harold Dumbroski's Astron 2; he had not seen it. Thomas Witt came away from Electronics Warehouse with a mixed impression. The non-establishment Stanley Pitts had obviously irritated Foster, but he had overcome his image and sold Electronic Warehouse one hundred circuit boards.

When Witt reported his findings to his boss, Moss to his surprise reacted with interest. Something about Stanley reminded him of other valley entrepreneurs. He used Atari as an example, and the founder was now worth millions. Moss opined that the immediate future for Astron Computers depended on its product. He directed Witt to check out the inventor. Witt talked with several engineers at HP and was surprised to learn that Harold Dumbroski was a valued employee held in high regard as an engineer by peers and supervisors alike despite the fact he did not hold a degree. One engineer stated flatly that if Harold built a piece of equipment, it would work. Moss directed Thomas Witt to arrange an appointment for him to visit Astron Computers and talk with the two principles himself.

With some difficulty, Witt set the meeting for ten o'clock on Saturday. Moss preferred a weekday time, but Dumbroski who worked full time at HP, could only join them evenings or weekends. When the two Moss Agency men arrived at the Pitts garage, they found Harold and Stanley waiting. Hoping to better impress Moss, Witt had advised Stanley to "dress up" for the meeting. To Witt's dismay, Stanley greeted them with his ponytail, T-shirt, jeans and bare feet in sandals. Although his visitors did not recognize it, Stanley had followed Witt's advice. He had put on a clean T-shirt and jeans for this important meeting.

Stanley had not told Witt that he had difficulty persuading Harold to participate. Dum did not like business meetings and in fact had still not agreed to include his Astron 2 in Stanley's plans for Astron Computers' future. For these reasons, Harold was not in a good mood when the Moss Agency men arrived.

Stanley greeted Moss and Witt at the garage door.

"Hi Witts. Good morning, sir," Stanley offered his hand. "This is my partner Harold Dumbroski."

Harold who sat on his stool in front of his workbench waved but did not offer his hand.

"Please have a seat," Stanley said. He pointed at the grimy overstuffed chair, the place of honor, and a straight chair from his mother's dining room. Stanley was too nervous to sit himself.

"Is that the Astron 2?" Moss asked the engineer, ignoring his initial surliness.

Harold nodded affirmatively.

"Yes sir," Stanley said.

"I understand you've got color and sound," Moss said.

"Don't worry about the case," Stanley said. "I've decided to do it in plastic. It will look great."

"May I see it run?" Moss asked Harold, ignoring Stanley.

Harold begrudgingly demonstrated the Astron 2. He only smiled once. That was when the bright red letters spelling "ASTRON COMPUTERS" and "READY" appeared on the monitor.

Following the demonstration, Moss sat himself in the place of honor. Witt selected the straight chair. Harold remained on the stool, only partially turned towards the guests. Stanley took a standing position near Dum's chair, leaning awkwardly against the bench. Astron 2 glowed in the background.

Stanley considered elbowing his friend. Harold when he was not working was usually a genial, polite guy. In an effort to break the ice, Stanley said: "Dum's working on an article for a magazine."

"Really," Moss smiled. "May I take a look at it?"

"No," Harold replied bluntly.

"You should know, Mr. Dumbroski, that if we decide to represent your company we would ask that we be consulted on everything you submit for publication."

"We understand," Stanley said.

"Fat chance," Dum said. "I don't need a flake to tell me how to describe my work."

"That's the last thing I want to do," Moss spoke smoothly. "You don't want to make it too technical. All the readers may not understand."

"The people I'm writing for will," Harold insisted.

"I'm sure we can reach an understanding," Stanley said. "I appreciate all the help we can get." Stanley turned towards Harold. "As we discussed, Dum, we need publicity to sell the Astron 2. We learned that in Atlantic City. You're building the world's best computer. If the world doesn't know that, they can't buy it."

"Our only goal is to help you meet your objectives," Witt said.

"We have but one," Stanley said. "To start a revolution."

Witt grimaced. He wished Stanley had used a different term.

"Technical revolution not world," Moss smiled.

"Exactly," Stanley agreed.

"Well, we have decided," Moss said, surprising his colleague. "We'll take you on."

Given Dumbroski's attitude, Witt would have declined.

Moss outlined the terms of their relationship. Stanley nodded his head in agreement and was particularly pleased when Moss described the financial terms. Stanley did not have to come up with an immediate payment. The Moss Agency agreed to take one percent of the profits.

After the visitors departed, Harold turned to Stanley. "I hope you know what you're doing."

"Don't worry. I'm in control," Stanley replied.

"You just gave away one percent of our company."

"You still have 49.5 percent, and I have 49.5 percent. This Moss Agency is going to make us rich."

"How?" Harold asked. "Are they going to buy our computers?"

"Hey. One percent of nothing is nothing. That's where we are now."

"We've got money in the bank."

"Dum. We've got exactly one thousand dollars, and we owe ten times that."

"Then maybe your advertising friends will help make up the deficit."

"It doesn't work that way."

"You never said anything about a plastic case," Harold changed the subject. Thinking about money really did not interest him.

"I'm working on it. I wanted to surprise you," Stanley said.

Monday morning Stanley woke in buoyant mood. He had decided that now that Astron Computer had a first class advertising agency they needed a new home, one that would live up to their enhanced billing. As his first task this morning, he had to acquire the money that would allow them to expand. From past experience, he knew the banks were out. His former employer at Atari had already turned him down. What he needed now was some venture capital. Stanley did not know any venture capitalists, but he had read about them in the valley newspapers. There were guys out there willing to bet cash on promising startups. In Stanley's mind, Astron Computers was exactly that. He pondered how to find a venture capitalist when the thought struck him. Surely, the Moss Agency should be able to advise him.

"Moss Agency," the familiar, sexy female voice answered.

"Stanley Pitts here," Stanley said.

"Oh Mr. Pitts. Please hang on," the voice said. "Mr. Witt wants to talk with you."

Stanley intended to ask for Moss, but he decided Witt would do.

"Witt here," Witt picked up the phone.

"Pitts here," Stanley mimicked.

"I hope we are not going to have to go through this routine every time we talk Stanley," Witt laughed. "It would go better if you dropped your s."

"You mean Pitt and Witt," Stanley asked.

"A good name for an agency," Witt said. "I'm glad you called. Mr. Moss would like to talk to you."

"Here I am," Stanley said.

"No. I mean at the office. Could you come in this morning about eleven?"

"Let me check my schedule," Stanley said. He rattled his newspaper. "OK, I can make eleven. I've got no time for lunch," he added.

"That's good," Witt said.

At exactly eleven o'clock, Stanley with sandals flopping entered the Moss Agency. Although he hid it, he was impressed. Walnut paneling, thick green carpets, and heavy wood furniture gave the agency a prosperous image.

"Mr. Pitts," the attractive brunette receptionist greeted him.

She looked every bit as good as Stanley had visualized. Before he could decide which chair would give him the best view, the girl continued. "Mr. Moss is waiting for you."

She rose and led Stanley down a long hallway to a corner office. Stanley watched the swaying hips as he counted at least ten offices.

Moss was on the phone when Stanley entered. He waved, covered the mouthpiece, told Stanley to take a seat, and asked the girl to have Witt join them. Moss returned his attention to the telephone. Stanley looked out the window then turned and appraised the office. It was twice as large as the reception area and was decorated in dark wood and leather. Stanley decided he would have to have one just like it soon.

"Stanley, good of you to join us," Moss said as soon as he hung up. "Another client," he smiled.

"We have a standard contract for you to sign," Moss nodded toward Witt.

"It's in my office, sir," Witt said.

"I want to say Stanley, it's great to have you on board," Moss said.

"We're pleased to have the opportunity to employ your agency," Stanley countered.

"There's just one thing, Stanley," Moss ignored Stanley's response. "Before we formalize things, I must discuss something personal."

Stanley wondered what this was about but nodded affirmation.

"There's only one way I can tell you, bluntly."

"You have to overlook Dum," Stanley assumed Moss was concerned about Dum's behavior on Saturday. "He's only interested in his computers. He leaves the business up to me, so you don't have to worry on that score."

"It's not Mr. Dumbroski. I understand engineers. It's you."

The words surprised Stanley. He said nothing.

"You are going to become the image of Astron Computers. By the time we're through, when anyone mentions the name Astron Computers, they will visualize your face."

"I can handle that," Stanley smiled.

"Yes? Well. We are going to have to change a few things."

"Such as?" Stanley was not sure he liked where the conversation was going.

"When you look in the mirror, as I am sure you do, we all do, what do you see?"

Stanley thought for a few seconds. Suddenly, he realized Moss was talking about his ponytail." I guess I'll have to give this up," Stanley waggled his ponytail. He was rather proud of it.

"What you see in that mirror, Stanley, is a handsome young man. Your appearance is perfect for your age, but that is not what we want your customers to see."

"And that is?" Stanley tried to keep the irritation out of his voice but failed. Moss was becoming too personal.

"You have a long shot at making it big time," Moss paused, wondering if he was laying it on a little too heavy, but Stanley was smiling.

"No more beatnik look," a placated Stanley said. "I can handle that."

"The hair can be long, stylishly long, but no ponytail. For business, the jeans and T-shirt have to go, and the sandals."

"No suits," Stanley fought back.

"That's fine. Expensive sport coats and slacks."

"No white shirts and ties," Stanley looked meaningfully at Witt. "That's not me."

"Correct. You're the trend setter."

Stanley smiled.

"No starched white shirts and ties. You're not the head of IBM. Expensive turtle necks will do."

Stanley nodded agreement. "And my sandals?"

"Nikes. Or expensive loafers, preferably with tassels."

"Depending on the occasion," Stanley agreed. "No socks," Stanley decided to show his independence.

"Depends on the occasion," Moss agreed.

"We've got a deal," Stanley capitulated.

Moss stood up and shook Stanley's hand. "Now Mr. Witt will walk you through the contract," Moss said.

Stanley had turned to follow Witt out of the room when he remembered his purpose for visiting the Agency. "There is one thing, Richard," he called Moss by his first name.

"Yes Stanley."

"I need some names."

"I thought you would."

"Of venture capitalists."

"You work out the agreement with Mr. Witt, and I'll see if I can come up with a list of likely investors for you."

Stanley suddenly felt like a charismatic leader. He nodded and followed Witt out of the room.

Fifteen minutes later, in Witt's much smaller office, Stanley was intensely studying the three-page document he was expected to sign word by word. He knew from television that he should read the small print before he signed anything, and he was determined to ensure that The Moss Agency did not expect a single penny of advance cash which Stanley was determined not to pay. He had just finished the third page and had not understood much of what he had read when Richard Moss entered the room without knocking.

"You can sign it," Moss said.

Stanley decided to chance it. He held out his hand and Witt gave him a ballpoint pen. Stanley signed.

"Check this guy out."

Stanley took the folded paper from Moss and handed the signed contract to Witt. He opened the paper and discovered one familiar name: John Davis.

"He's one of the guys that bankrolled Atari," Stanley decided to show he was not an innocent.

"I talked with him. He'll give you a call," Moss smiled.

"This is only one name." Moss had promised him a list, and Stanley was determined to show who was boss.

"If that doesn't work out, let me know," Moss said. "I'll leave you two to your business."

Stanley refolded the paper and handed it to Witt. "I won't be needing that," Stanley said.

"Do you want help with your wardrobe?" Witt asked as Stanley rose to leave.

"I can handle it," Stanley smiled.

Three weeks later Stanley checked himself out in the mirror and liked what he saw. He had had his first hair styling and had shaved the beard himself. He wore a yellow turtle neck shirt, khaki slacks, white Nikes, and a blue blazer.

"I think he's here Stanley," his mother called.

Stanley peeked out the front window and saw a silver Mercedes in the driveway. As he watched, a man with wavy silver hair and wearing a dark blue business suit got out of the door. He paused long enough to scan the neighborhood. Stanley could detect no visible reaction. Stanley turned back to the mirror and checked himself again. He really liked his new image. He felt ready to take on Mr. John Davis. Stanley had met Davis once before at Atari. Stanley had been working on the shop floor and company brass had been conducting Davis on a VIP tour. They had shaken hands, but Stanley was confident that Davis would not recognize him.

"My God, Stanley," his mother greeted him at the bottom of the stairs. "You're beautiful."

Stanley, pleased, smiled. He winked at his mother then circled through the kitchen and the door that connected with the garage. Everything was as he had planned it. Dum sat at his workbench fiddling with his transistors. On the workbench to Dum's left set their last remaining Astron 1. To Dum's right in the place of honor rested the latest version of Dum's Astron 2. The difference was striking. The Astron 2 was enclosed in a beige plastic case that Stanley had designed. The case was the only one they had, and the circuit board it contained was Dum's latest effort. Dum still was not satisfied with it and had protested when Stanley had insisted that they prepare this one prototype for the meeting with Davis. Stanley wanted to knock the venture capitalist's socks off. He was proud of his and Dum's efforts. The Astron 2 had a keyboard in front, and it looked much like an IBM Selectric with a monitor attached.

Stanley looked towards the open garage door and found the natty Davis studying him. He looked like a millionaire. His face was smooth with a hawk nose. Stanley decided that the silver hair was tinted. He estimated that Davis was no more than forty-five. He too, Stanley decided, was dressed for a role.

"Mr. Davis," Stanley smiled. "Welcome to Astron Computers. We're small, but we won't be for long." Stanley crossed the garage and offered his hand.

Harold snorted. Stanley glanced to his left. Harold was sitting on his stool facing away from his bench. Stanley liked the surprise on Dum's face. This was the first time he had seen the new Stanley. Dum with his dirty deans, T-shirt and shaggy beard looked like an inventor.

"Pleased to meet you Mr. Pitts," Davis said. He took Stanley's hand in a firm grip.

Stanley savored the moment. This for him was the launch of something big. All he had to do was hustle the suave Davis into giving him a million dollars to grow with.

"Call me Stanley," Stanley said.

Davis turned to Dum. "And this is Mr. Dumbroski, the resident genius."

Dum grunted. Smooth talk did not impress him. He nodded, took Davis' hand without rising, then half turned towards his bench with his gaze on the Astron 2. He had to admit that Stanley had it looking quite professional.

"And this is the Astron 2," Davis said.

"Quite a difference from Astron 1," Stanley said. He pointed towards the other end of the bench at the rough wood case with the exposed circuit board.

"Will you demonstrate it for me," Davis spoke directly to Harold.

"The Astron 2?" Dum asked.

"Yes. I understand it has color," Davis said.

"This is only a prototype. It needs lots of work before we can go into production," Dum said.

"No problem. I've been through this before. It's my business."

"Dum designs the inside, and I take care of the outside plus all business matters," Stanley tried to get control of the meeting.

Both Dum and Davis ignored him. Dum turned on the Astron 2, then waited. After about two minutes of silence, a clicking began inside the plastic case and lights blinked along the base of the monitor.

"It takes a while to warm up," Stanley said. "We'll shorten that as we go along."

Dum glanced at Stanley, his look conveying disbelief. However, he said nothing.

Suddenly, the screen lighted up and "ASTRON COMPUTERS" and "READY" flashed in bright red letters.

"Impressive," Davis said. "Please elaborate."

"I don't know what you know about computers," Harold said.

"Not much. I'm familiar with them. That's all."

"I've added more ROM, more RAM and eight expansion slots," Dum said quickly, then waited for Davis to react.

"ROM, Reading Only Memory, holds the programs we install," Stanley said. "It's non-volatile. That means the programs don't disappear when you turn the computer off."

"And RAM?" Davis asked.

Stanley looked sharply at Davis, suspecting that he was being tested. "RAM, Random Access Memory, is volatile. It stores whatever the user imputs to the computer. If the text the user writes is not saved to storage, it disappears when the computer is turned off."

"What are expansion slots for?" Davis asked.

"To connect external devices the user may need," Dum said.

"Like printers," Stanley added. "The Astron 2 will also have sound, so it will need speakers."

"Can you play me some music?" Davis asked.

"No," Harold replied bluntly. "Not ready."

"Show me a program," Davis directed Harold.

"I only have a simple one installed. I wrote it myself," Dum said.

Davis nodded.

Harold demonstrated a game of Tic, Tac, Toe. Fortunately, Astron 2 performed flawlessly.

"Very good," Davis said. "Now let's talk about your plans. Obviously, you can't go into production here."

"Mr. Davis," Stanley took over.

"Please call me John."

"Will you have a seat?" Stanley indicated the one overstuffed chair.

Stanley fetched a chair from his mother's kitchen, and Dum focused his attention on the workbench. Astron 2 continued to flash "READY" in bright red letters. Stanley had told Dum to leave it turned on during their meeting.

"The Astron 2 can also be connected to the family TV set," Stanley said. To him, this was a selling point. Virtually every home in America had a TV set."

"No," Dum interjected before Stanley could continue.

Davis looked at Harold for explanation.

"We may have a problem with that," Harold said.

"Oh?" Davis waited. Harold was clearly the technical man, and he wanted to appraise him before he bet any of his money on these two kids.

"Technically, there's no problem. It's the government. The FCC may not approve of it."

"We don't have to get into that, Dum," Stanley said, trying to head off a technical monologue.

"Please explain," Davis said.

"The RF modulator gives off too much interference," Dum said.

"The Radio Frequency modulator allows the user to use his TV screen instead of a monitor," Stanley tried to put a good face on the problem. He was confident Dum could eventually solve their problem.

"Can you fix that?" Davis asked Harold.

"Maybe," Dum said.

"We can always get around it. Electronic stores sell modulator kits and users can hook up their own. That way Astron Computers would have no problem. The FCC is not going to chase individual TV owners. It's a minor problem."

Davis looked at Harold. He nodded agreement.

"All right, tell me about your company's plans." Although Harold and Stanley did not know it, this was the key part of the meeting for Davis. Davis had made his many millions by thinking big. A successful venture capitalist, who sat on the boards of companies, Davis invested when he found small companies, even as small as Astron Computers, with promising product and management with vision, extremely grand vision. He wanted ambitious startup managers that thought in terms of billions not millions. The answer to his question would determine the extent of his interest. Thus far he had not been impressed. Dumbroski clearly was a competent engineer, but Davis could find those by the hundreds, even thousands. Pitts seemed to be an eager kid, but shallow. His clothes appeared brand new. Davis figured he could still find price tags if he looked.

"We've sold a hundred of the Astron 1 and expect to sell at least another one hundred to one hundred fifty."

"At what price?" Davis was not interested in the answer. The kid was talking peanuts.

"Six hundred and fifty dollars," Stanley said proudly, giving the retail price not their profit line.

Davis made no comment.

"But we have big plans for the Astron 2. As you can see, it's a superior product."

"How many do you expect to build?"

"A couple h...." Dum started to answer, but Stanley interrupted him. "After we get our financing lined up, the sky's the limit." Stanley had already figured out that Davis did not want to discuss small figures.

"I'm not interested in the sky, Stanley. I want to know about your plans."

"There's at least ten thousand computer hobbyists in the United States. We know from Taurus sales that every one will want one."

"So you figure 10,000 sales. At what price."

"One thousand dollars each," Stanley said, selecting a figure off the top of his head. He and Dum had yet to figure out how much they could charge. Stanley knew that Dum would always argue for a lower figure.

"So you're talking ten million in sales at the high side."

"Christ, Stanley," Dum blurted.

"Something in that range," Stanley ignored Dum.

The figure did not impress Davis. He calculated the kids had not even thought about the subject.

"What about your financing?"

"I've been thinking in the million dollar range," Stanley said. He glanced sharply at Dum warning him to remain quiet.

"You're going to need a lot more than that, kid," Davis laughed.

"Call me Stanley," Stanley countered. Davis' arrogance was beginning to irritate him.

"All right, Stanley, how much does your company have in the bank now?"

"Something in the range of ten thou," Stanley bluffed. They had less than a thousand.

"I don't believe you," Davis decided to push Stanley to see how he would respond. It would take a strong man to lead this garage company to the level that would interest Davis. "Do you have a bookkeeper?"

"Certainly we do." Stanley did not intend to divulge it was his mother for now.

"May I see your books?"

"I'm sorry they are not available," Stanley said. "Our bookkeeper is out of town." Stanley could have said she was sitting in the kitchen listening to this conversation. "How much were you thinking of investing?"

"That's the wrong question, kid. How much do you need?"

Stanley hesitated. If he answered honestly, Davis would recognize they were broke. If he pretended they already had some support, Davis would lose interest. He decided to try another tack. "We're on the front edge of a revolution. In ten, maybe twenty years, there are going to be computers in every home."

"Doing what?" Davis laughed.

"People will use them to correspond with."

"How?"

"With a modem a computer can talk to another computer over a telephone line," Dum came to Stanley's defense. He did not like Davis' attitude. As far as Dum was concerned, they did not need him or his money.

"What else will they do?"

"They'll pay bills, store records, manage appliances."

"You mean turn on my coffee pot in the morning."

"And wake you up, and play your music, and raise and lower your thermostat."

"All that stuff is nice, if true, but I thought you said your market was the computer hobbyists, ten thousand of them."

Stanley did not know how to respond, and his expression showed it. Davis decided to end the agony.

"Look guys. You're on to something. You've got to think bigger. I don't know if your computer revolution is coming or not or if your Astron 2 will be a part of it. I'm not interested, but there are others who might be. I can give you a name or two, but you're not thinking big enough for me. Decide what you want to be. Calculate what it's going to take to get there, and be honest with whoever appears interested."

"Thank you for being honest," Stanley recovered. "If you can give us the names of venture capitalists you think might be interested, we would appreciate it."

Davis took out a note pad and wrote down one name: Stuart Miller.

Chapter 10

David Howard stood at the window of his small office on the IBM campus in Armonk and stared at the parking lot. Above, the gray sky filled with black clouds matched his mood. For about the thousandth time in the past few months, he asked himself what in God's name he was doing here. He did not fit in with the IBM culture as most of his classmates did. Those few like David who failed to adapt had departed. At a meeting the previous day, David in boredom had without thinking propped his new wingtip black shoe on the rung of the empty chair next to him. He had thought he was in uniform. Then, the junior executive in the chair in front of him turned, stared at David's shoe and exposed ankle clad in approved black silk and frowned. The man, no more than two or three years David's senior, moved his right ankle to the right and elevated his trouser leg slightly. He tapped an exposed black garter, then smiled at David who immediately got the message. He was out of uniform. IBMers were supposed to wear garters for Christ sake.

David supposed that the only reason he hung on waiting for things to improve was because of his boss Ernie Hendricks. For some reason the Chief of the General Product Division had cast his protective mantle over David. Presumably, Hendricks, an ex jock, saw a kindred spirit. David hoped it was more than that. David had left the football field behind upon graduation. Now, he wanted success based on something more than the ability to run with a football or hit a softball. The one thing that David was thankful for was that John Parsons, IBM's Chief of Sales and IBMer personified, had not yet risen to Chief Executive Officer as most anticipated. David had inadvertently acquired Parsons' enmity on the softball field, and he had not even known who Parsons was at the time. To him David Parsons had simply been an opposing second baseman standing in his way. By instinctively charging into the man with the ball David had successfully destroyed his IBM career before it began.

Today, David as Ernie Hendricks' special assistant was identified within the company as a pale shadow of his boss who was the company maverick. David, probably out of self-interest, he recognized, had become one of Hendricks most ardent supporters. Since Hendricks stood alone among the IBM hierarchy as the one man who argued that IBM's future rested with microcomputers, David was equally isolated. David still knew little about computers, a strange situation for a junior executive in a company whose very existence depended on the manufacture and sale of such things. David had not joined IBM with an interest in production. He had planned on becoming a supersalesman who acquired whatever expertise he needed as he winged his way through the ranks of IBM's cadre of loyal customers. The company dominated the market. There was no one bigger. David concluded that was where the problem originated. IBM had grown so big that it had become a world unto itself. That's why it could set the rules with the virtual force of law for its employees: how to dress, where to live, how to speak, even to wear garters and wingtips like a 19th century gentleman.

"David," Margaret Wilson, the division's senior secretary spoke from the doorway to David's office, "there's some kid in the lobby who wants to meet with the Chief of the General Products Division."

"Does he have an appointment?" David asked.

Margaret, who David dated on the sly—IBM frowned on office romances— smiled her indulgence. If the visitor had an appointment, she would not be consulting David. In fact, consulting was the wrong word. Margaret had already made her decision.

"How do you know he's a kid?" David made no attempt to cover himself. Margaret was the only IBMer who understood him.

"That's what the guard said. "There some kid down here who claims to be Chief Executive Officer of some software company."

"Aw, Margaret, just tell me," David pleaded.

"Ever heard of Digital Software?"

"Software. Why doesn't he talk with Production or Sales?"

"Ask him."

"Why me?"

"Because Ernie's busy, and you're just standing there mooning out the window."

"I wasn't mooning," David said.

Margaret giggled.

"I'll show you mooning," David said and unbuckled his belt.

Margaret waited.

"I will," David threatened but knew Margaret had called his bluff. "Tonight," David capitulated. He refastened his belt and pulled up his pants leg. "See gaiters."

"They're garters not gaiters. Gaiters are something else."

"What are they then?"

"Put on your coat Mr. Howard and look like an IBM executive. Mr. Swift is on his way up."

David, properly dressed, was sitting behind his desk shuffling a small collection of meaningless papers when Margaret appeared in the doorway. David admired her marvelous figure and decided he knew why he stayed at IBM.

"Mr. Howard. Mr. Swift."

Margaret stepped aside, and David found himself staring at a kid. He had long hair that ringed his face like a cap on an acorn, horn-rimmed glasses, and a broad smile. He wore a wrinkled white shirt, open at the collar, khakis, and Nikes. "He won't fit in here, not in a million years," David thought to himself. David silently considered saying: "Come on in, kid," but did not.

David glared at Margaret for getting him into this waste of time. To his visitor, he said: "Welcome to IBM Mr. Swift."

Margaret retreated and closed the door behind her, pausing only long enough to blow David a kiss. That only irritated him more. She was making fun of him.

"Please have a seat." David indicated the only chair in his office other than his own.

David decided to violate company policy and take off his suit coat in the presence of a client. David could by no stretch of the imagination picture his visitor as a client. He looked like a high school kid on a field trip.

"Thank you for granting me the time, sir," the kid remained standing, waiting for David to seat himself first. "I know you're busy, so I'll get right to the point. I am the chief operating officer for a small software company currently but temporarily based in Albuquerque, New Mexico."

For David, speaking of Albuquerque sounded like Siberia. He doubted he could even spell Albuquerque. So he said nothing, struggling to keep his face expressionless. He was amused by the kid's earnestness.

"We are responsible for the software for the Taurus."

"Really," David said. "I'm impressed." And he was. David knew all about the Taurus. He had thought of purchasing one for himself as a way of learning more about microcomputers since he was destroying his career by arguing about their potential. Only the fact that he was not sure he could assemble the kit held him back.

"We are currently expanding our client list. Since I was passing through Armonk, on impulse I decided to stop."

David wondered where Swift going. Armonk was not on a highway that connected New Mexico with anywhere. "It might be best if you talked with our software people."

"Oh, that would waste everybody's time," Charles said. "We know that IBM writes its own software."

"Then why are you here?" David blurted, regretting the harsh words as soon as they were out of his mouth.

139

"Because I wanted to meet the people in the General Products Division who are advocating that IBM get into the microprocessor field," Charles smiled.

"Mr. Swift. May I ask where you heard that?" David did not know that Ernie's advocacy had leaked beyond IBM's solid walls.

"The microprocessor world is a small one," Charles said. "We thrive on rumor and innuendo, and IBM is the Goliath. We little guys all have to keep one eye on IBM. If IBM were to enter the microprocessor business, our universe would be altered."

"And if IBM did begin to produce microprocessors?" David knew there was little chance of that, but he was interested in knowing what the tech community was thinking outside of Armonk. He might be able to give Ernie a nugget he could drop on his peers who denied a world outside IBM existed.

"Then, you will need an advanced software company to provide the operating systems and programs that are beneath your programmers who write for the 360's."

Swift's words surprised David who immediately realized he had let his visitor's appearance lead him into underestimating him. David hesitated before speaking. "I'm afraid you have been badly misinformed, Mr. Swift," David said. "IBM has no plans for entering the microprocessor world. The market is too small."

"Please call me Charles," Charles said.

"I'm David. Charles, may I ask you a personal question?"

"Please David."

"How old are you?"

"I know. My appearance. I look like a kid."

David waited for an answer.

"I'm twenty-one, almost."

"I'll bet they ask you for proof of age," David referred to the mandatory drinking age laws.

Charles frowned. "There's a revolution under way, sir, and IBM will either lead it or be buried."

"I hope you're not threatening us." David was impressed with Charles' nerve.

"Certainly not. I'm just stating a fact."

"Tell me more about your work on the Taurus."

"My partner and I wrote the BASIC."

"I've heard that story." David recalled Ernie recounting how a kid from Harvard had written the software virtually overnight. "You're the kid... the programmer from Harvard."

"That's right. Digital Software is now writing programs for a number of Silicon Valley companies."

Although David suspected this was an exaggeration, his intuition told him that this kid and his company were on to something and might prove to be invaluable contacts in the future. "And you're here in Armonk showing the flag."

"That's right, sir."

"Let me tell you something in confidence, Charles."

"Yes sir. I can keep a business secret despite my appearance."

"I think you're right. Microcomputers are going to be big. Have you heard of Astron Computers?"

"Yes sir."

"Keep your eye on them." Ernie had returned from a computer fair in New Jersey with some funny stories about a couple of kids with a garage operation who had caught the attention of several big timers.

Charles made a mental note to check out Astron Computers.

"Where are you going now?" David asked.

"Cambridge," Charles answered without explaining he was returning to Harvard for the second semester on the direct orders of his mother.

"They have some interesting little companies in Boston," David emphasized the word little. He suddenly recognized that he sounded like a regular IBMer. To cut the sting of his words, David blurted: "Do you have time for lunch?"

Charles checked his watch. Eleven-thirty. He had time. He had promised he would be in Cambridge today at the latest. He could have lunch, drive to Cambridge, and call his mother before she launched the inevitable search.

"Yes sir."

"Good." David picked up the phone and pushed the intercom button. "Ms. Wilson," he said formally. "Mr. Swift and I are going out to lunch. Hold all my calls."

"Yes, Mr. Howard," Margaret said formally before giggling. "Are you taking the kid to MacDonalds?"

"Sure why not?" David said and hung up. Margaret was right. The kid was not dressed appropriately for the IBM Formal Dining Room. "You don't mind if we go out?" David asked as put on his suit coat.

Bill Oldham shoved his chair back from his desk and stared at the blank sheet of paper in his typewriter. He was supposed to be drafting a short story for his creative writing class. Usually, he could mull a situation over in his mind for a few minutes while walking around campus then return to his typewriter and bash out a quick story in under an hour. Tonight, the words were not coming. He did not like the topic: A man faces a life shaping decision; he wants to discuss it with his closest friend; he cannot, because…

Bill found the scenario trite. He turned to his roommate Mike Hamilton who lounged on his bed idly leafing through an advanced economics theory textbook. "How can you read that stuff?" Bill asked. Economics were the ultimate in boredom. His lowest grade at Harvard, a C, had come in Economics 1. He had never taken another economics course.

"Do you think Charles will make it?" Mike asked.

Bill shrugged. He walked to the window. Snow had begun to fall again. They already had three new inches on the ground. "If he doesn't kill himself in that damned Mustang."

Charles had phoned three days earlier to announce his return to Harvard. He asked if he could temporarily join Mike and Bill. As upperclassmen, they shared a small two-room suite, two beds in one room, with two desks and a battered couch in the other. "You get the couch," Bill had warned. Charles agreed and said he expected to be there three days later, following a brief stop in Armonk.

"What the hell is in Armonk?" Bill finally asked a question that had nagged him since Charles' call.

"IBM," Mike said.

Bill laughed. "Charles can't be thinking about going to work for IBM." Trying to picture the highly intelligent and independent Charles as an organization man was ludicrous.

"Maybe he's going to buy them out," Mike said.

Bill did not bother to respond. He, like Mike, had high respect for their friend's abilities, but IBM was too long a reach even for Charles. "Maybe he wants to sell them software,"

Bill was a little more knowledgeable about computers since his summer in Albuquerque. At least he knew the jargon. He still had trouble understanding Charles' obsession with microprocessors.

"If he wants to, he will."

The phone rang. Bill tried to ignore it. "Will you please answer that. You're closer," Mike said.

"It's Charles' mother," Bill said.

He still remembered his last conversation with Charles' mother. It had occurred the day after his return from Albuquerque. Mrs. Swift in her most imperious voice had asked to speak with Charles, thus informing Bill she expected her son to be at Harvard not in Albuquerque. Bill had tried to equivocate, but Mrs. Swift would have none of that. "Is he there or in Albuquerque?" she had demanded.

Bill had not known what to respond. He knew Charles did not plan on returning to Harvard.

"Your silence tells me Albuquerque," Mrs. Swift had declared. "Thank you for your help, William," Mrs. Swift had declared and hung up the phone.

"Let's not answer," Bill said. "Charles can talk with her himself."

"Maybe it's Charles on the phone. He's had an accident or something."
Bill picked up the phone.

"Hi Bill, this is Charles' mother. May I speak with my son?"

"Hello Mrs. Swift. I'm sorry but he is not here yet."

"But he's coming. Right?"

"You know Charles better than I. It's possible he got held up in Armonk."
Bill considered telling her that it was snowing again, but decided against it.

"Armonk! What's Armonk?"

"I understand that's where IBM is."

"In what state?"

"New York."

"Thank God. At least he's getting closer. Please have him call me as soon as he gets in."

"Yes ma'am."

"Do you no why he stopped in Armonk?"

"No ma'am."

"More of that damned computer business. I wish I had not gotten him started on that."

"Yes ma'am."

"Thank you Bill. Have him call."

Bill hung up and turned to find Mike smiling. "Somebody I know?" Mike asked. Bill ignored him.

Charles finally arrived at eleven-thirty just as Bill and Mike were preparing for bed.

"Damned snow," were Charles first words.

"Call your mother, now," Bill responded.

Charles dropped his suitcase and parka on the floor and phoned Seattle. Charles' side of the conversation consisted of a series of "yes mothers and "don't worry I am here now."

Charles registered for a math course, an advanced business course, European History, and Philosophy 300. Following his usual practice, he dutifully attended the first lecture of each course, learned the names of the textbooks and reading materials, then avoided class thereafter. He devoted his days to long telephone conversations with Clap in Albuquerque, writing code, Extended and Disk Basic, and marathon poker games with greatly elevated stakes. Charles continued to win moderately and consistently while some players suffered losses that could reach as high as one thousand dollars a day.

In February, Charles, following a lengthy conversation with Clapper, wrote a letter called "An Open Letter to Hobbyists." He dashed it off in an hour and gave it to Bill to proof read. In it Charles complained about hobbyists who were stealing software. He cited the fact that less than ten percent of all Taurus owners had purchased BASIC, but all seemed to have it. He concluded that the hobbyists were copying the software and passing it along without paying the programmer who wrote it. He concluded by asking that anyone who wanted to pay up to write him at an Albuquerque address. He signed the letter: "Charles Swift, General Partner, Digital Software."

Bill corrected a couple obvious grammatical errors in the text then joined Mike in an effort to dissuade Charles from sending it to Albuquerque for dispatch to PAMS' customer list. Charles listened impatiently to the arguments that ran the gamut from "nobody's going to pay for something they can get for free" to "they will just laugh at your naiveté." Charles stubbornly insisted that the users were thieves stealing from hardworking programmers like himself who deserved a just reward for their toil. He mailed the letter to the PAMS secretary who in turn sent it to every computer publication.

The reaction in the home computer community was immediate and strong. Many seized the opportunity to criticize PAMS for its buggy software, security agreements, misleading advertising and many other complaints that purchasers had about the Taurus. Few offered to pay up as Charles had challenged. Software piracy was not a new issue. Charles was not the first programmer to insist that the product of his intellectual efforts was protected by copyright, but no one could suggest how to end the piracy. Many respondents attacked Charles personally. One letter, from a lawyer, called the open message defamatory and lectured Charles on copyright law.

The reaction did not amuse Charles. He had been deadly serious. Software piracy posed a threat to the company he was trying to create. However, Charles did not desist from his efforts. In March, without first informing his mother, Charles flew to Albuquerque to attend the First Annual World Taurus Computer Convention. The income from the Taurus contract freed Charles from the tethers that were his parents' purse strings.

Clapper met Charles at the airport and drove immediately to the Airport Marina Hotel where the gala was set to open the next day.

"You can't believe the response," Clap gushed.

Charles, who had not been privy to the planning, looked skeptically at his friend. He could not bring himself to admit that business in Albuquerque was progressing without him. Damn Harvard.

"How is our booth?" Charles asked.

"It's got a lot of bullshit stuff," Clap said.

Charles called the games, the chess, backgammon and music programs, "bullshit stuff." Clap did not agree. Digital Software had to provide the products the users wanted. Clap humored his friend. He understood Charles' frustration over his isolation in Harvard while business was progressing without him. Clap had persuaded Ed Johns to feature Charles as the fair's leadoff speaker. Charles' confrontational approach on software piracy could generate considerable interest.

"But it looks good," Clap continued. We've got reporters from all the publications."

They toured the large room where the booths had been set up. Of course, the Taurus was the star of the show. It was the First Taurus Convention. They met Ed Johns who entered as they were leaving.

"How's it look James?" Ed greeted Clapper.

"Great, Ed," Clap said.

"And how's our lead speaker?" Johns acknowledged Charles.

Charles nodded and held out his hand. His relationship with Ed Johns had been cool since Charles' threat to leave if Johns did not sign the Digital Software contract. Johns crushed Charles' hand, then ignored him.

"Some of the dealers are complaining," Johns spoke directly to Clap.

"About our resistance to diversification?"

Johns nodded assent.

"Some of the Taurus dealers want to handle other brands," Clap explained to Charles.

"Makes sense," Charles said. He had read that other small companies were beginning to produce microcomputers that were superior to the Taurus.

"Not in my stores they won't," Johns stared at Charles. He decided he really did not like the kid, boy genius or not.

"Charles just got in," Clap changed the subject. "I'm taking him over to the house." During Charles' absence, Clapper had moved the four employees of Digital Software into a rundown house where they lived and worked.

Ed Johns grunted and turned to conduct his own inspection of the fair displays.

The next day twenty-year-old Charles Swift opened the fair with his first public chat about the future of the industry and software. He stressed that the microcomputers were the tools of the future. Every home will have one, and people will use them to think with just as today they use typewriters to write with. Despite his youthful appearance, his squeaky voice and his controversial stand on software piracy, Charles' depiction of the technological revolution appealed to his audience. They were the convinced. Charles to his surprise received an enthusiastic applause when he finished.

"Great job, Charles," Clap shook Charles' hand when he left the podium. Although he had worried about how Charles who lacked charisma would do, he had not mentioned his concern.

"We've got to sign a new contract," Charles said.

The timing of Charles' announcement surprised Clap, but the fact that Charles was not satisfied with their deal with PAMS was not news. They had given PAMS exclusive worldwide license for the software for a ten year period. They had agreed that PAMS would sublicense the software to all comers, thus giving Digital Software royalties on a broad scale. The catch turned out to be that Ed Johns made no effort to sublicense their software. His interest was in selling Tauruses, and he appeared to be indifferent that others were freely pirating their BASIC programs.

"The old contract only covers BASIC not 6800 BASIC," Charles whispered as he took a chair on the platform next to Clap. The next speaker was Ed Johns.

6800 BASIC was Charles' software for the new chip.

They listened to Ed's uninspired tribute to his Taurus, then left the platform. Charles ignored the displays.

"We'll license the 6800 BASIC on a nonexclusive basis for a flat fee to be paid monthly for the next two years."

"Ed will insist on exclusivity," Clapper said. "Why do that?"

"So we keep control of our software. Ed hasn't sold a single sublicense. We know there are other customers out there, so we can sell it ourselves."

Clapper nodded agreement.

"And we won't have to worry about the pirates stealing our software from the Taurus like they do now."

Clapper understood then. Currently they received royalties only when PAMS sold the BASIC with the Taurus. Most users were buying the Taurus without the software then making pirate copies at their clubs or borrowing it from friends. With Charles' new proposal, they would receive a flat fee and it would not matter if PAMS marketed their software or not. And, they could sell their new software themselves to the new computer makers.

Before he left Albuquerque, Charles informed Ed Johns of their decision. Johns did not like the loss of exclusivity, but he recognized he had no choice. The kid had offered a deal, take it or leave it. Johns took it. That step freed Digital Software from its dependence on PAMS. They were free to grow as they could.

The First World Taurus Trade Fair provided Digital Software with a platform, and Charles hoped to use it to free them from their dependence on

PAMS. Although he was back in Harvard, Charles continued to focus on Albuquerque. Joy Wilson supported Paul Harbinson, Charles' friend from high school who worked full time, and Dave Marks, who moonlighted while attending the University of New Mexico. The one full timer, two part timers and Clap kept DS going while Charles chaffed time away in Cambridge. Joy weekly sent Charles her clerical contributions to what passed for Digital Software's books. Charles managed the payroll, conspired over taxes, studied contracts, and searched for new customers. Charles relied heavily on Mike Hamilton' advice. Charles was determined to establish DS on a firm administrative footing, and Mike's growing expertise as an upper class Business Administration major helped.

Customers, however, not business savvy, were what DS needed most. The income from PAMS barely met their expenses. Some three weeks after Charles' return to Harvard, two clients appeared; both were electronics firms with growing reputations, Motorola and Intel. Their purchases of BASIC were small. Both insisted on buying in single unit quantities, hardly a relationship designed to expand DS' coffers, but Charles was pleased with the exposure. Then, just as Charles began to worry about cash flow, NCR, the cash register company, asked Digital Software to adapt BASIC for its data entry terminal. NCR, one of the so-called seven dwarfs competing with IBM, agreed to a contract, and Charles assigned Paul to work on it full time.

The NCR contract represented only a temporary life ring for DS. Having learned from the Taurus Fair that the relatively new phenomenon of electronics fairs offered opportunity to hunt for customers as well as the chance of viewing the latest technological developments, Charles decided to have Paul join him in New York for the National Computer Conference. Charles drafted a flyer that touted Digital Software's small portfolio of products. Bill Oldham rewrote it at no charge, and Charles had two hundred copies printed.

Paul and Charles worked the New York fair diligently. They passed out their pamphlets from the popular Taurus exhibition and assiduously assessed each display for potential customers for their BASIC. Having learned from their National Cash Register contract, they did not limit their search to the growing number of microprocessor companies. Charles returned to Harvard with a list of thirty prospective customers. To each of them he wrote and introduced himself and his company. He included a copy of his convention flyer just in case the customer had missed it. The New York trip produced an entry for the debit side of DS's ledger, but unfortunately that was all. Undeterred, Charles continued to insist that DS if it hoped to survive had to continue to spread word of its existence. To that end, he sent letters to potential customers, and in this effort he was joined by his New Mexico staff, particularly Paul, who concentrated on the electronic community now proliferating in what was becoming to be called the Silicon Valley.

Paul one day phoned a startup outfit called Astron Computers. The only number for Astron Computers that Paul could find produced a female who called herself Mrs. Pitts. Paul asked to speak to the head of the company, but he was not available. Paul left his number, but Mr. Stanley Pitts never returned his call. After several abortive efforts, Paul finally reached Mr. Pitts.

"Mr. Stanley Pitts?" Paul asked.

"Yes."

The way Pitts spoke that one word irritated Paul. The tone implied that Paul was intruding on the time of a busy man.

"I represent Digital Software," Paul said.

"So?"

"You may have received one of our flyers."

Silence ensued.

"We are based in Albuquerque where we provide BASIC for the Taurus," Paul tried again. Everyone had heard of the Taurus.

"That doesn't impress me."

Paul wondered who in the Hell Pitts thought he was. Astron Computers was just another startup. His information indicated they still worked out of a garage. Paul decided to ignore the man's attitude.

"We're interested in expanding…"

"Who isn't?"

"If you should need BASIC or any other software for your Astron, we can provide it."

"I doubt it."

"I assure you…"

"The Taurus is a piece of garbage, and your software is buggy," Pitts interrupted.

This time Paul was on familiar ground. "I assume you have one of Taurus's early products. It has been greatly improved."

"I wouldn't own that junk."

"Well, in any case, please keep Digital Software in mind." Paul decided to surrender. Pitts was a waste of time.

"If we need software, we'll write our own over the weekend. Don't call us, we'll call you." Pitts hung up.

Chapter 11

Stanley slammed down the phone, and his mother shook her forefinger in disapproval.

"Don't be rude, Stanley," she chastised.

"Sorry mom, but he's selling junk. I don't have time for it."

Stanley checked the sleeves of his blue blazer for wrinkles, then he smoothed his yellow turtleneck shirt over his khaki slacks. He stroked his Clark Gable mustache that his mother labeled "interesting" and preened, waiting for approval.

His mother ignored Stanley's posturing and stuck to her subject. "He seemed like a nice man."

"That's what salesmen do. They pretend, then they try to hook you. I'm expecting Mr. Miller. Now that's important."

"Oh? He's here now."

"Why didn't you tell me," Stanley complained.

"I was going to, but then that nice man called. You've been fussing at me since, so I didn't have a chance."

"Thanks mom."

Stanley hurried to the front of the house and peeked through the curtains at the driveway. He saw a gold Corvette convertible with the top down parked in front of the open garage doors. Stanley hurried out the front door, turned, then slowed his pace to a saunter, determined to make an entrance. As he approached the garage, he spotted a tall, slender man with thin, moderate length, dirty blond hair standing beside Dum at the workbench. An aging playboy in appearance, he looked like Stanley's research had led him to believe he would. That aside, Miller was a minor valley legend. A graduate of the University of California, Miller had started at Hughes Aircraft, moved on to Fairchild then Intel. While still a nobody, Miller had borrowed and invested big in Intel stock when few others were interested. He was thirty when Intel went public and made him a millionaire. Three years later Miller retired a multimillionaire. John Davis had given Stanley Miller's name as a venture capitalist who might be interested in Astron Computers; this was Miller's first visit, and Stanley was determined to impress him.

"Hi," Stanley announced his entrance.

Neither Dum nor Miller looked up. Both were studying the open wood case of the Astron 1. That wasn't what Stanley wanted. He had told Dum they had to impress Miller with the Astron 2. Stanley was proud of the classy plastic case. It looked like Stanley thought a computer should look. Instead, Dum was showing off his damned circuit board.

"That is a very professional piece of work," Miller patted Dum on the shoulder. Dum, who rarely showed emotion, was smiling broadly.

Stanley was pissed but tried not to show it. The two were acting as if he didn't exist. Dum, ignoring everything Stanley had told him, appeared to think he was at a computer club meeting of enthusiasts. Miller was not another hobbyist; he was their potential benefactor.

"Did you do it all yourself?" Miller asked.

"Yes."

Miller turned suddenly and nodded at Stanley, catching him with a frown on his face while he nervously tapped his right thigh. The fact that Dum had blatantly taken full credit for Astron 1, acting as if Stanley did not exist, truly irritated Stanley. Stanley forced a smile and offered his hand. Miller returned the smile and gave Stanley's hand a quick squeeze.

"Your partner is a talented engineer," Miller said.

"He is that," Stanley seized his opening. "I'm Stanley Pitts," Stanley said. "Dum handles the engineering, with a little help, and I manage the company."

"Pleased to meet you Stanley. Is this it?" Miller's gaze roamed over the garage.

In Stanley's opinion, it did not look bad. In a burst of enthusiasm, Stanley's father had lined the walls with paneling, built another workbench along the back wall, ran electricity, and constructed cabinets on the third wall. Stanley made a mental note to thank his father for the help. Miller seemed impressed.

"Would you like to take a seat?" Stanley indicated the old upholstered chair, the obvious place of honor, usually Stanley's throne when he talked to Dum's back as he worked.

"If you don't mind, I would like to study Mr. Dumbroski's work a little more closely. I'm an electrical engineer," Miller said.

"Please," Stanley capitulated. Dum sat on his stool behind Miller with a broad smile on his face. Stanley lowered himself into the easy chair and watched like an outcast as Miller and Stanley examined the damned circuit board.

"We've sold a hundred of those and are ready to ink a contract for another one hundred and fifty," Stanley tried to recapture Miller's attention.

"I've talked with Jake Foster," Miller smiled.

Stanley could just imagine what Foster had said. Most of the original circuit boards remained on the shelves of his stores. Miller silently studied the Astron 1 for ten minutes. It seemed like two hours to Stanley. He was convinced Dum had destroyed their opportunity. Finally, Miller stepped around Dum and approached the Astron 2.

"Now that is a real computer," Stanley began, assuming he now had the opportunity for his pitch. "It has color..." Stanley began then stopped. Miller did not appear to be listening.

"I don't see any screws in the case," Miller spoke directly to Dum.

"Stanley designed the case. He thought it looked better with the screws underneath," Dum threw Stanley a bone.

"It does," Miller nodded at Stanley. "And I like the molded plastic. Appearance alone should sell it."

"I tried to incorporate a modern look, just like HP uses for its calculators," Stanley tried again to guide the discussion.

"Can we open it up?" Miller addressed Dum who responded by handing Miller a screwdriver.

As Stanley watched, Miller tilted the monitor backward exposing the underside of the Astron 2's case. Miller deftly turned the ten screws and exposed Dum's handiwork. Again, Miller silently studied the circuitry. "I truly am impressed, Mr. Dumbroski. You are a talented engineer."

Stanley, frustrated, uncharacteristicly listened as Dum and Miller traded technical comments. After about half an hour, Stanley retreated to the kitchen where he brewed himself a cup of tea.

"How's it going?" Stanley's mother asked.

"We're dead," Stanley said. "I haven't been able to get a word in edgewise."

"Be patient," His mother counseled.

Stanley turned and started back to the garage, tea in hand.

"Are you going to offer your guest refreshments?" She asked.

"Let Dum do it," Stanley frowned.

Stanley returned to the garage where he found Miller now sitting on his mother's kitchen chair and Dum busily re-inserting the screws in Astron 2's bottom.

"Mr. Miller would you like to join us in tea?" Stanley's mother asked from the doorway.

"I would enjoy that," Miller replied.

Mrs. Pitts disappeared. Stanley sat down in the easy chair and sipped his tea.

"She's one of our employees," Dum announced.

Stanley choked, splashing tea on the front of his yellow turtleneck. He silently vowed to throttle Dumbroski. It was bad enough to have to admit that Astron Computers worked out of the Pitts garage. Now, Dum identified Stanley's mother as an employee. It made them look like two kids playing business.

Dum finished his work, righted the Astron 2, then turned it on. After a couple of minutes, the bright red announcement "ASTRON COMPUTERS" and

"READY" appeared on the screen. Dum nonchalantly hit a few keys and was soon playing a game of Tic, Tac, Toe. Mrs. Pitts served tea to Dum and Miller, then retreated to the kitchen.

Stanley morosely wiped at the spots on his shirt with his handkerchief. When he looked up, he found Miller staring at him.

"All right Stanley, tell me about your business plan."

"We are on the threshold of a technological revolution..." Stanley began. He had been working for a week on a presentation that let him appear as a visionary, a role he played well.

"No, I mean your business plan," Miller interrupted. "Don't forget. You're talking to the converted. I worked at Intel."

Miller recognized that his cool treatment of Stanley was irritating him, but Miller acted with deliberation. That's why he shut Stanley out of the conversation with Dumbroski. It was clear that Dumbroski was a qualified engineer, some of his colleagues described him as a genius, but Miller had downplayed that. Having seen Dum's careful work, he now suspected that the praise had been deserved. Now, Miller wanted to test Stanley. He had been described by some as an arrogant hustler who exploited his friend's talent. Few seemed to genuinely like Pitts, but most had grudgingly admitted that he had some talent. At least he was a pusher. Miller knew that momma's boys did not build companies, not the kind that Miller wished to use to enhance his already considerable reputation and fortune.

"Astron Computers is at a key juncture, Mr. Miller. We have a good product in the Astron 2. You know the potential market as well as I do. There is a future for the microprocessor. Some visionaries believe, as I do, that the future of home computers is virtually unlimited. The world is a large place. Imagine every home needing a computer." With these words, Stanley tried to slip back into his prepared presentation.

"How are you going to make Astron Computers a part of this revolution?" Miller asked. "Certainly, you aren't going to produce anything more than prototypes here."

"Of course we need space to grow," Stanley's irritation at being interrupted disclosed his natural arrogance. "I need capital. With that, I will hire the employees and build the plants and develop the advertising that will sell our products. The Astron 2 is only the beginning."

"Let's talk specifics, Stanley," Miller interrupted again.

"We have a public relations agency," Stanley began again.

"The Moss Agency. A wise choice," Miller said.

"We are going to produce the best," Stanley tried again.

"I believe you have the engineering talent to accomplish that," Miller smiled at Dum. "How are you going to become the best."

"By hiring only the brightest."

"Are there many of those around?"

"I'll find them."

Miller liked the way Stanley did not back down. It was going to require more than bluster to survive in the world that young Pitts intended to conquer. Miller checked his watch. He had learned enough to think about. He saw no need to torment Stanley further at this point.

"Gentlemen, I have another appointment," Miller said as he checked his watch. "I hope you will excuse me."

"Are you going to join us?" Stanley demanded.

"It's too soon to answer that question."

The frank response raised Stanley hopes. He smiled. "I'll start work immediately on a business plan which will outline exactly how we will use your investment."

"That sounds like a good idea," Miller agreed. "But remember, everything is still hypothetical at this point. I have much more homework to do myself."

Stanley recognized that meant evaluating him and Dum. Stanley hoped Miller stayed away from the high school. He did not have much of a reputation there. Only one teacher would speak for him. Actually, old Mrs. Connel had understood him.

Dum did not respond. He turned off the Astron 2 and started work again. He wanted to improve the power supply that tended to overheat.

Stanley followed Miller to his Corvette. "That's a great car," Stanley said.

"You'll have one someday to replace the old van you had to sell," Miller said, demonstrating he had already done some investigating.

Stanley did not know how to respond. He found this intrusion into his personal life unsettling and did not like it. He assumed that if their places were switched he would be investigating Miller in the same way, so he had to live with it. Still, he did not have to like it.

"When shall we meet again?" Stanley asked.

"Tomorrow too soon?" Miller smiled.

Stanley smiled back. He would meet with the millionaire as often as he wanted just as long as Stanley got his hands on some of his money when they finished.

To Stanley's dismay, and encouragement, Stuart Miller became a regular at the Pitts garage, dismay because he made no commitments, encouragement because he did not say no. He usually appeared two to three times a week and talked with Stanley about his plans while Harold worked. Periodically, Miller would examine Harold's efforts and complement him. He remained noncommittal about Stanley's views. Eventually, Stanley and Harold accepted Stuart as a regular, someone whose company they missed when he did not appear. Stanley had difficulty judging Stuart. He had expected a quick yes or no from the venture capitalist and did not get it. Stanley concluded that Stuart was too cautious to be a venture capitalist. What Stanley did not understand was that Stuart was not sure he wanted to get back in business. The hesitation did not come from doubt about Stanley or Harold. Stuart liked them both and was confident their nascent business had potential. Stuart had enough money to last him and his family a lifetime, maybe two. He lacked, however, respect. He knew his peers at Fairchild considered him an average engineer. They thought he had been lucky with his stock option speculations. Stuart knew better and tried to ignore these negative opinions, but deep inside him still existed a need for recognition that had not been sated.

Finally, after two months of careful study, appraisal and introspection, Stuart decided.

Stanley was lounging in the garage while Harold worked when he saw the familiar Corvette slide to a halt in the Pitts driveway. "Here's Stuart," he announced.

Harold did not respond. He liked Stuart, a fellow engineer, and did not find his visits a distraction. Stanley, however, still irritated him at times.

"Hi fellows," Stuart said.

"Hi Stuart," Stanley said.

Harold waved with a free hand.

Stuart retrieved a chair from the Pitts kitchen and placed it in the center of the open garage door where he could face Stanley and Harold equally. "I've made a decision," Stuart announced.

Stanley sat up straight. Harold continued to work.

"You should listen to this carefully, Harold," Stuart said.

Harold raised his hands from the bench and turned to face Stuart.

"You need working capital to expand. $250,000 should get things started," Stuart said.

"A new plant, personnel, advertising," Stanley said. He tried to hide his excitement, but a break in his voice when he spoke unmasked him. "Damn," Stanley said as he leaped to his feet and offered Stuart his hand.

Stuart smiled. He liked the enthusiasm. "But first we have much to discuss."

"We've already talked ourselves to death," Harold said. His sour expression signaled his concern. A quarter of a million dollars was a lot to risk on his inventions.

"We can do it, Harold," Stanley patted his friend on the back. "We'll build and sell your computers. Look out IBM," Stanley enthused and began pacing.

"I don't need any help," Harold said. Privately, he liked things the way they were.

"Please sit down, Stanley," Stuart said. "You're making me nervous."

"I can't help it," Stanley continued pacing. "When can we have the money?"

"I'll give you $250,000," Stuart began again.

"What do you get out of it?" Harold asked. Nobody just gave people money for nothing.

Stanley stopped pacing and waited for Stuart's answer. To get the money they needed, he would give Stuart anything he wanted.

"For the $250,000, nothing. It's for the company. However, I'll pay an additional $67,000 for one third interest in Astron Computers."

"Done," Stanley said.

"Wait," Harold said.

"One third of $317,000 is over $105,000," Stanley said. "Stuart just gave you $105,000."

The thought made Harold pause.

"The three of us are partners," Stanley said. "Plus one percent for the Moss Agency."

"We can negotiate that," Stuart said.

"My word to them is good," Stanley said. He did not want to upset the Moss Agency, at least not soon. They were preparing a press release featuring Stanley and Astron Computers. Stanley liked the idea of seeing himself described in print as the leader of a promising startup company. Who knows where that could lead?

"That's a minor problem, Stanley," Stuart said.

"Where does that leave me?" Harold asked.

"You are our chief engineer," Stanley decided.

"You will have to leave HP and devote your full attention to Astron Computers," Stuart ignored Stanley's comment.

"I can't," Harold said. Harold liked his job at HP. He was a married man and it gave him security.

"You have to," Stanley said. Stanley decided that as company president he would have to be firm with Harold.

"I haven't decided to let him have my computer," Harold nodded at the Astron 2.

"That's part of the deal," Stuart said.

"Do we need him?" Harold looked at Stanley.

"We need Stanley. Without him, you don't have a deal."

Stuart's strong defense of Stanley surprised Harold.

"Hey, I'm not the bad guy. These are my inventions," Harold blurted. "What will we pay him?" Harold again stared at Stanley who had stopped pacing and was studying his friend with evident disbelief.

"Now Dum..." Stanley began but Stuart interrupted him.

"We'll pay Stanley $20,000 and you $20,000 a year, just to begin," Stuart said.

"That's a lot of money Dum," Stanley said. "More than fair."

"What will he do to earn $20,000?" Harold persisted.

"I'll be chief operating officer," Stanley declared.

"Now that's something else we have to discuss," Stuart said. This time his words shocked Stanley who sat down quickly. "We must bring in a professional businessman to set up Astron Computers properly and to manage us efficiently."

Harold smiled. He turned back to his workbench and picked up his soldering iron. As far as he was concerned, the discussion was over. Stanley would never accept the deal, and he would not have to worry about quitting his job.

The color drained from Stanley's face. "What will I do?" Stanley pleaded. He had difficulty comprehending what was happening. Stuart had offered the key to open the door that could lead to the attainment of his every dream and now they were telling him he could not be a part of it.

"Stanley," Stuart spoke evenly. "You know nothing about running a business. We'll get a professional and build Astron Computers the way we should. We're on the threshold of something very, very big. You convinced me."

Stanley did not know how to respond, so he sat quietly with his mind churning. He had to be boss. It was all his idea. Without him, Dum's computers would just be toys.

"You must decide what you want, Stanley. Power, money, status? You can have these if we do things right. Or, you can have this." With a wave of his hand, Stuart indicated the Pitts garage.

"Power," Stanley spoke softly.

"Take money first," Stuart advised. "Then everything else will follow."

"OK. Then tell me this," Stanley continued to search for a way. "What will I do? Stay home? If that's all I do, I already have that." Stanley mocked Stuart's wave at the garage.

"You won't stay home. You will be one-third owner of an important company. You'll be able to do anything you want. You will have the power."

Stanley smiled thinly. "We have a professional manager who runs things, but I can do what I want?"

"On the condition you do not undermine the president's authority. He's the boss in administrative matters, personnel, bookkeeping, sales, all logistics."

"What about public relations?" Stanley asked.

"We'll work that out."

Stanley was torn. He was being offered money or nothing. He decided he would take the money then see if they could keep him from running the company. He did not need the title. Not at first. After they became a going concern, he could persuade Harold to vote with him and they could select whomever they wanted as chief executive officer. "I agree," Stanley said.

"That leaves it up to you," Stuart turned to Harold who had once again turned his back on his workbench. Stanley's answer had surprised him.

"What's up to me?" Harold asked. He now knew what Stanley had felt when Stuart's cold eyes had focused on him.

"We need your full time commitment," Stuart said.

"I can't leave HP," Harold said.

"Then we don't have a deal," Stuart said, standing.

"Wait," Stanley said.

Harold and Stuart waited for Stanley to continue. He did not know what to say. Everything now depended on Harold.

"Give us a little time," Stanley spoke to Stuart. "This is a big step. Let us discuss it."

Both Harold and Stuart smiled at Stanley. Their impulsive friend was the last person either expected to plead for time. Harold nodded. His expression indicated that nothing Stanley could say would influence him.

"Very well," Stuart agreed. "Take as long as you need. It's an important decision."

Stuart departed, leaving two very confused young men behind.

As soon as the gold Corvette backed out of the driveway, Stanley spoke: "Harold we can have one of those," he referred to the car.

"Fine, but I'm not quitting HP."

"What do you want in life Harold?"

"To be an engineer. I don't need to be rich, and I certainly don't want to run a company." The last words were intended to bring Stanley down to earth.

"You can do that," Stanley said. "You come to work at Astron Computers," in his mind it was already a big company, "and I guarantee it. You can be an engineer for the rest of your life. We will own two thirds of the stock. If we vote together, we can do whatever we want, no matter what Miller says."

Harold did not respond.

"You can't turn down one hundred and five thousand dollars. That's what he's offering us."

Harold turned back to the bench. He refused to discuss the matter further. Stanley dropped into his chair and tried to think. He realized that nothing he could say to Dum now would persuade him. Too excited to just sit there, Stanley without a word started out on a long walk.

Over the next two weeks Stanley avoided raising the subject again with his friend and partner. Instead, he tried subterfuge. First, he visited Dum's parents to enlist their assistance in persuading their son to be reasonable. To Stanley's surprise, Dum's father turned Stanley's appeal into an attack on Stanley, accusing him of exploiting his son. This unwarranted assault forced Stanley to retreat, but he did not surrender. He approached every mutual acquaintance that he and Harold had in common and enlisted their support in a campaign to persuade Harold to accept the simple premise that he had to seize this chance of a lifetime. For two weeks, colleagues at HP, old high school acquaintances, even a few from grammar school, besieged Harold with advice, but to no avail. He refused to commit himself. Finally, while Harold was at work at HP, Stanley visited Helen at their small apartment.

"Stanley, don't say it," Helen greeted Stanley at the door. "Dum's going to do what he wants to do. He doesn't listen to me."

"But Helen," Stanley began.

"If he listened to me, he wouldn't spend every night at your damned garage."

Stanley saw his opportunity. "That's it, Helen. With only one job, he'll be able to spend more time with you. I guarantee it."

"How can you guarantee anything? Want some coffee?"

"Do you have any tea?"

Stanley followed Helen into the tiny kitchen. "Would you like to have a bigger kitchen?" he asked. "New appliances? A freezer?"

"Don't tell me you're going to buy them for me," Helen laughed. She often wondered about her husband's strange choice of friends. Stanley was just a kid.

"Did he tell you about the money?"

"What money?"

"If he refuses Mr. Miller's offer, he's turning down one hundred and five thousand dollars."

"You're kidding me," Helen laughed uncertainly.

Stanley explained and stressed the $20,000 salary. "All Harold has to do is work full time at Astron Computers."

"In your father's garage?"

"Mr. Miller is a venture capitalist. He's a multimillionaire." Stanley did not know how rich Stuart was but knew he had a lot. "We're opening a factory."

"And if it goes belly up what happens to me?"

"Dum will get his job back with HP."

"You guarantee that?"

"You know I can't. Dum's a good engineer. They'd jump at the chance to get him back."

Helen sipped her coffee. Stanley drank his tea. He knew enough to wait. He was playing his last card.

Finally, Helen sighed. "What do you want me to do?"

"Tell him to take Mr. Miller's offer. He can be an engineer and have money too."

"I'll talk to him tonight at dinner."

That evening Stanley waited anxiously in his room for Harold to arrive. It was long after seven when Harold finally appeared, almost a full hour late. Stanley had planned to let Harold work alone for an hour. He hoped that Helen had not told Dum about his visit. However, five minutes after Dum's arrival, Stanley lost his resolve to be patent and hurried to the garage.

"Hi Dum," he tried to be nonchalant.

"Stanley," Dum said.

Stanley took his chair. He sat in silence while Dum worked for about fifteen minutes. Finally, Stanley could not stand the tension. He joined Dum at the workbench.

"Anything I can do to help?" Stanley asked.

Dum snorted. He made no secret of his disdain for Stanley's technical prowess. Dum never let him work on anything important. Sometimes, he allowed Stanley to install components on a circuit board, the same as Stanley's sister did. "You can hold this," Dum indicated the power supply he was trying to install in the back of the Astron 2.

"What about a fan?" Stanley asked. He knew Dum was worried about the heat buildup in the plastic case, but Stanley did not like the noise a fan made.

"Maybe we have enough room that the heat will dissipate through the vents without one," Dum decided to tease Stanley. He knew he had to install a fan. It made good engineering sense.

Their relationship had been troubled since the talk with Miller. Dum recognized it was his fault and that he had said some things that had insulted Stanley. He had said the same things in private and Stanley had not reacted. This time, however, Harold had spoken to Miller who Stanley always tried to impress.

"I'll do it," Harold said.

Stanley felt like jumping in the air and shouting. Helen had done it.

"Good," Stanley said calmly. "I'll tell Stuart."

Three weeks later Stanley and Harold dissolved their partnership. On January 3, Miller, Stanley and Harold formed the Astron Computer Company. Miller produced a bottle of champagne, and Mrs. Pitts joined them in the garage. They drank to the company's future, and Stuart gave Harold his first assignment as chief of Astron Computers R & D unit. "We have to have a floppy disk drive in time for the First West Coast Computer Faire in April."

Chapter 12

In December, Charles, still exiled to Harvard, encountered his first commercial crisis.

Faced with the demands of new business and his mandated return to Harvard, Charles in September had hired two Stanford programmers to join Paul Harbinson, Dave Marks and Joy, increasing DS' employee payroll to seven when the two partners were included. Somebody had to take up the slack created by Charles' absence and Clapper's full time preoccupation with PAMS. To accommodate the new staff, Charles moved Digital Software to offices on the eighth floor of the Park Towers.

Unfortunately, the new employees proved a mixed blessing. They fulfilled DS's contracts, but programmers did not bring in new contracts, and neither did the partners who were preoccupied elsewhere. Overhead, particularly the payroll and the new offices, ate deeply into the DS profit line. The situation was not helped when PAMS failed to license DS software to others as the contract required. The cause was not difficult to discern. Ed Johns, who was struggling to keep his own expanding company afloat, saw no merit in helping his competitors to share his market. Then, in early December, without a word of warning to his young associates, Ed Johns signed an agreement to sell PAMS to the Perfect Computer Corporation. Neither Charles nor Clapper could forecast with accuracy what this would mean to DS; they were not privy to Perfect's plans, and neither Ed Johns nor the Perfect representatives would speculate about the future. A worried Charles discussed his problems at length with his mother and father. Consequently, his concerned parents, still desirous of keeping their entrepreneur son in Harvard, offered a loan that Charles declined.

"What in the hell am I doing here studying Victorian history?" A desperate Charles asked.

Bill Oldham and Mike Hamilton exchanged exasperated looks.

"Charles, make up your mind. Be a businessman or a student. You can't do both," Bill said. He was tired of Charles' constant harping.

"He's right, Charles," Mike said. "Nobody can manage a company from one thousand miles away."

"I'm out of here," Charles said and started for the door.

Worried that he had offended his friend, Bill asked: "Want company?"

"No. Keep at the books. I'm just going upstairs."

"Poker won't solve anything," Mike persisted.

"Maybe I can win enough to keep DS afloat."

"When you need it, is when you lose it," Bill warned.

Charles closed the door behind him.

"He should fold this in and go to Albuquerque," Mike said.

"Here we are," Bill said. "Studying at the world's greatest university. Our roommate is one of the brightest young men on the planet. It should be a match made in heaven."

"Charles doesn't care about Harvard. It has nothing to offer him."

"I'll bet high school was the same way. He was too far ahead of everybody else."

"Well that's tough shit," Mike let a little jealousy show. "He's got to grow up and decide what he wants. Please his parents, or run his business."

"If he does," Bill said. "We'll need a new roommate."

Charles made his decision in February. The new semester started just like all the others. He attended the first day of class, then retreated to his shared apartment. The news from New Mexico grew more ominous. Suddenly, on a gray, overcast typically New England day, Charles packed his bags, loaded them in the Mustang, said goodbye to his friends, and departed Harvard, for good.

Two days later a haggard, road weary Charles parked the Mustang behind the Park Towers and took the elevator to the eighth floor. As the elevator rose, Charles chastised himself for agreeing to move to this building. He knew at the time that it was beyond Digital Software's grasp, but Clapper and Paul had argued they needed the space for the new programmers. Too late, Charles realized he like the others had been seduced by the fatal attraction of nice offices. Everyone wants to be considered a success.

Charles paused to inspect the new sign that marked their offices: "World Headquarters, Digital Software, Albuquerque."

"Whose idea was that?" Charles hollered as soon as he opened the door.

Joy frowned at the interruption then smiled when she saw Charles. "Welcome home boss."

Charles offered a brief handshake, then hurried past Dave and a young man Charles did not know. "Hi Dave," Charles waved. "I'm Charles," he spoke to the strange face.

"I'm Horace," the new kid called to Charles' back. "I'm the new part timer."

Charles went down the short hallway and stopped at the first open door. Inside, the two new programmers from Stanford were working. Neither looked up. "Hi guys, keep at it," Charles said and continued down the hall.

He opened the closed door and stopped. "Christ. What's all this?" Charles asked.

Paul Harbinson sat behind a large wood desk. Facing him was a Danish Modern couch and chair set. Drapes hung on the window.

"Have to work somewhere, boss," Paul smiled.

"No wonder we're going broke," Bill said.

"We got a good deal on the furniture," Paul said.

"I'm here to stay," Charles said. "This is my office. I'll be back in an hour. Be moved."

Charles returned to his car and drove to PAMS. He found his partner James Clapper at his Director of Software desk.

"Charles," Clap spoke first.

"What are you still doing here?" Charles demanded.

"This is my last week," Clap said.

"Christ, Clap, we're going bankrupt, and you're still playing here at PAMS. What does Perfect plan to do?"

"I don't know, and I don't think they do. I'm showing the new Perfect guy the ropes," Clap said.

"We don't have time for that. Are they going to pay us what they owe us?" PAMS had not paid a license fee in two months.

"I don't know," Clap said. "I'm still in a difficult position here. I'm a PAMS/Perfect employee until Friday."

"Clap," Charles said. "Take advantage of your position and make sure they pay us."

Charles left PAMS without saying hello to Ed Johns or the new owners' representative from Perfect.

****

The New Year began for Astron Computers with staggering change.

Despite his lack of formal title, Stanley took charge. He found them space in a concrete block building on Steven Creek Boulevard in Cupertino. They had one large room with a plasterboard divider that let Stanley establish a small office in front. This he commandeered for himself and the assistant he planned to hire. The move from the garage went smoothly, and Harold quickly established a workstation in the back right hand corner of the bigger room as his engineering section. Stanley purchased the benches and equipment he needed to establish the remainder of the room as their production line. A large sliding door in the rear opened on to a loading dock.

Stanley signed the lease for the building on his own authority. Fortunately, both Harold and Stuart Miller approved. Neither objected when Stanley moved a desk into the area he had appropriated as his office. With that tacit approval of his management authority, Stanley arranged for three telephone lines, one his private number, and began to hire employees. By the end of their second week, he had ten new employees, four engineers, two programmers, two utility workers who would man the loading dock and the secondhand van Stanley purchased, and one clerical who was to serve as Stanley's secretary. Stanley left his mother and sister behind.

Harold served notice at HP and took charge of the engineering department. He put the engineers to work on another one hundred Astron 1 boards. Harold concentrated on the Astron 2. Stanley wanted to have at least three computers to display at the April San Francisco Faire. Without consulting his partners, Stanley reserved a large booth early and as a consequence obtained one of the most favorable locations in the huge auditorium.

Stanley's enthusiasm and unilateral actions caught Stuart Miller by surprise. They should not have, but they did. He had no choice but to grant Stanley his head. Stuart needed time to devise a solid business plan for the growing company, but he postponed even that important project in order to hire the company president. Stuart knew that if he did not act quickly, Stanley would pre-empt the position despite the fact he had reluctantly acceded to the conditions imposed by Stuart.

Miller immediately turned to an old friend from Fairchild who was now employed by National Semiconductor. Willard Temple had worked in the electronics industry for ten years and now at thirty-one was in charge of a semiconductor assembly line that grossed twenty million dollars annually. Miller knew Temple as a conscientious manager who had the respect of his subordinates, had management experience, and rated high in cost consciousness. National Semiconductor valued his services, but Stuart knew Temple was bored. He was eleven years older than Stanley, experienced, an engineer, and had a strong personality. Stuart hoped he was young enough to work evenly with Stanley and stubborn enough to control the aggressive one-third owner when needed.

Temple did not hesitate when Stuart broached the subject of the presidency of Astron Computers. Temple respected his friend who had made a fortune in stock options, and he was young and ambitious enough to want his turn. He said yes. Then, Stuart faced the delicate job of obtaining the consent of his two partners. Stuart assumed Harold would agree as long as Temple understood that engineering was Dum's prerogative. The problem would be Stanley who already had the bit in his teeth. Stuart scheduled a meeting for ten AM on Monday to introduce Stanley and Harold to Temple and gauge their reactions.

At exactly ten o'clock, Stuart and Willard Temple arrived in Stuart's Corvette. Stuart had staged their arrival in this manner to project his sponsorship of his friend. They entered the Astron Computer premises and found the company office empty.

"They're around somewhere," Stuart said. "You wait here, and I'll round them up."

Harold was easy. Stuart found him alone at his workbench. "Can you join us?" Stuart asked.

Harold nodded. Stuart decided that Harold had to be the most laconic man alive. "Have you seen Stanley?" Stuart asked.

"I think he's out back."

I'll get him," Stuart said.

He went out on the loading dock and found Stanley helping their two loaders stack boxes in the van.

"Stanley, can you join us? Willard's here," Stuart said.

"Be right there," Stanley said. "As soon as we get this stack loaded."

Stuart assumed Stanley was going to be difficult. He considered ordering Stanley to come immediately then decided against it. "We'll be waiting."

"Start without me," Stanley countered. "This shipment's important."

Stuart surrendered. He returned to the company offices and found only Willard Temple waiting. "Harold's not here?" Stuart asked.

"What you see is what you get," Willard smiled.

"Let's get a coffee," Stuart said. "Sometimes my partners require patience. Harold's on his way, and Stanley is helping load a shipment into the van."

"Will that be in my job description too?" Willard asked.

"You're the CEO. Write your own job description," Stuart said more sharply than he intended. "They'll be here shortly."

Stuart introduced Willard to Jan Potter, Stanley's recently hired office assistant who they met at the coffeepot. When they returned to the office, they found Harold waiting. Stuart made the introductions. Harold took the chair, and Stuart and Willard sat side by side on the couch. Stuart and Willard chatted while Harold listened silently. After about five minutes, Harold stood up.

"I'm wasting time," he declared. "Just call me when you're ready."

"No, please," Stuart said. "Let's get his over with. We'll start without Stanley."

"Don't think that will do much good," Harold said.

"All right," Stuart said. "I'll get Stanley. He knows we're waiting. I'm tired of the games."

"Please excuse me," Stuart said to Willard who was smiling indulgently.

Stuart returned to the loading dock where he found Stanley smoking a cigarette and chatting with one of the loaders. The van had departed.

"Stanley," Stuart said firmly. "We're waiting."

"I'll be right there," Stanley said.

Stuart waited. Finally, Stanley appeared to capitulate. He tossed what remained of his cigarette into the alley and followed Stuart through the shop. All ten employees, including Jan and the engineers, openly watched. They knew the company's new president was waiting in the office, and they were all curious about how Stanley who had been acting as boss would react. Not one of them thought it would be well. Most assumed it would be entertaining.

"Willard, this is Stanley," Stuart said as he entered the office.

Willard did not respond, and Stuart looked behind him to find Stanley had turned left at the coffeepot and was now chatting with Jan. Stuart decided to ignore the challenge. He sat down on the couch next to Willard and waited.

After several silent minutes, Stanley, carrying a cup of tea, entered the room and sat down behind the desk he had appropriated as his own. Stanley ignored the other three and carefully raised and lowered the tea bag. "I prefer to brew my own," Stanley announced as he tossed the dripping bag into the wastebasket.

"Now, where are we?" Stanley asked, immediately assuming control of the meeting.

"This is Willard," an irritated Stuart said.

"Pleased to meet you," Stanley said, not rising or offering his hand.

Willard, not about to let the arrogant young man intimidate him, merely nodded.

"Willard can start..."Stuart began.

"Tell us something about yourself Temple," Stanley interrupted.

"Stuart has my resume," Willard responded.

"I gave you a copy, Stanley," Stuart said.

"Haven't read it," Stanley said blithely. "It must be here somewhere." He made a show of looking through the small stack of invoices on his desk.

"I'm a graduate engineer," Willard decided to stop the games. "I've been in business for ten years, at Fairchild as an engineer, and for the past two I've worked at National Semiconductor where I..."

"You'll find us small potatoes after National Semiconductor," Stanley said, pronouncing the words "National Semiconductor" as if they were dirty words.

"From what Stuart has told me, I doubt that Astron Computers will remain small potatoes for long," Willard countered.

Stuart, seeing that Willard could hold his own with the rude Stanley, leaned back to watch the sparring.

"How do you see your duties here?" Stanley asked.

"As chief executive officer I will run the company," Willard said.

"Except for engineering," Harold spoke for the first time.

"Engineering is your exclusive department, Harold," Willard said. "I stand by to support you, nothing more."

"Even though you're an engineer," Harold said.

"You are chief engineer, period," Willard said.

Harold leaned back, satisfied.

"And for the rest of the company?" Stanley asked.

"I will delegate as needed. However, as chief executive officer I reserve the right to step in when necessary."

"You won't have to worry about that," Stanley said. "You just worry about the big picture."

"Willard will start a week from today," Stuart tried to head Stanley off.

"That sounds good," Stanley said. "Tell us a little about your company plan," Stanley said.

"At this point, I don't have a company plan," Willard said. "That will come later after I get to know you all better."

"You're not going to have time to develop a plan later," Stanley challenged.

"Stanley, you know I have been working on a company plan," Stuart intervened.

"Have anything to share with your partners?" Stanley smiled thinly.

"As a matter of fact I do," Stuart accepted the challenge. "I've discussed some of this with Willard."

"That's neat," Stanley said, his expression showing he meant the opposite.

"We've got to expand our target customer base beyond the hobbyists."

"Great."

"I've been giving thought to the professionals."

"Working girls?"

Stuart frowned. "Doctors, dentists, tax preparers, professionals who run small offices or operate out of their homes. They all can benefit from a computer's word processing and spreadsheet capabilities."

Stanley did not react.

"There is a huge market out there for that kind of thing," Willard said. "However, for them we must provide a complete, reliable product."

Harold sat straighter, listening carefully. He was not prepared to accept any criticism of the Astron 2.

"Mind explaining that?" Stanley asked, obviously sensing Dum's intense interest. Stanley needed Dum's support to override anything the other two proposed.

"Hobbyists buy kits. Part of the challenge is building their own machines. Professionals will expect computers that they can unpack, plug in and use. They don't have the time to troubleshoot bugs," Temple said.

"Then, you had better forget that market, right Dum?" Stanley said.

Harold did not rise to the bait, so Stanley continued. "Computers are marvelous machines. In ten years they will control our lives, but they are complicated machines, and they are still in the Model T generation."

"We still have technical challenges," Harold agreed.

"We have to make our product as reliable as we can," Stuart, also an engineer, understood this. He and Stanley had discussed this situation for hours. He understood that Stanley was simply trying to establish his position in the company.

"That's a given," Willard agreed. He also was reading Stanley accurately. He felt it important, however, that he not let the young man undermine his authority from the very beginning.

"Have you thought about schools?" Stanley took off on a tangent.

"Not for a decade," Willard tried to make the question into a joke.

"As a market?" Stuart asked.

"Certainly," Stanley said. "There is none bigger. Computers are a natural for college students. Term papers, notes, research," Stanley said. "We teach kids to use computers in grade school and high school and then they belong to us."

Harold, who had not heard Stanley take this line before, was impressed. He realized that if students learned to use Astron computers in school, those would be the computers they would buy as adults.

"The smartest thing we could do would be to give every student an Astron 2," Stanley said.

"And the dumbest," Willard laughed. "Do you have any idea how much that would cost? There isn't a company in the world, not even IBM that could afford to do such a thing. Be realistic, Stanley."

Stanley glared at Willard but did not replay. He had really liked the idea.

"Stanley's right, in a sense," Stuart intervened. "If we can get the students of the world started on Astron 2s, they will be our customers for life. However, that's something for the future. We must concentrate on the attainable markets for the near term."

"Then that's the hobbyists," Harold said. He knew full well that it would be years before computers could be made reliable. Until then, the hobbyists who liked nothing better than solving problems would be a preferable market.

"We can save this discussion until I am able to table a company plan," Stuart said. "Any ideas you might have, Stanley, and Harold, and Willard are welcome."

"We've got to work as a team," Willard said.

"Right," Stuart agreed.

Harold nodded agreement, and Stanley smiled.

"Do we have any other questions?" Stuart decided to hurry this meeting to its conclusion. It was clear that Willard and Stanley were going to have problems, and that was in part his fault for putting Willard in over Stanley who had previously thought himself company president, but that was something Willard would have to work out.

"Yes," Stanley said. "I have a question."

"Let's hear it," Stuart said.

"What does he get paid?" Stanley looked directly at Stuart.

"$20,000.01," Stuart said, suppressing a smile. Stanley and Harold had agreed to salaries of $20,000 even. The one penny more for Willard established who was boss.

Stanley frowned. Harold smiled. He like his father doubted that Stanley had the experience to manage a complicated company. Stanley immediately recognized that Dum's reaction meant he would lose any debate. He could either pull out or stay. Stanley, despite his scarcely hidden hostility, had been impressed by their new chief executive officer's confidence. Stanley reached into his pocket, took out two cents, and tossed them to Dum. "There, Dum. You're now the highest paid employee here. That makes you boss."

Both Stuart and Willard recognized the challenge in Stanley's gesture but decided to ignore it as a demonstration of adolescent bravado.

Willard Temple reported for duty on the following Monday. Deliberately, he arrived at eight o'clock, one half hour before the company's declared opening. To his surprise, he found Stanley, Dum and a majority of the other employees present and working.

"Good morning boss," Stanley greeted Willard from his position behind the company's only executive desk. Jan had a secretarial desk which she had located near the front door where she could act as receptionist and control the company's second telephone. The first sat on Stanley's desk.

Willard had already thought about this problem.

"Sorry, Stanley," Willard said, loud enough for Jan and nearby employees to hear, "I'll need to claim my office."

"Sure thing Mr. President," Stanley said.

Stanley stood up and gestured toward the chair. "The throne is yours."

"Please take all your personal items," Willard said. "The desk will be exclusively mine from now on."

"And the telephone?"

"We'll have another line put in for you," Willard said.

While Stanley sorted through the desk drawers, Willard called Jan into the room.

"Please bring your pad," Willard said.

Willard took the chair and Jan the couch. "I want to have the carpenters come in and build partitions around my office," Willard said. "We need to be able to have confidential talks."

"Yes, Mr. Temple," Jan said, secretly pleased at Stanley's discomfort. She liked Stanley but felt he needed to be taken down a few pegs. He had a tendency to holler at whomever he pleased.

"Also, we need to order some safes to store company documents, and I want to install some security procedures. Check with the office supply stores and find out how we go about getting company badges."

"The kind you wear on a chain around your neck?"

"Exactly."

Stanley laughed.

"You don't agree Stanley?"

"What are we protecting ourselves from?"

"Corporate espionage. Dum's plans for the Astron 2 are priceless."

The idea made sense to Stanley. He wished he had thought of it. Since he had not, he felt it imperative that he denigrate it.

Three days later, badges were issued to every employee. Stanley deliberately held back. Finally, when he was the only one without a badge around his neck, he approached Jan's desk. "Here's yours, Stanley. I like the picture. It's you."

Stanley took the badge and studied it. The picture was a perfect likeness of a frowning Stanley, but that was not what irritated him. His badge was labeled Employee 00002.

"I'm number two?" Stanley blurted.

"Don't blame me," Jan said. "Mr. Temple assigned the numbers."

Stanley pivoted and hurried past the carpenters who were building the plasterboard partitions around his former office. He found Willard at his desk smiling. He clearly had been listening to the exchange between Stanley and Jan.

"Who is number 1?" Stanley demanded loudly.

The carpenters stopped hammering and listened with amused looks on their faces. They had been working at Astron Computers for two days and had already developed a dislike for the arrogant young man who was shouting at the mild tempered company president.

"Harold," Willard said calmly.

The answer surprised Stanley, and for a moment he was speechless. "What's your number?" Stanley demanded.

"Fifteen. We are assigning numbers according to date of employment. "I'm last, so far," Willard laughed.

"And Dum's first?"

"He invented the machines."

"I don't want to be number 2," Stanley said petulantly.

"What do you want to be Stanley?" Willard assumed he had Stanley outmaneuvered.

"I want zero."

Willard had no choice. "All right, Stanley you are Employee Number 00000."

One of the carpenters clapped his hands lightly. Stanley smiled victoriously and returned his badge to Jan for reprocessing.

Despite the angry storm clouds that hovered over Astron Computers, every employee arrived early, and most departed late, all afraid they would miss some of the excitement generated by the one employee with no assigned duties. Stanley discovered he thrived on challenge. He did not linger long enough in one place to require a chair let alone a desk. The newly installed partitions required three coats before the tint matched Stanley's articulated palette. All sighed with relief when finally the painters departed, their spackled white clothes smeared with three new colors. Then came the new telephone lines and extensions, and Stanley faced indecision—he could not choose among, the red, the olive or the tan phone receivers. He rejected three separate and distinct types of business cards that he, Willard Temple and Jeremy Thompson, the newly hired Sales Manager would carry.

When he was not occupied by these issues, Stanley visited each and every company employee with the exceptions of Willard Temple and Dum. The latter Stanley spared out of respect; the former Stanley simply avoided. The others viewed Stanley's stopovers with very mixed emotions. Inevitably, he would have suggestions to change whatever they were doing. Some of the time his observations were helpful. Occasionally, his advice would be so bad that disaster loomed. At these times, the wise employee discreetly sought the assistance of Willard, and did so at his or her own risk. At least three times a day Willard and Stanley crossed swords, not in gentle, polite fencing, but in angry denouements that bordered on fisticuffs. When this happened, the innocents retreated and waited for the heat to dissipate.

One of the longest and angriest conflagrations erupted in Willard's office when Stanley rejected the CEO's personal recommendation for a company logo as "garbage."

"What can possibly be wrong with a simple picture of a star?" An exasperated Willard demanded.

"Just that, it's simple," Stanley raised his voice. "It doesn't say anything. I selected the name 'Astron' because it's Greek for star."

"Exactly my point," Willard kept his voice low. He was aware that he and Stanley were considered a spectacle.

"Not any plain ordinary simple asshole of a star," Stanley shouted, reaching for his audience. "That's what this is, an asshole of a star."

"Shall I ask Moss Advertising to send over their artist so you can tell her that her star is an asshole of a star."

"Yes," Stanley smiled thinly. "You better."

Someone in the shop groaned loudly. Stanley stepped to the shop entrance and shouted: Do you want an asshole of a star as your company's logo?"

"Why not? Our managers are," someone shouted.

The distracted employees applauded.

"See," Stanley returned to Willard's office. "They agree with me."

"What would you suggest, Stanley?" An exasperated Willard asked.

"I see a stream of shooting stars, rising in an arc, starting small and growing large, with the top star glowing brightly."

"Now what does that signify?"

"That Astron Computers has a heaven full of stars, with the arc of stars representing the computers we are producing for the world."

"We still have a few to sell to live up to that image," Willard said.

"Think small, be small," Stanley challenged.

At this point, Temple made a mistake. "Stanley," he said. "I wash my hands of it. You are responsible for the logo."

"Thank you," Stanley said and marched triumphantly from the office. He dropped the rejected logo in Jan's wastepaper basket.

An hour later, Willard emerged from his office to find Stanley and a deliveryman dressed in neat brown standing near Jan's desk. He immediately perceived that something was amiss when he saw the frown on Jan's pretty face. Stanley held a clipboard and appeared to be signing a document.

"What's that?" Willard asked.

"A voucher," Jan said.

"May I see it?" Willard asked.

Stanley ignored him and passed the clipboard to the uniformed man. Willard held out his hand, and the deliveryman gave the board to Willard. Stanley turned, but Willard spoke: "Just a minute, please, Stanley."

Willard scanned the document. "You signed a voucher for $1,000," Willard said.

"I know it," Stanley said.

"You are not authorized to sign any vouchers. Only Dave or I have the authority to sign vouchers, and I do not wish to exercise my authority except in an emergency." David Barker was the recent hired Finance Officer who had been installed in the new offices that Willard had rented on the second floor.

"Who says I can't sign vouchers for a measly thousand dollars?" Stanley countered.

"I told him," Jan, who was caught in the middle, a not infrequent occurrence, said.

Willard quickly scratched out Stanley's signature, and scribbled his own. He handed the clipboard to the waiting deliveryman.

"Stanley, could I see you in my office," Willard said.

"Not on your life," Stanley said and strolled back into the shop.

"This place is a zoo," the uniformed man mumbled as he headed for the front door.

By the end of March, such heated confrontations had become a staple of Astron Computer life, and most employees learned to ignore them. Despite it all, morale remained high. Most bought into Stanley's frequently overstated image of Astron Computers. The closer they got to the April opening of the First West Coast Computer Faire sponsored by the HomeBrew Computer Club, the higher the tension became. While Dum and the engineers labored to prepare three Astron 2's for exhibition—Stanley had decreed that Astron Computers would simultaneously demonstrate three state of the art computers not a simple one that competitors would dub a prototype—Stanley took charge of the preparations. He consulted with the Moss Agency and worked with their designers to make the Astron Computer display professional and dramatic.

Stanley suggested the addition of a large television display that could be seen from a distance throughout the hall. He insisted that large silk screen versions of the company logo flank the television. The dramatic blue background with a stream of ever larger and brighter stars streaking deep into the universe was indeed magnetic. Even workmen assembling competing displays paused to examine Stanley's logo.

Determined to personally manage the creation of Astron Computers' image, Stanley decided to remake his own persona. He surprised Stuart Miller with a calm smile when the latter suggested that if Stanley wished to represent the company at the fair that he attempt to appear more cosmopolitan by wearing a suit. Without informing his colleagues of his plans, Stanley purchased a suit from the valley's most reputed tailor. Stanley, ever a stickler for detail and a slave to his own taste, paid four hundred dollars for a discreet black suit with conservative lines. The tailor suffered through four fittings before Stanley declared himself satisfied. When the tailor suggested a fitted linen dress shirt, Stanley declined, sarcastically noting he was not ready to sell out completely. Instead, from a catalogue of designs, Stanley chose a shiny black silk shirt intended to be worn without tie. The shirt required three fittings and cost half as much as the suit itself. Stanley completed his ensemble by purchasing a two hundred-dollar pair of black, Italian tassel loafers.

On the day before the opening of the Faire, Stanley truly indulged himself. He bought a green Corvette convertible from a local General Motors dealer. Stanley chose green because he did not want to create the impression of imitating Stuart Miller and because green was the color of the only slightly used Corvette on the lot. Stanley bargained aggressively and succeeded in reducing the price by one thousand dollars and limiting the monthly payments to four hundred dollars, down from the five hundred suggested by the salesman. The deal required two hours and ended with the salesman promising to throw in a hard wax detailing while Stanley waited.

At nine o'clock on Wednesday morning, exactly one hour before the West Coast Computer Faire was scheduled to open, Ann Page climbed out of her taxi. She reached through the window and handed the driver a five-dollar bill. He started the pretense of searching for change, but Ann dismissed the effort with a casual wave. She was on expense account. Two months previously, Ann had parted from Commodore and joined a small market research agency in San Francisco. Since her agency hoped to sell its services to the valley's burgeoning electronics industry, Ann had been deployed to scout for clients. This was her alleged area of expertise. Anxious, Ann had deliberately arrived early. She had a small but expensive Japanese camera in her bag and planned on taking pictures of the most promising booths before the crowds appeared.

Ann wore a tight white linen dress with a mid thigh hemline that she hoped was not too provocative. She had given more than the usual careful thought to her appearance and had dressed up for the occasion. She knew from experience that an attractive woman could get virtually any question answered.

Ann started up the stairs.

"Look at that," one of two workmen standing to her left said.

Ann had expected positive reactions, but not quite that quickly. She turned her head and smiled brightly. To her surprise, neither man was looking at her. Instead, they were staring wide mouthed at a glistening green Corvette. Ann, too, stared. The driver was dressed in all black. He had styled hair, a trim mustache, and a deep tan on his handsome face. He wheeled the car into a reserved parking place in front of the building. Like the others, Ann watched as the man got out of his car and casually strolled toward the building. He seemed vaguely familiar. Ann, unable to place him, decided he must be a second level movie star hired by one of the major companies to showcase their products. The black suit was obviously expensive, and she knew the black silk shirt had to be.

Suddenly, she felt underdressed in her favorite linen dress. Ann turned and continued toward the entrance.

"Nice wheels," one of the workers called behind her.

Ann did not hear the answer because the uniformed guard at the door held up his hand to halt her.

"Do you have a pass lady?" He asked.

Ann was prepared. She opened her purse and took out the press pant that the Faire credentials committee had sent the agency.

The man in black nudged her slightly as he moved past her.

"Delivering groceries?" he said before he nodded to the guard who stepped back and let him through.

"Good morning Stanley," the guard said. The man pushed past and entered the hall.

Ann almost dropped her purse. "Was that Stanley Pitts?" She asked the guard.

The last time Ann had seen Stanley he had worn a dirty T-shirt and jeans with bare feet in sandals, and his ponytail had not seen shampoo in months. He had taken her to a filthy storefront restaurant and had made her walk for miles from her hotel. She had cut him without bothering to check out his garage-based company. Ann's scathing but humorous account of her visit had amused coworkers for days, particularly her description of the erstwhile CEO. Ann tried to recall the name of Stanley's company.

"He's Astron Computers," the guard said as he stepped back to let Ann enter.

"Why it's Ms. Grocery Store herself," Stanley greeted her.

He was standing just inside the entrance. Embarrassed, Ann did not know what to say as she tried to decide which was the real Stanley, this pseudo movie star or the barefoot hippie.

"Slumming in Cupertino again?"

Ann thought she heard a slight sneer in his voice. "Hi Stanley. How are you?"

Ann offered her hand. Stanley took it and held it firmly.

"Nice suit," Ann said as she jerked her hand back.

Stanley laughed. "Would you like lunch?"

"No thank you," Ann said, remembering Stanley's veggie burger and her tuna salad.

"See you around," Stanley said, wheeling about, leaving Ann speechless. She was not accustomed to being dismissed so quickly and gracelessly. She decided only Stanley's appearance had changed. As she watched Stanley walk across the large room, she found herself thinking: "rude but handsome with a new look."

Ann took out her camera, checked the settings, then moved from booth to booth, taking pictures as the workers applied the finishing touches to their displays. At the Astron Computers booth, Ann paused and watched as Stanley directed the workers as they moved their computers an inch to the right or left. Stanley appeared to be a perfectionist, but the others accepted his finicky adjustments in good grace. They talked and joked with Stanley as they worked. Ann stepped back and took several pictures of the logos and the large television screen they flanked. As she watched, Stanley turned on one of the computers. After a brief wait, large red letters spelling "ASTRON COMPUTERS" flashed on the huge screen. Ann took another picture.

"Pretty impressive, huh?"

Ann looked up to find Stanley standing next to her. She really did find him handsome. "I like your mustache better than I did your ponytail," Ann said, surprising herself.

"You really must come out and see my garage this time," Stanley said, referring to Ann's last visit when she had declined to visit his company. "I have some splendid etchings."

Ann laughed easily at the cliché.

"Let me get your picture standing next to the television screen," Ann said impulsively. Stanley was standing too close to her, invading her space. Stanley posed as Ann asked: She snapped two frames, then retreated, pleased when a backward glance disclosed dismay on Stanley's handsome face. She rather liked the Roman nose with the Clark Gable mustache.

After the doors opened admitting a sizable crowd, Ann was surprised to see that Astron Computers and Stanley attracted a disproportionate share of chattering viewers, many of whom were not hobbyists. Ann recognized several of the important representatives of the larger electronic companies carefully appraising the Astron 2 computers. She overheard several discussing the Astron 2 engineering, praising both the styling and the electronics. The hobbyists and the competitors were not the only visitors attracted to the Astron display. Many of the bored wives who had accompanied their husbands to the Faire also gathered before the streaming stars logos, and they weren't appraising the computers. These women, young and old alike, were admiring Stanley. Ann heard more than one attractive young lady praise his movie star looks.

By noon, Ann had seen all she had to see. She made one last round then started for the door. She was half way there when she felt a tug on her purse strap. She turned quickly, determined to confront a clumsy thief, and found a smiling Stanley standing slightly behind her.

"Ready for our luncheon date?" He asked.

"I don't recall that we had one." Ann was flattered that Stanley had singled her out for attention.

"Indian OK?"

"French," Ann decided. She was unwilling to risk Stanley's choice of restaurants again.

Stanley hesitated.

"Café de Paris," Ann said.

"Don't believe I know it," Stanley said.

"Half a block from my hotel."

"Which is?"

"The Hilton."

"Need a lift?"

"I'm not going to walk," Ann said. "I learned my lesson last time."

"Think they'll have Niciose Salad?" Stanley asked, demonstrating that he remembered.

"I'm certain of it," Ann said.

Stanley took Ann by the arm and guided her through the crowd. Ann sensed the admiring glances that trailed in their wake.

"That's my car over there," Stanley pointed at the Corvette.

"Pretty," Ann said. She thought of saying more but decided against it. Instead, she wondered where Stanley had suddenly found this new style. Even more, she wondered where he had gotten the money. "Business must be good," she said, trolling for information.

"Not bad," Stanley held the Corvette door for her.

Ann slid on to the black leather seat. It was uncomfortably hot, but she said nothing.

Stanley got in the driver's side. As soon as his backside touched the hot leather, he lifted up slightly and looked at Ann. "Guess I should have leaned the seats forward out of the sun."

"Yes."

Stanley surprised Ann again. She expected him to burn rubber. He drove sedately, carefully obeying the speed limits.

"This is a beautiful machine, Stanley," Ann said. "Had it long?"

"Long enough. I'm thinking of trading it in."

"Oh don't," Ann said. "Let me know if you decide to sell it. Was it expensive?"

"Not bad," Stanley said. It had only cost him a year's salary.

"Does Astron Computers need any market researchers?"

Stanley hesitated. He had not given market research much thought. He had left that up to Stuart and Willard.

"Who designed your logo?"

"I did. Like it?"

"It's perfect. Who handles your advertising?"

"The Moss Agency. What do you do in marketing research?"

Ann sensed that Stanley was asking what is market research. "We try to find out about our customers. What kind of machines they prefer. For what reasons."

"Polling."

"Customer interviews. Also feedback on the warranty slips."

"We need to do more of that," Stanley said. "We're still learning."

Stanley drove past the Hilton and parked a short distance from the Cafe de Paris. Stanley had noticed the restaurant but had avoided it because of a reputation for high prices. The interior was cool and dark. As best as Stanley could tell, most of the tables were full. An attractive young lady met them near the entrance.

'Table for two," Stanley said, feeling like a phony out of the movies.

"Do you have a reservation sir?" She squinted. Stanley in his black suit and shirt was almost invisible.

Ann giggled.

Stanley glared at her. "No."

"We have a table sir, but I must apologize. It is a small table in the rear."

"That'll be fine," Ann said. She felt a need to apologize to Stanley. He thought she was laughing at him.

"The waiter will be right with you," the hostess said after depositing them at a small table near the busy kitchen door.

"She had trouble seeing you," Ann said.

"Why is that?"

"You're dressed in black, just like Lamont Cranston."

"Who is Lamont Cranston?"

"The Shadow."

Stanley thought for a moment. "Not only am I invisible, I have the power to cloud men's minds."

"And women?"

"Those too. But I promise not to use my power on you."

"I hope so."

The waiter approached.

"Would you like something to drink?" Stanley asked.

"A glass of white wine."

"I'll have the same," Stanley said.

"Do you have a preference?" The waiter asked Stanley.

He hesitated.

"House wine will do," Ann said.

The waiter looked at Stanley who nodded.

"Could we order now?" Ann asked. "We have to get back." She was speaking more for Stanley than herself.

"Certainly." The waiter produced two menus.

Ann handed her menu to Stanley. "You order for both of us."

Stanley took her menu, handed it to the waiter, and opened his. Everything was in French. Stanley pretended to study it, then looked at the waiter. "We want something light and quick. Do you have Nicoise Salad?"

"Certainly sir."

Stanley looked at Ann. "Exactly what I wanted," she said.

"And bring some rolls," Stanley ordered.

"We serve it with French bread," the waiter said. "Would you like..."

"The bread will do," Stanley said.

After the waiter departed, Ann decided to learn a little more about Stanley. "Would it be rude of me to ask how old you are Stanley?"

"Twenty-one. Old enough to order alcohol," Stanley said. The answer was true, but he did not take advantage of the opportunity often. He did not like alcoholic drinks. He preferred tea.

"That's not why I asked," Ann said. "Everything I say to you seems to come out wrong."

"It's my fault," Stanley said honestly. "I'm not very good with girls."

"Don't tell me that," Ann said, speaking more sharply than she intended. "I'll bet you have lots of girls."

"I don't have time."

"What about in school?"

"College you mean? I dropped out after one year. I hung around a while then went to India."

"No girls there?"

"In India. No. Not my kind."

"No I meant in college."

"I dated a couple of times but girls didn't like me. They were always dumping me."

"I don't understand that."

"We weren't interested in the same things. I like philosophy, electronics, things like that."

"So do I." Ann hadn't intended to say that.

The waiter served the wine.

"Your salads will be ready shortly. They're a house specialty. You made a wise choice, sir."

After the waiter had again retreated to the kitchen, Stanley picked up his wineglass and raised it in toast. "To many more chance meetings."

"They don't have to be by chance," Ann said.

"You just like my suit and car," Stanley laughed. "And the fact you can't see me in the dark."

Ann sipped her wine and wondered if that were so.

Stanley, feeling the lifting effects of three unaccustomed glasses of wine and the exclusive attention of a female companion, was in a good mood until he was stopped two blocks from the Civic Auditorium by a traffic policeman who insisted with his aggravating whistle that Stanley turn right.

"I'm an exhibitor," Stanley stopped in mid turn to appeal the decision.

"Keep Moving, Mister," the harassed policeman shouted.

Stanley drove around the block two additional times, confronting the unrelenting cop with each rotation. Finally, to Stanley's and the irritated cop's relief, a car abandoned a curbside parking place just as Stanley approached. Stanley hit his breaks and waited, ignoring the honking vehicles behind him. With little difficulty, Stanley backed his car into the space. He carefully opened his door, gave a finger to an irate driver who paused long enough to glare at Stanley, then reached in his pocket for change for the meter. After dropping several coins into the slot, Stanley turned and for the first time noticed the car parked just ahead of his Corvette. Stanley studied the unusual design, A Porsche 911. It looked like a real bomb. Stanley stepped back and compared the two cars. Both were green. Stanley decided he preferred the Corvette.

"What's the problem officer?" Stanley called to the red-faced policeman as they waited their turn to cross the intersection.

"Civic Auditorium," The cop growled.

Stanley checked his watch as they strolled along. Three o'clock. Lunch had gone on longer that he had anticipated. Stanley smiled. The others could do some of the work for a change.

When they reached the intersection across from the Auditorium, Stanley was surprised to see a long double line stretching from the entrance doors, around the corner, and down the block for as far as he could see. Stanley took his nametag that identified him as a display sponsor and pressed the adhesive back against his suit pocket. He hoped the glue would not damage the material. They made their way along the line, ignoring the waiting audience. Only a few frowned. Most seemed to be in a jubilant mood.

Stanley pointed to his card when he approached the door. The guard stepped back and waved them through.

"Good luck," the guard laughed.

Inside, the large exhibition hall was jammed. Stanley and Ann parted.

"Thank you for a very enjoyable lunch," Ann said.

"We'll do it again," Stanley smiled.

Stanley slowly made his way to the Astron Computer display in the center position Stanley had selected. The crowd circling the large television display was larger than most. It stretched almost twenty deep. Dum, Stuart, Willard and Jan were all busy responding to questions.

"Where have you been Stanley?" Jan spotted him first.

"Lunch with a customer," Stanley said.

"I saw her. The white linen suit."

"Way to go Stanley," a stranger in the front row called.

Stanley smiled.

"I haven't had lunch yet," Jan protested.

"Be my guest," Stanley replied nonchalantly. "Try the Nicoise Salad."

"Whatever that is," Jan said.

"Frog salad. With anchovies, slices of egg, and tuna," a well-dressed businessman said.

"Stanley," Dum called.

"Yo," Stanley said. He still felt the high from his lunch.

Dum irritably motioned for Stanley to join him.

Stanley side stepped behind the tables holding the Astron 2s and joined Dum.

"Thanks for joining us Stanley," Dum grumbled.

"No problem."

"This is Charles Swift," Dum said, consulting a business card he held in his hand.

"Hi," Stanley said. He glanced at the kid who stood in front of Dum. With his shock of unruly sandy hair, horn-rimmed glasses, white shirt, rolled up sleeves and khakis, he appeared to be about sixteen.

"What?" Stanley dismissed the kid and addressed Dum.

"Charles is president of Digital Software," Dum said.

Dum seemed to be impressed so Stanley took another look at the kid. Nothing had changed. Stanley waited for Dum to explain.

"He wrote the BASIC for the Taurus," Dum said.

Stanley shrugged. The Taurus was history. "Too bad they couldn't get a booth," Stanley said. The fact that Taurus had not sent a demonstration team had caused some talk among the other exhibitors.

"Pleased to meet you Mr. Pitts," Charles said. Charles estimated that he and Pitts were about the same age, but Charles had no difficulty being polite when it served his interests. He was trolling for customers. Many of the Faire attendees had been impressed with the Astron Computer. It could be the next big thing.

"You from around here?" Stanley asked.

"Albuquerque."

The answer did it for Stanley. He had no interest in Albuquerque. "Well, I've got to get working," Stanley said.

"Wait," Dum grabbed Stanley's arm. "Stanley's in charge of our subcontracting," Dum spoke to Charles.

"Not any more. You want to talk with our President." Stanley nodded in the direction of Willard Temple.

Stanley noticed two pretty young ladies standing at the other end of the display staring at him. Immediately, Stanley decided to test his newly found attractiveness to women.

"Digital is interested in helping us with our software problems," Dum said.

"We don't have any software problems," Stanley said sharply. Dum's deference to the kid irritated him.

"Please take our card," Charles slid a card into Stanley's hand.

Stanley glanced at the card, then handed it to Dum. "Mr. Dumbroski writes our software," Stanley said.

"I don't now," Dum said.

Stanley recalled that Dum had begun to complain that he did not have the time to write all the software for the Astron 2. Stanley was not a programmer, did not really understand software, thus had no interest in what was a technical problem. "You're in charge of the engineering department. Talk to Temple."

As he turned away, Stanley decided that having a chief operating officer had its advantages. Some one else could help solve problems.

Stanley smiled as he approached the two girls. "Interested in computers, ladies?"

"Don't mind him," Dum said to Charles. "His new suit has gone to his head."

"Should I talk with Mr. Temple?" Charles asked. He needed the business.

"I've got your card. We'll be in touch. You wrote the BASIC for the Taurus?"

"Yes."

"I don't mean your company. You."

"Yes."

"I read about you," Dum said.

"Really?"

"Yes. You're the Harvard kid mentioned in the fliers."

After the guards closed the front door to the Civic Auditorium, and the crowds of excited hobbyists and company representatives dissipated, Stanley departed ahead of the rest of his Astron Computer colleagues. Two attractive young ladies accompanied the dapper Stanley who was more worried about how he would fit his two companions into the Corvette than he was about his coworkers who he was abandoning. The cop was gone when Stanley and friends reached the block where he had parked his car. Most of the other vehicles had departed, but the two green sports cars, the Porsche and the Corvette remained.

"Oh what a pretty car," the brunette said as they approached the Porsche.

"That's mine, the Corvette," Stanley said, irritated.

"It's pretty too," the brunette said.

"I like the Corvette best," the blonde cooed.

Stanley began to speculate how they could get rid of the brunette.

After they had crowded into the Corvette, the two girls sharing one seat, Stanley started his engine.

"Listen to that baby purr," Stanley said.

While he waited for traffic to pass, Stanley noticed the kid from the software company approaching. Stanley really liked the image he presented: two girls, a snappy sports car, top down, and his new black suit.

"Hi kid," Stanley waved. "Sorry, we don't have room." Stanley smiled as he indicated his two passengers.

"That's OK," Charles said as he unlocked the Porsche.

Stanley thought about offering the kid the brunette, but decided against it. The kid probably wouldn't know what to do with her anyway.

Chapter 13

As the southwestern spring merged into summer, Charles in a brief moment of passive contemplation decided he definitely hated Albuquerque. The barren countryside and the mountains had their grotesque charm, but the heat and the lack of green made him miss Seattle. Nature did not preoccupy him much, however. He spent so much time traveling, looking for customers, working, working, working, and plotting against Perfect/PAMS whose problems he was forced to share that he existed oblivious to his environment.

After his return from the San Francisco Trade Faire, Charles drafted a letter formally protesting the overdue royalty payments and PAMS' clear policy of not implementing its agreement to license BASIC to others, thus denying Digital Software of overdue and needed revenue. Charles threatened to terminate the agreement. Perfect/PAMS reacted with a legal response designed to string out the disagreement. Their lawyer alleged that Perfect/PAMS was current in royalties owed and denied that the agreement required PAMS to sell licenses for BASIC to competitors. The lawyer observed that if DS did not concur in this interpretation it could submit any dispute to arbitration.

Before Charles could decide on his next step, Perfect/PAMS in an inconsistent lawyerly act sent Digital Systems a check covering delinquent royalties. Before Charles could complain that this check proved his previous point, he received a copy of a Request for Arbitration that his protagonist had sent to the American Arbitration Association. This was followed by a restraining order that prohibited Charles from licensing his code to anyone other than PAMS pending the arbitrator's determination.

"Damned lawyers," became Charles' most frequent utterance during this period. Digital Software was in jeopardy. Their main customer, PAMS, refused to meet the terms of their contract and perversely had succeeded in preventing Charles from dealing directly with other potential licensees for his BASIC.

Charles as senior partner—he now held the majority position of the sixty-four/thirty-six partnership—formally called for a meeting to determine what they should do. The two friends, each holding a can of diet cola, met in what passed for a living room in the rundown house that Charles and Clap shared with Paul

Harbinson and whoever needed a place to stay while working for Digital Software.

"We're in a real bind," Charles said.

"Ed did not do us any favors when he sold out to Perfect," Clap agreed. "I thought Ed was a weak manager, but he was basically a fair man. I don't know what to make of these Perfect people."

"Let's focus on us," Charles said. He was weary of the lament about PAMS. "We can't expect another penny from PAMS until the arbitration ruling." Charles was afraid to cash the last four-month royalty check that his letter had produced. Charles' local lawyer speculated that Perfect/PAMS might argue that the check satisfied all royalty payments.

"When do you expect the arbitrator to rule?"

"Who knows? I'm told it could drag on all summer." Charles was so angry with lawyers that he refused to cite the name of their lawyer,

"Can we get Texas Instruments to make an advance payment?" Digital Software had a $100,000 contract to provide a version of PLUS BASIC for one of TI's new machines. James was busy working on a simulator for the new chip.

"Not a chance," Charles, who handled the negotiations with TI, was adamant. "They won't pay until we deliver the code."

At the time Charles signed the contract, he had optimistically thought he could carry the overhead with the PAMS' royalties until DS met TI's rigid conditions. TI flatly admitted that they minimized their risks in contracting with a minor software company like Digital Software. Because the big companies like IBM and TI and Commodore preferred to write their own software in house, Charles had had to compromise to get the contract. By deliberately keeping his price low and offering code on a flat fee no royalty basis, he had persuaded TI that he could satisfy their requirements at a lower cost that they would incur in writing their own software. Now, he faced the consequences of his risk taking in interest of long-term benefits they might derive from a relationship with a major company.

"Are we in too deep to back out?" Clapper asked.

"Too deep. If they paid us for the work we've already done, they would give us $65,000."

James considered that dilemma. "What if we went to them and told them we're in a bind?"

"They would void the contract, and we'd be out $65,000. That would finish us."

"Then, we don't have any choice."

"We fill the TI contract, and we're flying. Screw PAMS and all the lawyers." Charles sipped on his coke.

"We'll tough it out," Clapper declared.

"Neither of us will be able to take any money out of Digital Software," Charles cautioned.

"Who cares," Clapper said.

"We started with nothing," Charles said.

"And we can start again."

"We have to take care of Paul and the others," Charles said. He knew their situation was too precarious to risk deluding themselves with false bravado.

"I'll talk frankly with them," Clapper said. "I'm sure they'll understand. This is a fling for them anyway. They know we're taking a high risk chance at making it big."

"And being part of something important," Charles added.

Digital Software limped into the summer. The arbiter finally commenced hearings at the end of June. Charles, Clapper, and Paul, who had negotiated with PAMS in Charles' absences and during Clapper's employment by them, testified. The arbiter appeared to find a journal Paul kept during that time to be of interest, but the hearings dragged on. The judge issued and reissued the restraining order prohibiting DS from selling its BASIC to other companies. Through it all, their attorney remained optimistic. He assured Charles that he had a good chance of winning. Charles hid his worry from the others as they labored on the Texas Instruments project. Charles sought distraction by racing his Porsche up the mountain highways at night, but this resulted in little more than the accumulation of speeding tickets. It reached the point where the State Police set up traps aimed at the green Porsche and its kid driver because they suspected he was engaged in illegal pursuits beyond speeding, possibly drug trafficking. Twice, Charles had to have Clapper bail him out of the local jail.

In July, Charles did something he had vowed he would never do. He accepted a $7,000 loan from an employee, Paul.

"I don't need it," Paul said. "What would I spend it for? I live in your house, drink your coke, eat your pizza and work twenty-four hours a day. You can repay me with stock options."

"Options are no good if the company goes under before we go public," Charles said.

"We'll make it," Paul said.

Charles' parents again offered a loan. Charles talked frequently with them on the phone and frankly discussed his problems. His father provided free legal advice, and his mother wanted to do more, but Charles was determined to show his independence. He could not bring himself to admit that he had been wrong. Just as Charles was about to surrender and accept his mother's assistance, an unexpected phone call provided salvation.

Joy answered the phone.

"I want to speak with..." the caller paused. "With...shit I can't find the kid's name," the voice said. "The one who's president of Digital Software. I got the number from information. Is this Digital Software?"

"Yes sir. Just a moment and I will put Mr. Swift on."

"Charles," Joy called. "Somebody's calling long distance and wants to talk with you."

"I hope it's from Texas," Charles said. "Charles Swift," Charles picked up the receiver.

"You the kid that wrote the BASIC for the Taurus?" The voice asked.

"Yes, sir," Charles wished the man would identify himself.

"This is Stanley A. Pitts, Astron Computers."

Charles smiled as he remembered the arrogant young man in the black suit and green Corvette. He restrained the urge to ask how he had made out with the two groupies he had collected at the Faire. "Yes, sir. I congratulate you on your success." Astron Computers had taken off since the April Faire. One magazine article had described its growth as miraculous.

"Yes," Stanley said. "We're expanding so fast we don't have time to write our own software. Dum tells me you have something called 6502 BASIC that he needs."

"Yes sir," Charles said.

"I know we could write it cheaper, but neither Dum nor I have the time," Stanley said.

"I'm sure we can accommodate you," Charles said.

By the time he hung up, Charles had Stanley's agreement to send a check for $10,000 as half payment of a flat fee license to use Digital Software's 6502 Basic in its Astron 2 computer. Stanley promised to send the check as soon as he received the tapes containing the 6502 BASIC that fortunately was not covered by the Arbitrators' inquiry. Charles immediately prepared the tapes and sent them off by insured mail return receipt requested. The signed check and a contract appeared in the return mail, and Charles had enough cash to carry Digital Software into September.

On the first day of September, Charles received a phone call from their lawyer: "Come over Charles, we have news."

"Good or bad?"

"We're opening the champagne."

Charles leaped into the air and slapped the ceiling with his palm for the first time in years of proud finger touching. "We won," Charles shouted, and the company's seven employees gathered around him to share.

"Tell us!" Paul demanded.

"That's all I know," Charles said. He turned to Clap. "Let's go partner."

On the way out, Charles stopped and turned. "Paul get the champagne."

"How many bottles?"

"Six. No get a case," Charles said.

Clapper drove, thankful that Charles' Porsche was in the garage. Charles had ignored a blinking oil warning light and the engine had seized. It was now undergoing repairs.

"Faster," Charles pleaded, but Clap carefully adhered to the speed limit.

As soon as Clapper parked in a near empty lot, Charles raced ahead.

The lead lawyer, his three partners and the firm's clerical staff were waiting. As soon as Charles opened the door, they all began to applaud. Charles hesitated at the door, waiting for Clapper to join him. After shaking hands around the room, pausing to kiss the women, even the matron with gray hair who served as private secretary to the chief lawyer, Charles stopped.

"Tell us everything," he demanded.

"The arbiter issued his ruling this afternoon. He virtually accused PAMS of corporate piracy. He found that PAMS violated the agreement. They can continue to sell your BASIC on its own machines, paying DS a royalty of course, and its exclusive license has been terminated. You can sell BASIC to anyone you wish without regard to PAMS."

Charles was so excited his knees shook. He accepted a glass of champagne and toasted his lawyers.

The Taurus had created Digital Software, and now, thanks to the Arbiter, the Taurus had liberated them. Digital Software was now free to pursue its own sales. Charles and Clap returned to their own offices and celebrated with their employees. Paul had purchased a case of champagne that he had cooling in a waste paper basket full of ice. They all took turns making speeches, Charles most of all. Later he remembered giving everyone the rest of the day off, then passing out on the mat behind his desk. When he woke several hours later, Charles stumbled to the bathroom, splashed water on his burning red face and doused his hair in an attempt to still the pain. When he returned to his office, he looked out the window. The black of night still cloaked the deserted streets. Charles heard someone snoring in the outer office. He turned off the ceiling light and lighted the small desk lamp. Then, he sat down and began to make a list of the customers he would call as soon as they opened for business. He divided the list into two columns, one labeled BASIC and one FORTRAN. At the top of the BASIC list he placed Radio Shack. Then, he added Intel, MECA, NCR and Commodore.

At seven-thirty Joy appeared. She stopped at the fridge, retrieved a can of diet coke, and silently delivered it to Charles. He smiled thinly, took the can, and held it against his forehead.

"We celebrated too much," Charles mumbled.

A disheveled Clapper appeared in the doorway. "Anyone alive in here?"

Joy fetched another coke.

Charles sipped his drink. "Now we have our chance," he declared.

As he and Clap talked, the other employees, one by one, submissively stumbled in to join them. By nine o'clock Charles was ready to begin. His first call was to Radio Shack.

Charles soon discovered to his delight that during the months since the April San Francisco Faire while he had been preoccupied with the Arbiter hearings the microprocessor market had caught fire. Led by phenomenal sales by Astron Computers, a number of startup companies had sprung up, all determined to share in meeting the unanticipated demand. Most were founded by engineers too busy creating and building their hardware to bother with writing their own software. Determined to dominate the market with his BASIC and FORTRAN, Charles priced his software low. His profit margin was small, but Charles counted on volume to carry his company. By the end of the year, Digital Software's books showed a $110,000 pretax profit line out of total sales of close to $400,000.

In Cupertino, the news was even better, or at least it appeared that way to the outside observer. The orders streamed in, and Astron Computer expanded and expanded again, but the growth carried with it the seeds of sometimes seemingly unsolveable problems. Willard Temple's experience came into play. The only way they could keep abreast with demand was to subcontract. They began with the circuit boards. There was no way that the perfectionist Harold and his three engineers could produce fifty boards a day, because that was the number the growing customer list wanted to buy. Willard contracted with Bank of America to handle the payroll. They were hiring new employees every day and Willard did not have time to calculate withholding and social security. To deal with the books, he brought in a professional accounting firm. All of the new employees and support services put extreme pressure on Astron Computers' cash flow. They filled individual orders, but demand soon outpaced their ability to keep up. They turned to distributors and retailers who were happy to stock up on Astron Computers. All of this was a blessing, but it was a mixed blessing, because the computers shipped much faster than the payments came back. It took time for the computers to reach the market, time for the retailers to stock their shelves, time for the customers to buy, and even longer for the retailers who liked the credit lag to pay their bills.

By November Astron Computers was so successful that only a $250,000 loan from Stuart and Willard enabled them to avoid bankruptcy.

Throughout the happy travail, Stanley was ecstatic. He thrived by letting Willard Temple cope with the problems while he pushed production. Stanley's personality continued to create problems. He was unreasonable, inconsistent, demanding, and rude but for some reason the others, mostly young, responded to him. They emulated his work habits. Stanley labored long into the night, oblivious to niceties such as eight-hour days, and the others worked with him. They learned to ride with his outbursts—"that's the poorest workmanship I have ever seen"—and to share his enthusiasm. Stanley did everything. He stacked boxes with the loaders, chastised the suppliers for the color of their boxes, insisted that a manual be provided with each computer, and made every employee's life miserable with his perfectionism.

Stanley's private life became a problem. Stanley decided it was inappropriate for a successful businessman of his stature to live at home. Over his mother's objection, Stanley joined Jeff Briggs, a new employee and former high school classmate, and Sally "Bubbles" Turner in a house-sharing endeavor. Bubbles was the blonde groupie that Stanley had picked up at the Faire. A local waitress, Bubbles was a flighty but ambitious girl. When she hooked Stanley, Bubbles imagined she had caught a rising star. In her imagination, she often pictured herself holding the reins to a soaring Stanley riding a bright star much like the one in the Astron Computers logo. Unfortunately, house sharing with Stanley did not turn out as Bubbles dreamed. Stanley devoted his entire attention to his company. He placed it above everything, family, Bubbles, sports, life itself. Stanley, an early riser, left the house at six o'clock. He snacked at the company and seldom ate at home or the diner where Bubbles worked. Since he rarely returned home before midnight, Bubbles saw little of him. On occasion he would slip beside her on her mat, but those proved to be very transitory experiences, slam, bang and not even a thank you ma'am.

This life, far different than Bubbles had anticipated, turned sour after six weeks. The fact that Bubbles found herself pregnant did not help. She assumed Stanley was the father, but she could not be sure.

One night, some two months after they had become housemates, Bubbles confronted Stanley. He had just rolled over after one of his slam bang sessions and was trying to sleep when Bubbles snuggled up against his backside.

"Stanley?" Bubbles whispered. She did not want Jeff who was sleeping on a mat nearby to hear.

Stanley, who wanted to sleep, ignored her.

"Stanley!" Bubbles said a little more loudly, pulling his hair in the back.

"What?" Stanley surrendered. He didn't know what Bubbles wanted, but he knew what he wanted: to go to sleep and not to have to listen to her complaints.

Bubbles had morphed into a whiner, and Stanley had decided to unload her. He was already looking for a replacement.

"Stanley, I need to talk."

"Not now. I need to sleep."

"We have a problem."

"No we don't. I do," Stanley resisted. "I need to sleep."

"I'm pregnant."

Stanley said nothing, refusing to respond.

"Did you hear me?"

"Yes."

"Oh for Christ sake, will you two shut up or go get married. I'm trying to sleep," Jeff said.

"Me too," Stanley said.

"This is none of your business," Bubbles said.

"How do I know that?" Stanley asked, unable to resist the opportunity.

Bubbles slapped Stanley on the back of the head, hard, and jumped up.

"Good," Stanley said.

"Good," Jeff echoed. He tended to pattern himself after Stanley.

Bubbles turned on the ceiling light.

"Turn that damned light off," Stanley ordered.

Jeff wisely stayed quiet.

Bubbles left the room with the light glaring. Both Jeff and Stanley heard water running in the kitchen. In a few minutes, Bubbles returned and dumped a bucket of cold water on Stanley.

"What did you do that for?" Stanley demanded. "You got the mat all wet."

Bubbles left. Stanley raised into a sitting position. He pulled the drenched T-shirt over his head, and used it to dry his face. Again, they heard running water.

"It's your turn," Stanley laughed.

Bubbles returned carrying the full bucket. She ignored Jeff and headed towards Stanley.

"Wait," Stanley said, waving his T-shirt over his head. "Truce."

"I want to talk," Bubbles said.

"You don't look pregnant," Jeff said.

"In the living room," Bubbles said.

Stanley thought about refusing but decided against it. He wanted to sleep but realized he would have to placate the stubborn Bubbles first.

Stanley pulled on his shorts and followed Bubbles into the living room. She picked up a marking pen that Stanley had brought home and began writing on the white wall in big red letters:

"I AM PREGNANT"

Stanley dropped into their faded upholstered chair and placed one leg over the arm. "OK. What do you want to talk about?"

Bubbles turned and threw the marking pen at Stanley who tried to catch it but failed.

"Damn you. I'm pregnant."

"So what is that to me?" Stanley asked.

"It's your fault."

"How do I know that?"

Bubbles walked to the couch with as much dignity as she could muster and began crying.

"If that's all you got to say, I'm going to bed," Stanley said.

Bubbles did not respond. Stanley was not reacting the way she thought he should, and she did not know what to do about it. Stanley stood up and returned to the bedroom, closing the door behind him. He turned off the ceiling light, pulled the mat away from the pool of water, and turned it over.

"What are you going to do about it?" Jeff asked.

"We'll have to get another roommate," Stanley said. "Now shut up. I want to go to sleep."

The next morning Stanley slipped past Bubbles, who was sleeping on the couch, and went to the shop. Later in the morning, he was berating an assembler who was having difficulty fitting a board into one of the new plastic cases Stanley had ordered when Stuart Miller approached. Stanley ignored him.

"They've got the measurements wrong," the assembler insisted.

"How could they? It looks fine to me. You're too dumb to get the board in," Stanley shouted.

Stuart grimaced. He spent too much time placating angry employees who had suffered Stanley's sharp tongue. "Why don't you measure it?" Stuart suggested.

The assembler grabbed a tape ruler from the workbench and handed it to Stanley. "You do it," he said. "I already have. It's a quarter of an inch short."

Stanley grabbed the tape and measured the case. It was a half-inch short. "I'll take care of this," Stanley said. He patted the assembler on the back. "The contractors screwed up."

"Are you sure you didn't?" The assembler challenged.

The nearby workers paused to listen. They always like to see someone get the best of Stanley who was usually right even if he didn't know how to act.

"I need to see you now," Stuart Miller interceded. "In the office."

Stanley obediently followed. "Don't worry. I'll take care of it," he smiled back at the assembler.

The listening employees applauded. Stanley assumed it was for him. In fact, the employees were facing the assembler who had a broad grin on his face.

Stuart led the way into Willard's office, waited for Stanley to follow, then closed the door. Stuart sat down behind Willard's desk, and Stanley slumped into the chair opposite.

"I had a phone call this morning," Stuart said.

Stanley said nothing.

"From Bubbles."

Stanley laughed.

"Couldn't you have found a girl friend with a more sophisticated name?" Stuart asked.

"She's not my girl friend," Stanley said.

"What is she?"

"Housemate. Just like Jeff."

"Is Jeff pregnant?"

Stanley did not trouble to answer the silly question.

"Bubbles says she is," Stuart ignored the non-response.

"Then she better get it taken care of," Stanley said.

"She says it's yours."

"Who knows?" Stanley shrugged.

"She claims she's going to file a paternity suit."

"She better move out," Stanley said. This was working out like he hoped.

"Why don't you give her a couple thousand and take care of her like you should?"

"It's not my business. Everybody at the diner knows she's available."

"Can you prove it?"

"If I need to. Ask Jeff." Stanley stood up. "Is that all Stuart?"

"It's none of my business," Stuart said.

"Right," Stanley said. "I've got to call that contractor."

After Stanley departed, Stuart leaned back in Willard's chair and studied the ceiling. It was things like this that made him happy he was retired. He could walk away from it. He tried to be an older brother to Stanley and Dum, but it was not always easy. Stanley was more predictable. Dum had his own problems. With the influx of business, Dum too was spending more time than he should at the plant. His wife, already feeling ignored, had given Dum a choice. Spend more time with her or divorce. Dum accepted divorce. Now his problem was the settlement. Stuart had counseled that he give his wife fifteen percent of his stock, but Dum did not like that idea. Stuart decided to keep his brotherly advice to himself.

Chapter 14

The intercom buzzed; David Howard stared at the blinking light; he checked his watch, four-thirty. It was a strange time for the intercom. Friday afternoons in Armonk usually just petered out. It was late summer, and most were looking forward to their tennis or golf. For David, it was different. Softball season was over, and he did not enjoy either tennis or golf. He considered them old men's games. David had no plans for the weekend; he just hoped the intercom did not mean work. He had had all the busy work for one week that he could stomach.

"Yes?" David's response was neutral. He did not know if Ernie Hendricks, his boss, or Ernie's secretary, Margaret, David's girl friend, was on the line. Of course he would respond differently to each.

"David," Margaret said. "Ernie wants to see you."

"On my way, sweets." Margaret definitely did not like to be called sweets, but it was Friday afternoon, and David felt a little cavalier as he anticipated the weekend.

Margaret had a worried expression on her face when David passed through her office. Noting Ernie's open door, David winked but said nothing. He held his hands palms up. "What's up?" His hands asked.

Margaret shrugged her shoulders. "He just got back from an unscheduled Management Committee meeting," she whispered.

"Shit," Ernie said softly. The summons meant busy work.

David entered Ernie's office and deliberately left the door open so Margaret could hear. They had a date scheduled for that night, and David wanted her to know the reason if he had to break it.

"Hi David," Ernie greeted him. He glanced at the open door to signal he detected David's maneuver. He smiled.

David balanced his notepad on his knee and waited with ballpoint pen poised.

"I've got news," Ernie paused. "I don't know whether it's good or bad."

David waited, nervously. He tapped the pen on the pad.

"I'm afraid I have to find a new third baseman for next year."

David relaxed. If the news related to softball, he could handle it. He could comfortably play any position but pitcher or catcher.

"We had a tense Management Committee meeting," Ernie said.

David could not tell if he were changing the subject or not. He sat up straighter in his chair. Something had obviously happened that affected him.

"Crane," Ernie referred to James Archibald Crane, IBM CEO, "was on his high horse. He kept asking 'where's my star?'"

"He can't blame you for that," David blurted. David recognized that Crane was referring to Astron Computers. "You've been advocating that we build a microcomputer since I got here."

"Even before that memorable event," Ernie smiled. "Crane was just setting us up."

"Us?"

"The Management Committee. Of course, the majority led by Parsons registered their immediate opposition."

"So what else is new?" Parsons, the Director of Sales and the heir apparent to the CEO, had become more outspoken of late. Rumors had it that he had the support of IBM's board and was merely marking time until Crane retired next year.

"They pissed Crane off. First time I really saw him really lose his temper. He usually just plays with the opposition."

David began to worry. It was unlike Ernie to discuss senior management politics with him.

"After the meeting, the CEO took me into his office and gave me my marching orders." Ernie paused and studied his assistant, looking for all the world like someone trying to decide how to convey bad news.

David could think of nothing that might have brought him to the attention of Crane.

"What I am about to tell you is just between us," Ernie said. "And of course Margaret who knows everything," Ernie spoke louder for the benefit of his listening secretary.

"I've got three months to come up with a plan for a microprocessor," Ernie paused for effect. "Not a proposal but a detailed plan."

"How do you do that without everyone knowing about it?" David asked. He did not know much, but he did realize that as soon as Ernie started levying engineering requirements the entire company would vibrate with the news.

"That's what we have to do," Ernie continued. "Crane wants to present the opposition with a fait accompli. Like in complete surprise."

"And how do you do that?" David asked, worried about the answer. He feared the direction the conversation was taking.

"That's where you come in."

"But I don't know anything about computers. Not even little tiny ones."

"The little ones can be as complex as big ones," Ernie laughed.

David waited.

"You are my secret weapon."

Margaret laughed.

David did not dare.

"Have you heard of Boca Raton?"

"In Florida?"

"The same. We, the General Products Division, have a small research facility down there."

David shook his head, indicating puzzlement. He had never heard of any IBM plant in Boca Raton.

"It's no big secret," Ernie said. "We don't talk about them because they don't do much. It's just a handful of engineers brainstorming new products."

"Like Xerox's place in Palo Alto?" That David had heard of. Xerox Corporation had set up a research center in Palo Alto, California to create the future. PARC, as the Palo Alto Research Center was known, functioned separately from Xerox proper. Their simple task was to keep the company on the cutting edge of technological research.

"Nothing quite as grand. That was our original intent, but we have never been able to bring ourselves to grant the resources, men or money, to do what we pretended we wanted to do. I've been thinking about shutting it down."

"That's what you want me to do," David relaxed. A trip to Florida was just about what he needed to break the boredom.

"Not quite, again." Ernie smiled. "I want you to move to Boca Raton and come up with a plan for me."

"A plan for what?"

"A microprocessor."

"I don't know shit about microprocessors," David protested. "I'm no engineer."

"I know that. As pitiful a thing as you are, you are all I have."

"Then you're in trouble."

"We, you and me, are going to do something revolutionary."

"I'll bet."

"We're going to buy ourselves a microcomputer."

David did not understand. IBM prided itself on the fact that it built everything it needed itself. Hardware, software, IBM designed their own. To buy a component from others was heresy. David owed his career at IBM, as miserable as it was, to Ernie. He waited for an explanation, but Ernie said nothing, waiting for David to speak.

"Why not go to Radio Shack and buy one?"

"That's not what I mean. We are going to do a little market research first. Identify potential customer markets, determine what kind of computer they need, what software, then buy the components, work IBM magic on them, then assemble our IBM home computer."

"If you can't sell the concept of building an IBM microcomputer to the Management Committee, how will you sell a hybrid?"

"That's Crane's problem. He's given us the go ahead."

"In secret."

"In secret."

"And he expects me, the third baseman, to come up with it?" David was not only skeptical, he was afraid. He had picked up the talk at IBM. He could walk and talk like an IBMer, but he did not know squat about computers. He had bought his own Astron 2 on the sly and was currently trying to figure out how to operate it. It had taken him three hours just to hook the damned thing up.

"No. That's me. I am the one who selected you."

David decided Ernie was serious. "I assume there are engineers at Boca Raton who can help me."

"Half right. There are engineers at Boca Raton, but they can't help you. Remember the project is secret. You are to tell no one down there what you are doing."

"What will I say?"

"That's up to you. Special Projects. They won't care. I'll tell them to give you an office a desk, a safe—all special projects require a safe to hide secrets in—and a telephone. Then, you'll be on your own."

"I'll need a secretary." David decided he might as well build his own little empire. "I'll need Margaret."

A cheer echoed from the outer office.

"Nice try. No secretary. You're on your own."

"How long? A month. Two months?"

"As long as it takes."

"What about expenses?"

"We'll give you an advance. You account for it through Margaret."

"I repeat I know nothing about computers."

"It's time you learned."

"When do I leave?" David capitulated. It might be fun to be on his own with a blank check.

"What about Monday?"

This time a hand smacked hard on a desk in the outer office.

"I need more time than that."

"No you don't. Tell your apartment mates that they will have to pick up the slack in the rent. It's an IBM special project. They'll buy it."

"I'll drive down."

"Will the Volkswagon make it?"

"Certainly."

"The Management Committee will be relieved to get it out of the employee parking lot. It's an embarrassment." Ernie smiled.

The evening began disastrously with Margaret blaming him for the abrupt transfer.

"If you had tried to fit in," she accused, "this would not be happening."

"I tried."

"No you didn't. You let Ernie use you."

"How?"

Margaret refused to answer. "What am I supposed to do while you lounge on the beach in Florida?"

"I don't like the beach," David said, only half truthfully. "Florida is dreadful in the summer."

"How do you know?"

"I read it."

"Do you really think you can build a computer?" Margaret lowered her voice.

"I don't have to."

"That's what you think."

"Ernie said..."

"I know what Ernie said," Margaret said. "He's determined to build a small computer."

"Tell me something. You've worked for him longer than I have. Why?"

"Why what?"

"Why a small computer? Why me?"

"Because IBM already dominates the big computer market. Ernie seriously believes that the microcomputer threatens that."

"Is he right?"

"The others don't think so."

"They why does he?"

"If he saves the company, it'll belong to him."

"The stockholders might have something to say about that."

"Don't be obtuse. He'll be CEO."

This all sounded logical to David. Margaret confirmed what he had already deduced on his own. "Why me?"

"You belong to him, that's why. You're the fair haired boy."

David wondered if modesty demanded that he deny the claim. He liked hearing Margaret say it, however. He smiled. "What about us?"

"Why don't you start answering the questions," Margaret countered

By the end of the evening they had worked out a modus vivendi. David agreed to call Margaret frequently, at IBM expense of course, and to return to Armonk as soon as he could. Although he made the commitments, David was not sure about his ability to fulfill them. Escape from Armonk sounded good to him. Working independently sounded even better. He would have to travel to California and New Mexico and anywhere else he could find viable startup computer companies. With the contacts that resulted, he might even be able to devise an escape from IBM for himself, and Margaret, of course.

One week later, David, two suitcases and the weary Volkswagon found themselves at a Holiday Inn on the outskirts of Boca Raton. As he shaved, David made his plans. He would locate the IBM Boca Raton Laboratory—he had an address—and introduce himself. Ernie had promised to pave the way with a phone call to the Laboratory director, Jeffery James Madison, III, an exiled IBM engineer with a faded reputation for weird and occasionally useful ideas. Ernie had assured him that Madison, a most distant relative to the former president, was a long time colleague and friend who had fallen on hard times and did not know it. Madison, an absent-minded engineer, was apparently better suited to the halls of ivy than the corporate world. Early in his career, he had contributed a few theoretical innovations that attracted the blessing of Tom Watson Jr., the son of IBM's founder, and had guaranteed Madison a career, which had topped out at a sub mid grade level. Margaret had confided that when Ernie took over the General Products Division, he had dispatched Madison to a safe haven with the Boca Raton Laboratory where he was expected to contribute little to the company cause other than congenial silence.

Today being Friday, David decided to make an appearance at the Laboratory then devote himself to finding less transitory and less expensive lodgings. Margaret had assured him that IBM would pay the bills, David being on temporary duty. David decided that discretion required an attention to economy. He had asked the friendly front desk manager for advice, and he had suggested that if David planned only to be in Boca Raton for two months, he might find a splendid deal by subleasing a condominium owned by one of the snow birds, residents who spent the summers in the north and their winters in southern Florida.

David carefully followed the street instructions provided by the helpful manager and arrived at the IBM Boca Raton Laboratory at exactly ten o'clock. He had chosen that time carefully, not wanting to draw attention to himself by arriving before the exiled employees had time to settle in for their long boring day. To his surprise, the Laboratory did IBM proud. David had expected to find the Laboratory housed in a long neglected building not far from the ocean. Instead, he found an immaculate two-story stucco building with a red tile roof.

In fact, a team of painters was busy applying a fresh coat of white paint. The building looked more like a millionaire's residence than a laboratory. David followed the drive that circled the building until he arrived at a parking lot in the rear. A tidy lawn of Bermuda grass dotted with tall palms flanked the building and included a glistening Olympic size swimming pool and two well-maintained tennis courts. "Not a bad deal," David decided. Exile to Boca Raton obviously had its advantages.

After parking, David made his way around the building, pausing to admire the painters' work, then entered the front door. An attractive brunette with a deep tan and wearing a bright yellow sundress to advantage greeted him with a smile.

"Good morning, Mr. Howard."

Surprised, David hesitated. The brunette smiled. "Dr. Madison is expecting you."

David stared, liking what he saw. "How did you know my name?"

"We don't get many visitors, and Armonk alerted us to expect you at ten sharp. And there you are."

David assumed Armonk meant either Ernie or Margaret. "I don't usually get such treatment," David said. "I'm not a VIP."

"Anyone from headquarters is a VIP to us. I'm Terry."

David offered his hand, and she gripped it firmly, holding it a little longer than David expected. He wondered if she was flirting with him. He felt a little guilty, but he liked it. "Call me David," he smiled.

"Play tennis?" Terry asked.

"Not well."

"You're built for it. We play doubles at noon."

"Lunch time," David smiled.

"Sort of."

"Got a racquet with you?"

David shook his head negatively. "I brought a swim suit."

"You'll need both," Terry said.

She leaned back in her chair and smiled invitingly at David. "You're the youngest visitor we have ever had, at least in my two years here. Usually, they're old goats who come down for a few days in the sun in January or February.

"Oh, I'm not a visitor," David said.

"I know. You must have done something really bad to be sent here at your age."

David did not know what to say.

"I'm not complaining mind you, but the old boys are not very good tennis players."

"I'll bet you are."

Terry smiled. "Dr. Madison's office is the last on the right." She pointed to her left.

"Facing the ocean," David said.

"See you at lunch," Terry said.

David turned to his right. He was struck by the silence. His leather heels made the only noise on the highly polished wood floor. David looked back over his shoulder and found Terry watching. Embarrassed, he waved. She flicked her fingers back. He wondered what she had meant when she had said "see you at lunch."

David hesitated when he came to the dark wood door on the right. He debated whether to knock or not, decided that he couldn't go wrong by tapping lightly, did so and entered. David found himself in an office half again as large as Ernie's in Armonk. Somebody had not spared any expense. The dark wood floor was highly polished. The walls were paneled, a huge desk faced the door, and leather furniture lined the walls with a conference table on the right and a sofa and chairs forming a chat area on the left. A middle-aged man with pure white hair and a red face sat behind the desk talking on the phone. He looked like a southern version of a friendly family doctor. He smiled at David and indicated he should take a seat. David chose the couch.

"That's all right dear, I'll be sure to remember," the man said.

He listened patiently.

"I will dear," he said.

David thought he sounded sincere.

"I'm making a note of it right now." He seized a gold pen and indeed wrote something on the pad in front of him.

"My wife," he mouthed silently to David.

"Yes, dear. I will. You must forgive me, I must go."

"I know dear," he said.

"Yes dear," he said again.

When he finally hung up, he shook his head. "She's afraid I will forget. I'm supposed to buy some cucumbers, two cantaloupes, and some firm tomatoes on the way home. I buy the produce. She thinks I have a nose for it."

David stood up. "Please excuse me for interrupting, sir..." he began.

"That's quite all right Mr. Howard," Madison said as he rose from the chair and circled his desk. The man appeared quite trim, in good shape for his age. Madison offered his hand to David. Madison's firm grip surprised him.

"It's not often that we get a special projects officer from headquarters," Madison said. He indicated that David should sit down on the sofa, and Madison politely waited for him to do so before sitting in a nearby chair himself.

"I hope no one misunderstands," David said. "I'm nobody important."

"You don't have to worry about that, Mr. Howard. Ernie said. "We will all respect your privacy, secrecy. We know about commercial espionage."

"I'm not a spy," David said.

"Oh my, I misled you. I meant by our competitors. We all have a need to protect privileged information."

"Thank you for understanding," David said.

"I hope you find Boca Raton hospitable," Madison said. "I offered to have Terry find a decent accommodation for you, but Ernie said you had already departed. Did you have a difficult drive?"

"No sir. I enjoyed it. I'm from the northeast and have never been to Florida before."

"A football player I understand."

"I played a little."

"A lot, Ernie says. We don't have a softball league. Not enough of us. But we do have tennis and a pool. For employee recreation."

"I appreciate that. I'm afraid I won't have time to play. I have much to learn."

"We all do," Madison smiled. "Now how can I help you?"

"I will need an office," David began.

"Please wait a minute while I get my pad."

"That won't be necessary sir."

"Believe me it is. I tend to forget things if I don't write them down. I'm not doddering or anything like that. I've always been that way."

"I'm sorry sir. I just meant that all I'll need is an office."

"We have one of those. Just like mine in the other wing. All by yourself."

"All I need is a desk and a telephone. Nothing grand like this, sir."

"I'm afraid these are all we have. Our engineers spend their time in the laboratory on the second floor. Will you be needing tools or workspace?"

"Just a desk and a telephone."

"Do you need a secretary? I'm sure Terry..."

"No sir. I expect to spend much time traveling."

"I see. Is there nothing I can do for you?"

"Thank you sir. No."

"Well," Madison sounded distracted. "I'm sure Terry can introduce you to the others. If not..."

David recognized that he was being dismissed. He stood. "Thank you sir."

He shook Madison's hand again. As he turned for the door he realized why Ernie had sent him down here. He was sure there would be no prying eyes.

"By the way," Madison, who stood by his chair watching David, called after him. "We play tennis at noon. Doubles if you're interested. Just friendly exercise."

"Thank you sir," David said. The words "friendly" always made him suspicious when connected to an athletic contest.

Chapter 15

Bill Oldham decided not to wait for the graduation ceremony. He had more important things to do. The day he departed Cambridge his roommate, Mike Hamilton, was still debating his options, take an executive trainee position with Procter and Gamble in Cincinnati, or attend the Stanford Business School where he had been accepted. Bill, unfortunately, had no options to ponder; he did not have the luxury of choice. Not even a Harvard liberal arts graduate had drawing power in Jimmy Carter's declining job market. Bill had known all along that an English major was not a ticket to future employment, but he had deluded himself into thinking that the magical name Harvard would open doors. It did for some—those with wealthy and influential parents, and those who had selected practical degrees like Business Administration—but not for scholarship students with English majors.

Bill returned to Clarksburg for rations and quarters while he conducted his job search. He had no money to support inquiries in New York, or Boston, or Washington. Those Meccas did not appeal to him in any case. He sensed a need to return to the mountains. This was strange because four years previously he had suffered from an urge to escape. Now, the image of a more relaxed lifestyle appealed to him; Bill had a modest portion of ambition, but he was not work driven like Charles. He did not want to return to the shoe store; he needed to write. Maybe, he reasoned, he could find an undemanding employment that would support him independent of his parents while he wrote his first novel. He had close to one hundred pages written, a good start. He had surprised himself with his choice of locale, the mountains, of hero, a one room school teacher, of heroines, a prototype Daisy May, short skirts and all.

"Well, Bill, what now?" His father greeted him on arrival.

Bill did not know how to answer. The simple words had considerable depth behind them. His father was a simple businessman who had difficulty comprehending Harvard; he believed the university in Morgantown was good enough for a simple mountain boy. Bill's selection of an English major had caused his father problems. "What in God's name can you do with that?" He had asked, unable to conceal his irritation with his son's choice of venue and subject. At the time, Bill had not tried to explain. When Charles had dropped out to work his dream, Bill's father had insisted: "I hope Charles has not infected you with his West Coast ideas." Anything west of Ohio was peculiar in his father's outlook. Bill could not answer honestly, that he wanted to write. His father would not understand that either.

"I thought I would take my time in deciding what I want to do with the rest of my life," Bill equivocated. He respected his father but had long ago concluded that he lived in a different world. He held values no longer pertinent in a rapidly changing world. Bill theorized that the mountains kept outside influences at bay.

"Don't take too long about it," his father cautioned.

His mother smiled and patted Bill's shoulder, softening his father's harsher views, as she always did. "I'm sure Bill knows what he wants," she said.

"I sure to God hope so," his father muttered and turned the page of the Clarksburg Dominion, his primary source of information about the world. Flatland sources were not allowed to intrude in the Oldham house.

Bill drifted for a week, sleeping late, rising after his father's departure for the shoe store, and always being sure he was out of the house before his father's evening return. Bill simply had nothing to report, and he understood an answer would be expected soon. He suspected the order to return to the store rested on the tip of his father's tongue. In his father's world, adult males did not languish around the house. That was women's work.

At his mother's insistence, Bill attended the annual family picnic. Every year his father's brother, a World War II veteran, hosted the Oldham clan at his cabin on Lake Floyd. Bill had a host of cousins, aunts and uncles, and they were a comfortable lot. He could count on them to congratulate him on his Harvard degree and not press on his prospects. At least, he hoped he could. After greeting the clan, accepting their sincere congratulations, Bill found refuge at the barbecue where he donned an apron and joined his Uncle David in turning hamburgers and hot dogs. After the first wave of chattering Oldhams had been served, Uncle David invited Bill to join him for a drink. A drink meant a frosty bottle of Fort Pitt.

"Your father tells me you're at loose ends," Uncle David moved right to the point, an Oldham trait.

"Yes," Bill said.

"Got anything in mind?"

"I know what I want."

"And that is?"

"I want to write," Bill said honestly. He had always been able to talk with Uncle David, a Morgantown graduate, unlike his father who went to work after high school, and an engineer who searched West Virginia mountains for natural gas deposits.

"And you need something to tide you over while you do," Uncle David said.

"Yes, but I'm not..."

"I'm not offering you a job, Bill," Uncle David said. "Not exactly."

Embarrassed that he had misread his uncle, Bill said nothing.

"My old roommate might need some help," Uncle David offered.

"Oh?"

"Old Hump is editor of a small daily in Virginia," Uncle David said.

Bill smiled at the name.

"His name is Humphrey Tydings, Jr. His family called him Junior but we went with Hump."

"That sounds interesting."

"I could give him a call. Maybe he needs a reporter or something."

Bill hesitated. He was desperate. A small town newspaper might be exactly what he needed, a modest income, undemanding assignments, time to write.

"Please do, Uncle David," Bill accepted the lifeline.

The next day, a Tuesday, Uncle David phoned while Bill was eating breakfast.

"Old Hump says come on down," Uncle David said.

"I'm on my way," Bill said. "Thanks Uncle David."

"That's what family is for."

Uncle David told Bill to go the Fairfax Journal office on Edsall Road in Alexandria and to ask for Hump. "They should know who he is. He's the boss."

"Did he say when?"

"Anytime. Now's OK. Hump's a very good friend I haven't seen in years, but he's there when you need him. Sounded the same to me."

Bill borrowed his mother's five year old Ford and headed for Alexandria. He had to check the map. He had heard of Alexandria, Virginia, but had never given it much thought. He knew it was across the Potomac River from Washington, that Alexandria was a southern city whose residents had sided with the Confederacy, but that's all. For Bill, a mountain boy, Alexandria represented Virginia, the state that West Virginia had broken from at the outbreak of the Civil War. Bill like his peers had always considered himself a northern boy. He had thought about a civil service job in Washington but had rejected it. The only thing that sounded interesting was the Foreign Service or the CIA, but overseas life held no attraction for him. The other things, Agriculture, or Veterans Affairs, the mainline domestic departments like Labor or Interior, all sounded too sterilely bureaucratic to interest him. He wondered what Alexandria had to do with a little paper called the "Fairfax Journal."

Bill calculated he could follow Route 50 to his destination. That appeared to be easy enough. According to the map, all he had to do to reach Alexandria was keep right at the last intersection before the bridge. He assumed he would spot Edsall Road on the way and then the Journal's building. He did not have a number but hoped it would be marked by a big sign.

Bill ended up half-right and helplessly lost. Some twenty miles from Washington, at a place called Fairfax City, the suburbs began. Bill had envisioned something like Clarksburg. Instead, he found himself on a four-lane highway jammed with speeding cars, all traveling bumper to bumper. He went under the beltway and stuck with Route 50. He tried to read the street signs as he passed, but it was difficult because he had to watch the cars and trucks which constantly jumped around him, changing lanes at will at high speeds. He failed to spot Edsall Road. As he neared the Potomac, he chose to turn right at the last opportunity before the Potomac bridge. The sign pointed toward Alexandria, but turning proved to be a tactical mistake. Bill did well, or at least thought he did, until he encountered the Pentagon and its massive parking lots on his right. He survived those with minimum difficulty then he encountered the maze of cloverleaf highways protecting the 14th Street Bridge. From that point on, he was hopelessly lost. He spent three hours visiting Washington, crossing back to Virginia, circling the Pentagon, touring Arlington and even passing Langley, Virginia and the beltway that threatened to take him to Maryland before he reversed directions and started over again. He finally located Alexandria by accident. That did not prove to be a blessing. He crossed Alexandria and met the beltway again. Still, he could find no Edsall Street. He paused at an Exxon station, learned he could find Edsall Street if he simply went south on the beltway and watched for the sign. Having learned his lesson, he researched the yellow pages at the Exxon station's phone booth and obtained a street number for the Journal, 6408.

Before venturing back out into the traffic, Bill paused behind the wheel of the Ford and debated whether he should go forward or not. He wished he had asked his Uncle David a little more about this little country newspaper. Bill had envisioned a sleepy little southern Virginia town where he might be tasked to cover the local high school football games. Northern Virginia was a huge metropolitan area. Bill was tempted to return to the sanctity of Clarksburg, but he did not know what he would say to his father and his Uncle David. He could hardly tell the truth—this big city bustle worried him. He decided that continuing was the least of two evils.

To his surprise, the Exxon attendant's terse directions proved amazingly accurate. He parked one block from the Journal's offices in twenty minutes. Bill combed his hair and checked his tie in the rearview mirror. He locked the doors, carefully, retrieved his sports coat from the trunk and turned to make his way to his future.

"Check de honkie," a deep voice said.

Bill turned and found two smiling, black teenagers appraising him. Bill nodded and walked quickly away. He worried about his mother's car, but he did not dare look back. He passed an Asian couple speaking a foreign language. He assumed it was Vietnamese but was not sure. He reached 6408 Edsall Road and entered.

"Can I help you son?"

Bill smiled at the elderly guard in the blue uniform who sat at a desk to the right of the door.

"I'm looking for the "Fairfax Journal.""

"You found it."

"I need to speak to," Bill hesitated. He almost said "Hump." "To Mr. Tydings."

"They're closed for the day."

"On Thursdays?" He checked his watch. Four o'clock.

"Newspapers keep strange hours."

"When can I find Mr. Tydings in his office?"

"He'll be in by six."

"Should I wait?"

"In the morning."

"I'll be back tomorrow."

"Good idea." The guard picked up a copy of the newspaper he had been reading.

"Could you recommend a decent, inexpensive motel nearby?" Bill asked.

"Beats me."

Bill retrieved his mother's car, made a U-turn, and followed Edsall Road back to the beltway. He recalled passing a motel on Route 1 near the beltway. Rather that risk getting lost again, he retraced his route. After a hair-raising ten-minute race on the beltway in thickening traffic, he found Route 1 and a Howard Johnson's Motel complete with restaurant. Pleased with himself, Bill decided Alexandria was not so difficult as he pulled into the motel parking lot. He negotiated a single for thirty-five dollars a night. He paused outside the office and purchased the Journal and the Post from distribution boxes.

The Journal proved to be about what he anticipated. It looked like a small town newspaper without the smattering of foreign and national news. Obviously, the Journal left that coverage to the Washington Post and the networks. He learned that the "Fairfax" in the masthead referred to Fairfax County. Bill decided that the Journal was a suburban newspaper, not a small town paper. Like its country cousins, the Journal covered community news, sports, schools, births, weddings, deaths and the like. The front page had an article about the governor failing to solve Northern Virginia's traffic congestion problems. Bill could sympathize with that complaint. The Fairfax Journal apparently had sister papers in Alexandria, Arlington, Prince William, Montgomery and Prince George's, so it was part of a much larger enterprise than Bill had anticipated. Being a small city boy himself, he might fit in. He had never acquired a taste for world events, a topic that seemed to obsess the Post.

Weary from his trip, Bill ate a cheeseburger at the Howard Johnson's restaurant, then retreated to his room where he watched television for a while then took a shower and went to bed.

He woke at eight, had two eggs over medium, homefries, and sausage links at Howard Johnson's again, found the three dollar and fifty tab high, vowed to find a cheaper motel and a MacDonalds if he decided to stay overnight again, then returned to his room. He brushed his teeth, put on a clean white shirt and tie, and headed for Edsall Street.

A different overweight, elderly white guard protected the door. This time, however, the lobby was filled with people. Most were young, in their twenties and thirties, and all seem to be preoccupied with matters of importance. At least that was the way it seemed to Bill. Harvard with its leisurely self administered pace was beginning to appear more attractive than it had at the time.

"Mr. Tydings office, please." Bill knew what to say this time and tried to act less like a new boy.

"Hump?" The guard smiled, apparently at his own humor. "Second floor. The fat guy at the big desk."

"Thank you." Bill concealed his surprise at the guard's lack of decorum.

"Take the stairs," the guard indicated the staircase to his right. "Never know when the elevators will break down."

Bill smiled, not knowing whether he should laugh or not.

He climbed the stairs and found himself in a hallway that ran the length of the building. Directly in front of him were double doors. A faded dirty sign on the right proclaimed: "City Room."

Inside, Bill encountered a large room with no partitions. Three rows of three littered gunmetal desks faced the door. Only three of the desks were occupied. At the middle desk closest to the door sat a mildly attractive brunette, probably in her mid thirties. She glanced up from her typewriter, then back down, dismissing Bill. A second desk was occupied by an older man, pudgy, with a receding hairline. He was talking on the phone and did not look up. Another older man sat at the desk in the far left corner. His back faced the door. At the rear of the room, sitting at a desk commanding a view of the others, Bill identified the most likely candidate to be Humphrey Tydings. The man looked like one of John Galsworthy's Forsytes. He had receding hairline, full red face, large nose, a busy white mustache and a pigeon "chest." He wore a wrinkled white shirt with open collar, rolled up sleeves and tie askew. A cigarette dangled from thick lips and an inch long ash threatened to join others on the shirtfront.

Bill discreetly circled the aligned desks and approached the latter day Forsyte. The man looked up, appraised Bill, then without speaking returned his attention to the copysheet in front of him. Bill held his ground and waited. The man read quickly, grunted, initialed the lower right hand corner of the page with a slanted T, then raised his eyes to study Bill.

"If you won't go away, state your business," he said as he tossed the page into his outbox. The ash fell from his cigarette, tumbled across his belly and crashed on the floor. The man did not seem to notice.

"My name is Oldham." Bill said.

"OK, Oldham, good work, you found me," the man canted his head, indicating that Bill should take the chair to the left of his desk.

"Your Uncle David called me." The man spoke softly, apparently so the others could not hear.

"Thank you for seeing me," Bill said, not knowing what else to reply. He found his uncle's old roommate intimidating. He had expected a southern gentleman and instead got a caricature of a city editor.

"Just graduated?"

"Yes sir."

"Harvard?"

"Yes sir."

"Relax. We won't hold that against you."

Bill smiled.

"Much. Major?"

"English, sir."

"Waste of time. No job offers?"

"No sir."

"What did you expect?"

Bill hesitated.

"What can you do?"

Again, Bill did not know how to answer. This interview was not going as he had anticipated. Bill decided he might not need to worry about moving to cheaper lodgings, Uncle David's old roommate or not.

"David said you're a writer."

"I want to be, sir."

"Drop the sir, crap," Tydings said gruffly. "Hump will do."

Bill wanted to turn to see if the others were watching his humiliation but did not dare.

"Don't worry about them," Tydings appeared to read Bill's mind.

"No sir."

Tydings laughed. "Work on the Crimson?"

"No sir."

"Christ. What are we going to do with you?"

Bill heard the woman giggle.

Tydings glared over Bill's shoulder, but he ended with a smile. Bill suddenly realized that Tydings was enjoying playing the gruff old newspaperman.

"Know anything about the "Journal?"

"You're a community newspaper," Bill responded quickly.

"At least you can read the masthead. Not bad for a Harvard man. All right, it just happens we're short a man."

The girl behind Bill coughed.

"Person," Tydings smiled.

"Did they teach you how to answer the telephone at Harvard?"

"I think I can handle it."

"Take the seat over there near the girl person. If the phone rings, say hello."

Bill wondered what he should say after hello but did not ask. He was tempted to ask about salary, hours, that kind of thing, but Tydings had returned his attention to another sheet of paper.

Bill meekly approached the desk next to the woman.

"Hi," she smiled. "I'm Clarice Taylor. CT will do."

"Bill Oldham. Mind if I sit here?"

"I'd be angry with you if you didn't."

Bill wondered if that were an invitation. CT was not bad looking but a little old for him.

"I'm not trying to seduce you," CT laughed.

Before Bill could respond, the phone on his desk rang. Bill looked at it.

"Don't be shy, answer it."

As Bill reached for the receiver, CT said: "You're our life and death reporter."

"Fairfax Journal," Bill said.

"I want to report an obituary," a mature female voice said. "My father died last night."

"Yes ma'am," Bill said. "I sympathize with your grief."

"Don't. He was a mean old bastard."

Bill did not know how to respond.

"But he lived in Fairfax County for eighty years. I want to let all those who hated him know they can relax. He's dead."

"Could you give me some details?" Bill asked. He opened the desk drawer and found a yellow legal pad and ballpoint pen.

"The old bastard wrote his own. Shall I read it to you?"

Bill took three pages of notes. When he finally hung up, CT, now typing, called over her shoulder. "You'll have to cut that to a paragraph."

During the course of the next two hours, Bill learned that he was in charge of births, weddings, deaths, and news tips. The later he reported to Hump who issued assignments to others. Bill wrote the announcements. He devoted a half-hour to polishing the "old bastard's" obituary.

"You're going to have to write faster than that," CT said when Bill asked her what he should do with it. "Hump's in box," she said.

With trepidation, Bill approached Hump's desk.

"First story?" Hump laughed and held out his hand.

He glanced through it, crossed out several words, then tossed it in his out box. "No three syllable words, complex sentences, adjectives, adverbs, or passive voice," he said. "Simple sentences, noun and verb, will do."

"Yes sir," Bill said.

"And learn to write in one draft. We're not looking for literature here. Just quick prose."

Bill's phone rang again. He rushed back and answered: "Fairfax Journal."

"Hi, I'm a father," a male caller reported. "Can you put it in the paper?"

Between calls CT informed Bill he should check in with the paper's administrative offices on the third floor. "They'll give you the forms to fill out for taxes and all that. Need money?"

Bill nodded twice.

"They'll give you an advance, but you'll have to ask for it. They don't volunteer anything around here."

Bill introduced himself to the two older males. One handled the local news stories, the Fairfax County Government, the police and fire, Fairfax City politics. The other was the business section editor. CT admitted she was the local features reporter. She covered local community and family news, food and home, garden, arts, the community college and the fledgling George Mason University.

"Will I get a chance to write something other than announcements?" Bill asked softly.

"We all start there," CT counseled. "Be patient. You've only been here an hour. You'll get your chance."

As the morning wore on, the frequency of Bill's phone calls increased. About noon the other reporters broke for lunch. "We're writing tomorrow's stories," CT confided. "We have a different timetable than the big papers." To

Bill's dismay, CT did not trouble to explain what the timetable was, so he worked through lunch. At three, the others stopped working for the day.

"You can work to five if you want," Hump stopped at Bill's desk after the others had silently departed. "Where are you staying?" He asked.

"At a motel. I have to find a room somewhere," Bill admitted. "If you are going to hire me."

"We'll give it a try," Hump said. "Despite Harvard."

Tydings turned to go then paused. "Find something out near Fairfax City," Hump ordered.

"Yes sir."

"Try the Carl M Freeman apartments on Route 236 just inside the beltway," Tydings said. "They'll have a decent one bedroom garden apartment for under one fifty."

"Yes sir." Bill wondered what a garden apartment was. It sounded good.

Tydings started to leave then stopped again. "You got transportation?"

"Yes sir." Bill did not say he had his mother's car.

"Spend the weekend learning the county. Ever heard of Tyson's Corners?"

"No sir."

"Check it out. I don't mean just the shopping center. Learn where the businesses are. Northern Virginia is becoming a commercial center. The big companies are locating here. Take a look at Reston, and learn your way around the government center at Fairfax City. Christ, you've got a lot to learn kid. You should've been born here if you want to work at the 'Journal.'"

Bill felt like asking how he should have arranged that but did not. "Yes sir. I'm a quick learner." He did not say he had been lost for four hours trying to get past the Pentagon complex.

Chapter 16

The Taurus was the electronic magnet that drew Charles and Clapper to Albuquerque, and the dispute with Perfect/PAMS negated the force that held them there. From the day that the Arbiter severed Digital Software's tie to the Taurus, Charles and his partner worried about their location. Charles, who handled most of the sales, finance and marketing, spent too much time on the road. Charles still wrote code, but selling BASIC to others became one of his primary responsibilities. A phone call from Japan greatly expanded Charles' horizon. The brief exchange led to a meeting in Albuquerque and the signing of a page and a half contract that gave a young Japanese hustler exclusive distribution rights for DS BASIC in East Asia. After one tedious trans-Pacific flight, Charles called a partnership meeting to discuss the future.

"There's no reason to stay here just because this is where we started," Clapper opened the discussion. "I hate Albuquerque."

"The only reason we came," Charles agreed, "was the Taurus. Now, our business center is elsewhere. Should we move to the Valley?" The Silicon Valley outside San Francisco had become the focus of the electronics industry, and it appealed to Charles.

"I miss home," Clapper said frankly. "I need the green."

"It would be nice to be able to drive to meet customers," Charles said. "Christ, I even have to fly to Texas to see Radio Shack. I used to think that flying was exciting. Now, it's worse than taking the bus. Maybe I should learn to moo. I feel like one of the herd."

"All of our employees except for Marion are from the northwest," Clapper persisted.

"We can't leave Marion behind," Charles said.

Clapper smiled. He remembered the day when Charles was out of town and he had impulsively hired Marion as office manager. On her third morning Marion had arrived early and discovered a teenager sitting at the desk in the president's office. Marion had rushed off to Clapper to report: "There's some kid sitting at Mr. Swift's desk. We have to get him out of there. Mr. Swift is due back today." Clapper had followed Marion into Charles office and found Charles with his feet on the desk. He had laughed: "Marion, I hate to tell you this, but that kid is your boss."

"We can pay to relocate Marion and her husband," Charles said.

With that commitment, the two partners agreed to move. "Where to" was a subject they deferred until later. They informed their eleven employees of their tentative plans, and all but Marion excitedly concurred. They, too, joined the "where to" debate. To Charles's surprise, Clap's suggestion that they relocate to Seattle was accepted almost unanimously, leaving him and Marion in the minority. Fond memories and his affection for his parents pulled him towards Seattle, but his practical business sense argued for the California Bay area where Intel, Motorola, and National Semiconductor along with upstart Astron Computers reigned. Sentiment and popular opinion won. They moved to Seattle.

Three months after settling into their new offices on the sixth floor of the First National building in Bellevue, Charles finally admitted that Clap had been right about returning home. Charles stood at the window of his corner office and admired the familiar blue waters of Lake Washington. He found the view quite relaxing. He could almost see his parents' home on the far shore, but not quite. Business had not fallen off as he feared it might. They now had forty-eight customers for BASIC and twenty-nine for FORTRAN, and Charles had just hired a new marketing director whose marching orders called for the sale of PASCAL. Charles smiled when he thought of the man's shocked reaction when he learned quite by accident that he was hawking a product that existed only in Charles' imagination.

Charles had patiently explained that DS could sell promises because it had the people with the intelligence and work ethic to produce the nonexistent when necessary. Charles did not like the phrase vaporware, but privately, he admitted that at times he sold promises.

Charles dismissed the thought and returned to a subject he had been debating with himself for several weeks. Should DS produce packaged software for the retail market? The hobbyists and small businessmen and professionals who had taken a chance on microprocessors purchased their software by mail. Charles was convinced that the market was now large enough to support specially packaged software—word processors, games, utilities—that could be retailed in electronic stores, maybe even the chains like Sears and Montgomery Wards. Charles had not originated the idea. Radio Shack had tentatively begun to test the demand by stocking selected items of software in its outlets.

"Charles," Pam, his office assistant, a replacement for Marion who declined to leave Albuquerque, interrupted Charles' musings. "You have less than an hour to catch your plane."

"I know," Charles agreed.

Charles hated waiting in airports and always pushed his luck by waiting to the last minute then rushing to be the last to board the plane. Usually, Charles was sanguine about missing flights. He knew there would always be another. This time, however, he had been invited to visit H. Ross Perot himself, the multimillionaire owner of Electronic Data who had made himself the king of software for mainframes. This was one appointment Charles did not want to miss. He looked forward to meeting Perot. The man was a legend. Charles was under no delusions about the value of the meeting for DS. Perot was not about to contract with DS to supply software. Perot's company manufactured its own. Charles assumed Perot wanted to take his measure. The big boys were always looking to merge with, that is to absorb, the little guys, particularly when they had something to offer, and DS had that. Maybe, Perot was thinking of entering the microcomputer software market. If so, that was something Charles needed to know. The competition would be brutal.

Charles grabbed his briefcase and headed for the door. He was wearing his customary open necked shirt, khaki slacks and Nikes, comfortable for travel, but out of deference to Perot he had a suit and dress shoes in his travel bag in the trunk of the Porsche. Charles planned to change clothes at the airport and arrive at Electronic Data System's headquarters neatly attired like any establishment executive. Charles had heard that Perot, an older man, judged adversaries by their style. Given his youthful appearance, Charles knew he had enough to overcome without having how he was attired work against him.

"Drive carefully," Pam called. Despite her age, Pam, who was younger than Charles, still tried to compete with the legend of her predecessor Marion, the company housemother.

"Yes mother," Charles winked.

"Where will your car be?" Pam asked.

"In the long term parking lot," Charles said.

The direct flight from Seattle to Dallas arrived a virtually unprecedented fifteen minutes early. Charles, with time to spare, decided to get a haircut at the airport barbershop. Unfortunately, he got a slow, chatty barber.

"Just a trim," Charles repeated several times in an effort to hurry the man along.

"What do you think about the Cowboys?" The barber asked.

"I'm a Redskins fan," Charles said, trying to shut the man's mouth and hurry his clippers.

"George Allen is destroying that team," the barber warmed to his topic.

Charles tuned him out. Finally, the man finished.

"Fine, Fine," Charles said as the man held up the mirror. Charles really did not care about hair. He just wanted not to offend Perot.

Charles tipped the barber two dollars and hurried off. He changed clothes in the men's restroom, then checked his bag in a locker. Pocketing the key and carrying his briefcase, Charles made his way to the taxi stand. Thoroughly familiar with the Dallas airport, Charles outraced several other similarly clad businessmen to the head of the line.

Charles need not have hurried. After announcing his arrival to the receptionist at Perot's plush headquarters, Charles was escorted to the anteroom of one of Perot's lieutenants. Charles knew the drill, so he relaxed with a two-month-old issue of <u>Time</u> magazine. After a fifteen-minute wait, the intercom on the secretary's desk buzzed. She picked up the phone, said, "Yes suh" twice in a flat Texas accent, then smiled at Charles. He dutifully admired her evenly spaced teeth. "You may go in now, suh."

Charles hesitated. He suddenly realized he had nothing to say to these vultures who were circling trying to decide whether the bones of his small company were worth the effort. "What am I doing here?" he asked himself. "To find out what these people are planning," he answered. He retrieved his briefcase and entered the inner office. Despite himself, he was impressed. The office was twice the size of Charles' own office. Rich wood paneling, walnut and leather furniture, and an immense desk set the stage.

"Mr. Swift, welcome to Dallas," a large, middle aged man greeted him. He stood and offered his hand. He towered over Charles who estimated that his protagonist stood at least six feet six.

"Good morning sir," Charles smiled.

The man's huge paw swallowed Charles' hand. He squeezed just hard enough to show who was bigger, then nodded at a younger man who remained seated on his left. "I'm John Wilson, Mr. Perot's executive vice president, and this is Harry Johnston."

Charles turned towards Johnston who responded with a thin smile and did not offer his hand.

"Please have a seat over here," Wilson led the way to a couch and two chairs that flanked a glass coffee table. An identical grouping set on the opposite side of the room. Charles wondered how Wilson used both at once. As Charles placed his briefcase next to his chair, he caught Wilson and Johnston exchanging amused looks. He knew immediately that they were reacting to his apparent age. Charles settled into his oversized chair and waited for Wilson to lead the conversation. The portly man in the black suit said nothing. Charles waited. Finally, Wilson broke the ice:

"May I offer you something? Coffee?"

"No thank you," Charles said.

"Are you sure? I think I'll have coffee," Wilson said.

Johnston leaped to his feet. "I'll take care of it."

While Johnston was conveying the order to the secretary, Wilson tried again.

"How are the Seahawks this year?" He asked.

"I really don't have a clue," Charles smiled. "I'm not a sports fan."

"Oh," Wilson said, his tone conveying disapproval.

Johnston rejoined them.

"Charles is not a sports fan," Wilson said.

"That's probably for the good," Johnston smiled. "The Cowboys are loaded this year."

The two Texans waited for Charles to respond. He said nothing. The secretary entered bearing a silver tray with two delicate cups. "May I get you something Mr. Swift?" She asked.

"No thank you," Charles smiled. He hoped they would soon get down to business.

"I understand you are in the software business," Wilson said.

"Yes," Charles said. "Digital Software."

"I'm not sure..." Wilson began.

"Software for microprocessors," Johnston intervened.

Charles concluded that Wilson who obviously had not done his homework did not know why Charles was there.

"Yes," Charles said. "We did the BASIC for the Taurus and sell FORTRAN and BASIC to a number of other companies. Japan is one of our growing markets. Our representative there..."

"For microcomputers?" Wilson interrupted.

"Yes. We..." Charles tried again.

Wilson glared at Johnston, obviously wondering why he was wasting his time with a kid.

"Mr. Perot invited Mr. Swift down for a meeting," Johnston tried to explain.

"I don't understand," Wilson said. "We're not interested in software for little computers. Our market is mainframes." Wilson looked at Charles, expecting an explanation.

Charles did not respond. Johnston said nothing. They sat in silence.

"I guess Ross has his reasons," Wilson said. "Well," he stood up. "I won't waste any more of your time, Mr. Swift."

"Escort him to Ross' office," Wilson glowered at Johnston, putting the blame for this travesty on him.

Charles was most unimpressed. He had obviously flown all the way to Dallas wasting a whole day for nothing. Charles followed Johnston into the hall.

A disconcerted Johnston made no attempt at small talk as he escorted Charles to the elevator. "Top floor, straight ahead," Johnston said as he pushed the up button. Without another word, Johnston left Charles standing in the hallway. Charles decided if he were Perot he would immediately fire those two.

The elevator arrived, and Charles pushed the button for the tenth floor. The elevator silently rose then opened directly into a large reception room. An American flag and a large eagle proclaimed the occupant's patriotism. Oil paintings with a western motif lined the walls. Two secretaries flanked the entrance.

"Good morning, Mr. Swift," Mr. Perot is expecting you," the attractive matron greeted Charles briskly. "You may go right in."

"She sounds like my dentist's receptionist," Charles thought.

"Mr. Perot, Mr. Swift," the secretary announced and stepped back to let Charles enter.

Perot's office was even larger. It was similarly furnished but had obviously expensive original paintings on the dark walls.

"Pleased to meet you, Charles," a smiling Perot met Charles half way into the room. He offered his hand. Charles noted he was at least two inches taller than the diminutive multimillionaire.

"I've been looking forward to meeting you, sir," Charles responded to the man's magnetism. Perot smiled broadly, but Charles saw the hard eyes coldly assessing him.

"Please sit down," Perot led the way to the corner conversation grouping of furniture. "You met Wilson?"

"Yes sir."

"Did he tell you why I wanted to talk with you?"

"No sir."

Perot smiled. "He didn't know. I'll bet he never heard of Digital Software."

Charles, not wanting to criticize Perot's executive vice president, did not respond.

"I thought so," Perot said. "Well, he will. You wrote the BASIC for the Taurus, didn't you?"

"Yes sir."

"Nice job. I tried it myself. The Taurus is a buggy little machine, but I like your BASIC. Where did you learn it? Harvard?"

"No sir. I taught myself. Several years ago."

Perot laughed. "You must have been in grade school."

"I wrote my first program when I was thirteen."

"Thought so. You have contracts with Radio Shack, Commodore, IMSAI, Ricoh and that little Cupertino company," Perot preened, proud of his knowledge.

"Yes sir. Astron Computers."

"What do you think of them?"

"Their Astron 2 is a good product. Needs a floppy disk drive. Dumbroski is a talented engineer."

"I agree. What about that fellow Pitts? I don't understand what he does. He's not an engineer is he?"

"Stanley has a difficult personality, but he has vision," Charles said. He was uncomfortable talking about customers.

"You moved out of Albuquerque."

"Yes sir. We're in Seattle now."

"Strange choice. Why not the Valley?"

"It was a close decision. Like all of my employees, I'm from Seattle. My partner James Clapper wanted to return home. The University of Washington has a good computer department."

"Right, a good source of employee talent."

"Yes sir. We try to hire only the brightest."

"Not always the easiest thing to do."

"We've been lucky so far. And we calculated we would have a better chance of keeping the good ones in Seattle than we would in the Valley."

"Fewer headhunters and temptations."

"Yes sir."

"Good thinking. What did you gross last year?"

Charles hesitated. Company finances were not a subject he wanted to discuss with Perot. The man had a reputation for eccentricity. "Over a million."

"Two million?"

"I'd prefer not saying," Charles said.

"Good. Keep your company's financial information between you and the damned IRS. What do you see as the future of the little computers?"

"There's going to be a computer on every desk and in every home. They will be used to conduct business, explore the world, provide entertainment, play games, communicate with friends, shop, send pictures, all from home or office."

"That's a little grand isn't it?"

"No sir. The day will come when there will be more than one hundred million computers in the world."

"Any room for mainframes?" Perot laughed.

"Certainly. But ten years from now microprocessors will have the computing power of today's mainframes."

"Are you sure of that?"

"Yes sir."

"And what makes you so certain."

"Products like the Astron 2."

"If the future for microprocessors is so bright, why doesn't IBM agree with you."

"They're too bureaucratic. IBM is blinded by their success. Today they own sixty-seven percent of the mainframe market. IBM is the first name big business thinks of when the word computer is mentioned."

"Why is that?"

"Quality. Reliability. Good customer service. If a customer has a problem, IBM representatives promptly dispatch representatives to solve it."

"What will happen when you have hundreds, thousands of little companies making little computers?"

"A few quality companies will survive."

"Who will they be?"

"I don't know, but IBM is the big boy. They'll wake up some day and dominate the market."

"And where does that leave you?"

"Digital Software will sell to everybody. One day we'll have our share of the market. There's nobody ahead of us."

"Not even Electronic Data Systems?"

"I'm sure you can speak for EDS better than I. Now, may I please ask a question?" Charles said. "How is EDS going to meet the microprocessor challenge? Will EDS write software for the microcomputers?"

"Good questions," Perot evaded answering.

Charles did not fill the silence. He had heard that Perot had been scouting the Valley startups looking for buys. It was rumored he had made an offer to Astron Computers.

"What do you reckon a merger is worth to Digital Software?"

"That's hard to say."

"If the future matches your vision, you're going to need a big partner if you hope to grow into it. IBM writes it own software, you know."

"They may find others can do it for them cheaper."

"Really?"

"Yes."

Perot focused his hard gray eyes on Charles. He expected the young man to bend, but he did not. Finally, Perot asked another question: "If a fellow were to buy a company like yours, would he make money?"

"If he could get a company like mine."

"Son, why don't you go back to Seattle and give this matter a lot of thought. If you can see yourself reaching the five/six million level, drop me a line."

The figure surprised Charles. With his contracts, he could realistically set his price at two million, but that did not take into account the future. If he believed his vision, and he did, six million would prove to be paltry within a year. Charles decided to leave matters as they stood.

"Thank you for your time, Mr. Perot."

"Thank you son. Keep me in mind." Perot was surprised that the young man had not jumped with joy at his figure. He made a mental note to keep an eye on Digital Software's youthful president. He had backbone and vision, a remarkable combination.

Charles returned to Seattle. He discussed his Dallas visit with Clap and his parents. Charles' mother, surprised at the magnitude of the offer for her son's small company, marshaled her boardroom personality and recommended that Charles and Clap sell. His father, the lawyer, came down on both sides of the issue, and Clap deferred to Charles. Charles wrote a polite no to the Texas millionaire.

Having made his decision, Charles decided he had to play the cards he had and back his vision to the hilt. In order to do so, he needed help. Digital Software had twenty-six employees, and they all reported directly to Charles. He no longer had time to write code or think about how to position their company for the anticipated technological advance. Clapper had made his niche working with the young programmers. In order to grow the company, they had to add muscle to the business side. Charles did not want to bring in a retread from another company. He wanted youth, smart youth. He immediately thought of his Harvard roommate, Mike Hamilton.

Mike was a hardhead but a smart guy. He had graduated from Harvard with a degree in mathematics and business administration and where Charles had lived on the fringes of university life Mike had joined. He worked on the Harvard Crimson, the university literary magazine, belonged to a club, and managed the football team. After graduation, he had sold cake mix for Procter & Gamble, tried to start at the bottom at Universal Studios, and enrolled at Stanford Business School. During his occasional visits to Seattle, Mike had kept up a good front, but Charles could tell he was still looking. It was only natural that his name popped into Charles' consciousness when he began searching for help.

Charles found Mike at his parents' home.

"Hey Mike," Charles said.

"Hey Charles," Mike countered. He and Charles always competed, even when talking on the phone. Mike still savored Charles' disappointment whenever Mike outscored him on a math test.

"What's up?"

"Killing time."

"When do you start Stanford?"

"September." Mike assumed Charles was calling for a purpose. "Just drifting for the summer." Mike gave Charles an opening.

"Mike, I'm really in trouble," Charles decided to tempt Mike with another opportunity to better him.

Mike did not react, waiting for Charles to drop the bait.

"Business is booming, and I can't handle it. We need someone with business experience to take over as general manager."

"That's great news."

"Interested?"

"Sure."

"You can help us out?"

"For the summer."

"We need someone for the long pull."

"Sorry, Charles. I can't help you. Let me check around. Maybe someone from Proctor & Gamble."

"I want you," Charles said. "You didn't like Universal?"

"They wanted me to start at the bottom."

"You can start at the top here."

"Sorry Charles."

"OK. How's life in Virginia."

"Dull. I saw Bill last week."

"Really? Where?"

"He's a newspaperman. The "Fairfax Journal.""

"Never heard of it."

"A suburban paper near D.C."

"How's he doing?"

"He loves it. The "Journal" is not a typical small town paper. Northern Virginia is a growing metropolis, and Bill's angling to break in on their business section."

"He learned something from us," Charles laughed. "If you see him, tell him to give me a call."

"OK."

"And the same applies to you. If you change your mind, let me know, but do it soon. I've got to get help."

"OK Charles."

Charles hung up the telephone, satisfied. He knew Mike well enough to realize he had taken the bait. Now, Charles had to set the hook. He briefed Clap. Two days later when Mike called back, Clapper answered the phone.

"Hi Clap," Mike greeted him.

"Hi Mike. When are you coming out to join us?" Clap acted as if it were a done deal.

"Nothing's certain," Mike replied. "I wanted to talk a little more with Charles."

"We really need you," Clap said.

"Is he there?"

"No. I think he's at the pad."

Mike hung up and dialed Charles' home, a hillside house with a dramatic view of Lake Washington.

"Hello." Susan, Charles' latest live-in housekeeper, answered.

"Hi Susan," Mike said, wishing he had a sexy young housekeeper of his own. "Is he there?"

"I hear you are coming to join us," Susan said.

"Only if you promise to leave Charles and assist me," Mike said. Charles insisted on calling Susan his personal assistant.

"Anything's possible," Susan flirted.

Mike was not certain if anything was going on between Susan and Charles. He knew there would be if he were in Charles' shoes. Susan was a perky twenty year old, an unusual housekeeper.

"Charles!" Susan called. "It's Mike."

"See you when you get here," Susan said.

"Is that a promise?"

Susan did not answer, and Mike was not sure she had heard him.

"Hi Mike," Charles, after a short delay, picked up the phone. "When do you get here?"

"Can we talk some more?" Mike said.

"Sure."

"You know, I don't know squat about computers."

"We've got technical people."

"What would I do?"

"Just about anything you want. We've got twenty-seven employees and need more. We still keep our books by hand, in a journal for Christ's sake. I've just had a small rebellion. The employees want overtime."

"You can pay the professionals on an annual salary basis no matter how many hours they work. But you have to pay the clericals time and a half for overtime. That's the law."

"So I'm told. The clericals filed a complaint."

"So pay them."

"I don't want to set a precedent."

"Fuck that Charles. You have to pay them for overtime."

"See, I need you."

"What precisely would be my title?"

"We could work that out."

"I would be in charge of administration, personnel, finance, operations?"

"Sure."

"Could you be a little more precise?"

"Mike. I need someone I can trust. I trust you, and you're a smart guy. You will have the authority to do anything you want."

"What about sales, marketing?"

"That too."

"But I can't program. I guess I could learn to talk the jargon."

"I guess you could."

"What salary?"

"Big."

"How much?"

"As much as me." Charles prided himself on taking out less in salary than some of his younger programmers.

"Could you be more precise?"

"Sure."

"Charles, cut the shit. How much?"

"How does thirty-five sound?"

"Too low."

"Let's think about it."

"You think about it. I'll be in touch." Mike hung up.

Charles smiled.

One week later, Mike accepted the position of Assistant to the President at a salary of $50,000 a year plus a chance to get percentage share of the company depending on growth.

Chapter 17

Success came rapidly to Astron Computers, but the two founders changed little. Harold grudgingly agreed to perfect the disk drive that the Astron 2 needed so badly. Without a wife to demand a share of his time, Harold dedicated himself exclusively to his project, sublimating the most fundamental of human activities, like eating and sleeping, and his workbench became his home. After a month of intensive effort, Harold completed his task and introduced his disk drive at the Consumer Electronics Show and the Second West Coast Computer Faire. The response was electric.

Disk drives are small metal boxes that hold disks covered with a special coating that permits a magnetized head to read and write data. The computer retains its vital instructions and information on these disks where it is available for recall on demand. Harold's belated contribution gave the Astron 2 the edge. Radio Shack, Commodore and other manufacturers had been in the chase, but Harold won. His design drew immediate praise, and the accumulated inventories that had troubled Stuart and Willard quickly dissipated. With a startling suddenness, Astron Computers pulled ahead in the chase. A common phenomenon came into play, the rolling bandwagon.

Stuart Miller's phone began to ring, and ring, and ring. The big names in the world of venture capitalism, the proverbial sharks, joined the chase. Stuart raised $500,000 in new capital in one day, and Astron Computers was revalued at five million dollars.

Timing is important in the world of business. Like fine wine, every product has its time, and this is as true in the marketplace as it is in the most carefully selected cellar. Astron Computers offered its Astron 2 in an American business climate that needed good news. Battered by Japanese industry, the country's manufacturing foundation had lost its confidence. Even mighty Detroit was in the doldrums. When cocky, little Astron Computers strode confidently into the shaken marketplace, the American people and Wall Street responded with enthusiasm for a new product that had not been seen for years.

The Moss Agency, sensing this opportunity, offered Astron as the symbol of the new American technology. Richard S. Moss highlighted the story of the two teenagers and the Pitts family garage, and the American media loved it. Stanley particularly thrived on the attention. Before long, the media forgot about Harold Dumbroski and began to depict the photogenic Stanley as the technical genius leading America's electronic renaissance. The publicity fanned the craving for America's newest electronic toy, and Stuart jacked Astron's advertising budget to $600,000 a year. The Moss Agency used the money to tout its growing client in newspapers and consumer magazines that reached every home.

In Cupertino, Astron Computers exploded. They hired new employees by the hundreds and acquired a fancy headquarters on a corporate campus with four new buildings. Stanley with his self-inflating ego rose to the occasion. Still without title, Stanley imposed his whimsy on every facet of the business. In his usual attire, sandals, T-shirt and jeans, Stanley led by intimidation.

"That's a piece of shit," was a favorite expression.

Stanley developed a love hate relationship with his colleagues. Somewhere deep in his genes, a little fellow responsible for charm kept those who felt his sharp tongue ambivalent. One moment Stanley would say something hateful, and the victim despised him; then, extravagant praise from the same source converted the recipient into a friend for life. In addition to having a total void of tact, Stanley was a perfectionist who had extreme confidence in his own taste and judgment. His disputes with Willard Temple, the company president, became legendary. On Stanley's birthday, Willard presented Stanley with a wreath of white roses with a black ribbon bearing the letters R.I.P. Stanley took the gift as a compliment.

Bubbles, Stanley's discarded housemate, refused to go away. Bubbles gave birth to a girl who she named Stella after Astron's top-secret new computer that was still under development. Stanley himself had named the new project; Stella in Latin meant star. Stanley tried to ignore Bubbles and her daughter, but the simple act of purloining the name Stella infuriated him. He recognized that Bubbles was vindictively getting her revenge; the electronic Stella was Stanley's adoptive brainchild while Bubble's daughter was the product of his loins. Bubbles sued Stanley for child support, and Stanley fought back with denial. Quite coincidentally, Personnel posted a picture of a baby in a wheelchair, an appeal from the United Way, on the employee bulletin board. One of Stanley's anonymous detractors promptly added a postscript:

"Hi, My name is Stella, and I can't even get child support."

The addition struck a chord with most employees, but an angry Stanley failed to grasp the dark humor. He tore the poster to shreds.

The next day, Stella posters appeared throughout the plant.

The Bubbles/Stella problem dragged on without resolution. Finally, Stanley agreed to a blood test. It came back negative for Stanley. The laboratory rated Stanley's chances of being Stella's father at 95%. Stanley chose to interpret this as a victory. He claimed that the tests showed that five percent of the men of the entire United States could be Stella's father. The court found otherwise and directed that Stanley pay $400 a month in child support. Stanley protested but paid.

The months passed and business boomed. At times it seemed like half the population of the United States wanted an Astron 2.

"The world's next," Stuart Miller confidently predicted.

Stanley purchased a new home in Los Altos and acquired a new housemate, Nancy, a junior account executive at the Moss Agency. Nancy moved into Stanley's new home and shared the sparse furnishings for a while. Money and fame carry their own negatives. Stanley had as much trouble decorating his home as he had selecting his women. With so many choices suddenly available, he could not decide what he wanted.

Monthly orders for the Astron 2 reached the one hundred thousand dollars plateau. The Astron 2 sold retail for $1,298 and came with two game paddles, a demo cassette and 4K of memory. Buyers could use their own TV set as a monitor. Harold's disk drive, an expensive extra, added to the bottom line. The number of dealers selling Astron 2's increased to 300. At Stanley's urging, Astron Computers created an Astron Computers Education Foundation whose goal was to provide Astron systems to schools. The engineers designed their first printer, and the software shop released a spreadsheet. To cope with the overwhelming expansion, Willard reorganized the company into divisions and created a management committee composed of the new chiefs. One man could not supervise all the current and anticipated projects. Only Stanley, who emerged without a position, protested the reorganization.

Stanley complained to Stuart Miller who consulted Willard.

"I can't make Stanley a manager," Willard said. "Nobody would work for him."

"We need you to maintain our public image," Stuart told Stanley.

"I've been stabbed in the back, by friends," Stanley angrily replied. With tears streaming down his face, Stanley rushed from the building and did not reappear for three days.

"They were the quietest three days in company history" was a common refrain throughout all four buildings.

For a week, corporate sanity returned to Astron Computers as the new division managers reorganized their baronies. Stanley presented a placid exterior while he seethed internally and plotted behind closed doors. Willard, pleased with himself, called his first Management Committee meeting. He invited his six division chiefs and Stuart Miller as an observer from the Board. He deliberately excluded Stanley, knowing full well that only chaos would result if Stanley attended. Willard had deliberately isolated Stanley from management and had no intention of retreating.

At the appointed time, the six division managers assembled. They aligned themselves in two groups of three on each side or the table with Willard at the head. Stuart did not sit at the table. He selected a chair along the wall, to the right and behind Willard. Stuart hoped his position would signal support to the company chief executive. The Chief of Production sat to Willard's right and the Director of Administration to his left. Harold, as Chief of R & D, sat at the foot of the table on the right. Harold had at first refused to attend because he had no interest in what he called "corporate bullshit," but Willard had pleaded with Harold:

"You must attend the first meeting, Harold, just to set a precedent. Then, you can send another engineer as your representative whenever you wish."

Over his better judgment, Harold capitulated, but not because he had any interest whatsoever in setting a precedent. Willard thought he had outmaneuvered Stanley, but Harold knew his old friend better than that. Harold went just to watch Stanley perform as Harold assumed he inevitably would. Harold seldom agreed with Stanley any more. He felt Stanley was publicly assuming credit where none was due, but Harold did not make an issue of it. He was determined to be an engineer and nothing more.

"Thank you for coming, gentlemen," Stuart opened the meeting with a smile.

Before he could say another word the door opened, and Stanley entered.

"Sorry I'm late, gentlemen," Stanley smiled and took the open chair at the opposite end of the table from Willard.

In Stanley's fertile mind the competing knights were in place, sitting face to face with only the long table separating them. Stanley with a smile on his face sat stiff and straight. His body language conveyed his message: "Let the games begin."

Willard turned to Stuart for support. Stuart shrugged his shoulders. Willard turned back and found each of his division managers watching him closely. Stuart had two choices: he could precipitate a corporate crisis with Stanley who owned one third of the stock, or he could capitulate and allow Stanley to participate in the meeting. Stuart looked at Harold who also held one third of the stock. If Harold voted with Stanley, Willard would be out of a job no matter what Stuart said. Harold smiled noncommittally and waited for Willard to respond to the challenge. When Harold declined to take a position, Willard could not judge whether that was good or bad.

"Thank you for joining us, Stanley," Willard capitulated, confirming for all present that the disruptive Stanley could do just about anything he wanted within the walls of Astron Computers.

Stanley smiled.

Willard opened with a short prepared speech describing his rationale for the reorganization. All of the new division chiefs smiled throughout. Harold listened without concealing his disinterest, and Stanley frowned. Willard then explained how they would conduct their future management meetings. He stressed his hope that they would be able to meld together as a team and guide the speeding express they were riding forward. Finally, Willard exhausted his fund of cliches and opened the agenda to questions.

"What is your business plan for the Astron 2?" Stanley asked loudly, ignoring the politely raised hands of the division chiefs.

Harold jerked erect in his chair. The Astron 2 was his baby.

"I don't have any specific plans that you are unaware of, as you well know Stanley," Willard countered.

The division chiefs turned to Stanley.

"Then we're fucked," Stanley said.

"We can expect to sell the Astron 2 for at least three more years," Willard said. "Or maybe five," Willard added when he saw the pained expression on Harold's face.

"We're fucked," Stanley repeated.

The division chiefs' heads as one turned back to Willard to see his response.

"Do you have something specific in mind, Stanley?" Stuart intervened. He was not sure what Stanley was up to, but as a matter of course he usually tried to moderate Stanley's outbursts.

"Certainly," Stanley said, taking the floor, which was his objective in the first place. "The Astron 2 is our Model T. In this business we either keep leaping forward or we fall back. There is no standing still. We're number one now. The Taurus was. Where is PAMS now? It no longer exists."

"Stanley, Astron Computers is on the rise. We have nothing to fear." Willard hoped to avoid a debate with Stanley, but he knew that he could not let Stanley's Chicken Little call stand. The allegations would leak, the media would feast, the market would react, the stockholders would clamor, and the Board would convene. "As you know we have our New Projects Division working on Stella and Asteroid." As it always did, the use of the word "Stella" brought smiles to the faces of the waiting division chiefs. They turned to Stanley to see his reaction. All knew that the engineers were in the initial stage of designing Stella as the microprocessor of the future. Asteroid was a gleam in Dick Ruffins' eye.

Stanley snorted. "Who is in charge of the New Project Division?"

Now, Willard saw where Stanley was headed. Stanley had set his sights on the vacant division chief position. "Dick Ruffin is our Acting Chief of New Projects."

"Why do we need a Division of New Projects and Harold's R & D?" Stanley slid sidewise. He smiled at Harold who reacted as Stanley anticipated he would.

"Because I don't want it," Harold exclaimed. "I'll run my own little R&D shop."

"We offered it to Harold. Now we have Personnel searching for a qualified engineer." Willard looked at the Director of Administration who nodded. "We believe the Director of New Projects should be an engineer," Willard spoke quickly to head off a windy comment by Administration while telling Stanley he did not have a chance at getting the post.

"Is Ruffin an engineer?" Stanley asked softly, smiling insincerely at Dick Ruffin who sat to Stanley's immediate right.

"Dick has a degree in computer science," Willard countered. Willard assumed that Stanley was in no position to question Dick's qualifications.

Everybody present was familiar with Ruffin's impressive academic background. He indeed had advanced degrees in computer science and medieval studies. A universal man, he was a scientist, an intellectual, and a practical engineer. Stanley himself had hired the overqualified Ruffin to write the manual for the Astron 2 while Astron Computers still operated out of the Pitts garage.

"Dick has a tendency to spread himself too thin," Stanley accepted the challenge. He smiled at Ruffin to blunt the brutal impact of his words, then continued. "I've told him that many times. We all have. He will waste everybody's time with this PARC bullshit project of his."

Ruffin had visited the Palo Alto Research Center, PARC, Xerox's independent research group that had been divorced from the main company to free the center's engineers and scientists to keep that company in the forefront of technical research. Intellectually challenged by what he saw there, Dick had returned to Astron Computers and had conceived his own computer of the future, the Asteroid. A jealous Stanley had promptly denounced Asteroid as a piece of shit.

"Let's not waste everybody's time with old news Stanley," Willard said before Stanley could continue. "We all know your views. If you insist, we can schedule Asteroid for a future meeting, and we can discuss it in detail then."

"I agree," Stuart added his weight to the effort.

Stanley glared at Stuart then Willard. "We need a Division Chief for New Products now," Stanley persisted.

Willard looked at Stuart who nodded. "I think we have devoted enough time to this meeting," Willard declared. He stood up, and the others, except for Stanley, followed his example.

"About time," Harold said, referring to his oft-stated conviction that all meetings were a bureaucratic waste of his time.

Stanley held his position. Willard turned to Stuart. "Stuart, could I see you for a minute in my office?"

"Certainly," Stuart replied and followed Willard out of the conference room.

"We're fucked," Stanley declared in a voice loud enough for the departing attendees and the clerical staff in the outer room to hear.

Willard stepped back and let Stuart precede him into his office. After Stuart entered, Willard turned to his secretary. "I don't want to be disturbed, by anyone," he said as he glanced in the direction of the conference room where Stanley still fumed.

As soon as he closed the door behind him, Willard spoke softly to Stuart. "What are we going to do about Stanley?"

"Harold's a problem too," Stuart said.

"I can live with Harold. Stanley could destroy this company."

"What do you suggest?"

That was the question Willard wanted. "We could go public."

Willard had given the matter considerable thought. Willard knew his suggestion was a solution that would appeal to Stuart who had invested cash in Astron Computers betting that the day would come when a public offering of stock would increase his millions many times over.

"Going public would make Stanley and Herald very rich men indeed," Stuart tried to deflect the discussion away from where he thought Willard was heading. Together, Stanley and Harold as shareholders controlled sixty-six percent of the company.

"And dilute their control of the company by putting it in the hands of the Board of Directors where you could influence decisions," Willard said.

"Money might divert Stanley," Stuart again chose to avoid Willard's implications of self-interest. "But I doubt it would mean much to Harold."

"We've got to get Stanley under control. Did you see how he attacked Dick?"

"That was uncalled for," Stuart said. "Do you need anything more from me?"

"No. Just give the matter of the public sale of stock some consideration. The time is ripe, and I'm sure many of the employees would appreciate it."

In order to keep personnel costs low, Astron Computers had adopted a policy of low salaries salted with stock incentives. Many of the longer-term employees made no secret of their belief that now was the time to cash in by taking the company public.

Whimsy and pride had drawn Stanley into the conference room to challenge Willard Temple and his new bureaucratic toy, the Management Committee. Temple had tried to neutralize Stanley by leaving him outside the management structure, but Stanley decided to teach the man who thought he was running Stanley's company a lesson. Stanley held enough stock to guarantee he could attend any meeting he wished and impose his views. The fencing with Temple had recharged Stanley's waning interest and given him a target, the New Projects Division. Stanley would have preferred to have a Division Chief in place; he needed a foil for his outlandish sallies. Ruffin, the acting chief, was no challenge. Stanley knew he could flatten this talented but mild man without effort. Stanley had decided to dismiss Ruffin and his Asteroid project out of hand and take on Stella as the replacement for the Astron 2. That Bubbles in a fit of pique had chosen the name Stella for her child irritated Stanley, but he was determined not to let that stand in his way. He assumed that if he ignored the public derision it would soon dissipate. The fact that Stella was the responsibility of Tom Williamson mattered not at all to Stanley.

Simple envy led Stanley to disregard Williamson. As a graduate student at the University of Washington, Williamson had distinguished himself by combining a double interest, computer science and medical radiology. His doctoral dissertation had been widely acclaimed. Upon graduation, Williamson had spurned medicine and had signed on to write applications for Astron 2. Willard Temple's adoption of Williamson as a protégé had led to his selection to lead the Stella research team. This simple act had of course spawned Stanley's disdain.

In the weeks following the initial Management Committee meeting, Stanley devoted himself to perfecting his campaign to take control of the Stella Project. He began by denouncing all of the work to date and insisting that the small team of engineers begin anew. Williamson protested but to no avail. Willard Temple, delighted that Stanley had found a project to distract himself, encouraged Williamson to humor Stanley. Williamson who had left academia because he had disliked faculty politics decided that capitulation was easier than conflict. He deferred to Stanley, assuming that eventually the volatile minority owner would grow bored and find something else to quibble over. Williamson misread his antagonist. Stanley who thrived on the conflict did not accept victory gracefully. His lack of technological knowledge did not protect the engineers because Stanley fancied himself an innovator.

Stanley's distraction proved a source of relief for Dick Ruffin and his Asteroid Project. Now that Stanley had his own hobbyhorse to ride, he almost left Dick in relative peace to pursue his own goals. Having repeatedly denounced the Asteroid as piece of shit, Stanley now ignored it. He did, however, seize on a proposal tabled by Ruffin as Acting Chief of New Projects, and took it as his own. Ruffin had suggested a second visit to PARC Laboratories, and Stanley arbitrarily decided to lead the Astron delegation himself. He wanted to see firsthand what all the talk was about.

The Palo Alto Research Center, PARC, is located on Coyote Hill Road near Route 280. In the early seventies, Xerox, the copier company, decided to establish a research center whose purpose would be to design the office of the future. At the time, Xerox sat on the summit of the duplicator industry, dominating it much as IBM did the mainframe business. A forward looking Xerox chief operating officer had decreed that PARC be established to keep Xerox in this favorable position. Consequently, Xerox collected some of the nation's best scientists and established them on the other side of the country from corporate headquarters. "Go and dream," Xerox management directed, and dream they did. The entire Valley was awash with rumors about PARC's successes.

Although Dick Ruffin failed in his initial efforts to persuade PARC to open its Pandora's Box of technological achievements to a group of engineers from what some were now describing as the number one microcomputer company in the nation, Stanley was not daunted.

In December, just before Christmas, Stanley worked out a deal. He had heard that Xerox corporate headquarters did not grasp what a resource they had at PARC. To the copier executives of the late seventies, PARC represented Xerox's gift to the world of research. The forward leaning chief operating officer had passed on, and less interested executives had replaced him. Somebody someplace told Stanley that Xerox was seeking to diversify its investments as a hedge to protect its copier domination. Since Astron Computers had attracted exploratory Wall Street interest with its hints about going public, Stanley contacted the Xerox office responsible for diversifying. Stanley suggested a deal: Astron Computers would allow Xerox to buy 100,000 shares of Astron at ten dollars a share. In return, PARC would allow a team from Astron Computers to view its reputed marvels.

Stanley did not realize what a Christmas gift this would be. He had a simple motivation: he needed some new ideas to bolster his status as an innovator. He led the Astron group with this selfish purpose in mind. Stanley graciously allowed his chief engineer for the Stella Project, Tom Williamson, to accompany him. Stanley needed someone to take notes, and Williamson, who had surrendered his status as project leader to Stanley, was a logical choice. Stanley assumed it would be a humbling experience for his more educated subordinate. Williamson pleaded that three of his senior engineers be allowed to accompany them; the opportunity was too great to ignore. Stanley agreed. Then, Dick Ruffin who had originally proposed the visit heard of Stanley's plans. Ruffin, as Acting New Projects Director, politely asked Stanley if he could join the delegation, and Stanley dismissed him out of hand. Persistent, Ruffin rushed to enlist company president Willard Temple's support. He was the only one who dared confront Stanley. Temple decided he too would visit PARC. He so informed Stanley and invited Ruffin to join them.

"Christ, they'll think we're invading," Stanley had protested.

They drove to PARC in three cars. Stanley led the way, alone in his new Mercedes. Willard and Ruffin followed in car two, and Williamson and his engineers brought up the rear in a company van.

Gene Vance, a PARC engineer, met the Astron Computer delegation at the door and conducted them to the demonstration room. Stanley, wearing a blue sports coats, a gray turtle neck shirt, beige slacks and tassel loafers without socks, marched in the forefront of the Astron contingent.

"I hope you have something worthwhile to show us," Stanley signaled that he was in charge.

Ruffin glanced at Willard who shook his head to indicate Stanley's presumptive behavior did not matter. Willard was prepared to announce his status as company president, later, if needed.

Vance, accustomed to briefing visitors, ignored Stanley's comment. This visit had been arranged by Xerox corporate headquarters. Vance did not know why corporate was catering to these valley upstarts but was prepared to play his role. Corporate did not often call upon PARC to engage in public relations, and the truth was that Vance did not mind showing off for the wide-eyed visitors. All he was going to do was demonstrate Smalltalk and the Alto. Smalltalk was the software program that managed the Alto's desktop. It provided icons and small menus that enable the user to choose the actions he wanted the computer to perform without typing in esoteric command instructions. The user activated the desired application by selecting an icon and then choosing from a drop down menu. A device called a mouse provided the buttons the user pushed to signal his selections. It all appeared quite simple, but technically it was quite complex. In computers like the Astron 2, the user typed instructions on a keyboard which informed computer memory what action it should generate on the screen. The Alto used something called bit mapping. Every pixel, which is single dot, on the monitor screen was available to be controlled by memory. This required tremendous amounts of memory as the Alto screen presented over 500,000 pixels. Vance liked to describe the screen as a desktop. Most visitors could imagine that. Using the Smalltalk software and the mouse, the Alto freed the user from the ubiquitous command line. Gene Vance assumed these technically literate engineers would grasp what this all meant. The Alto lifted the microcomputer from the domain of the hobbyists and put it into the hands of the common man.

Vance did not worry about the visitors stealing PARC's revolutionary secrets. He would only demonstrate the Alto and its software. That would be enough to impress them if they knew their business as he assumed they did. He would not open the Alto and would not respond to any engineering questions. He would merely smile when they asked, "how did you do that." In any case, corporate headquarters had authorized the briefing. Vance, like his fellow engineers and scientists at PARC, did not mind demonstrating their superiority on occasion, at least not as long as it was properly appreciated.

Stanley, the first Astron Computer employee to enter the briefing room following Vance, took the seat of honor in the center of the first row of seats facing the simple PARC display on table. Willard took the seat on Stanley's left and Bill Atkinson maneuvered into the seat on the right. Dick Ruffin joined the others in the second row. Gene Vance positioned himself to the right of the Alto. Stanley smiled. He had seen enough hardware to be unimpressed so far. The Alto had a monitor, an unimpressive case, a keyboard and an odd little object that Stanley identified as the famous mouse. Prior to departing their offices, Stanley and the others had done their homework. Stanley knew what the mouse was designed to do.

"Welcome to PARC gentlemen," Gene Vance said. "My name is Gene Vance, and I am one of the engineers working on Alto. I remind you, Alto is a work in progress."

Vance paused, but no one in the Astron delegation volunteered a comment.

"Now, let me introduce Alto." Vance keyed a switch on the front of the box. The monitor flickered, and Alto began to hum. "While we wait for Alto to get ready, let me say that the first view you see will be that of the desktop. That's what we call our graphical user interface."

"What the hell is that?" Stanley blurted.

"The graphical user interface, the gui, pronounced gooey, like gummy only with a goo, enables the user to click icons instead of typing text commands. It is a simple program."

"Not so simple," Tom Williamson said. As chief of the Stella Project he would be the one to conceive the programs that would give Stella a graphical user interface if they decided to go that route.

The Astron engineers laughed politely. They knew what a graphical user interface was and how to pronounce gui.

Stanley sat straight in his chair staring at the monitor as a series of icons arrayed themselves across the screen.

"I will continue with the desktop metaphor," Vance said. "Each application is represented by an icon. Look on them as files on your desk. When I click on an icon, a window will open, and the application symbolized by the icon will appear."

Vance, standing to one side of the monitor so all his visitors could view the screen, reached down and pushed a button on the mouse. Immediately, a word processor program opened."

Stanley leaned forward until his nose almost touched the monitor. "Open another program," he ordered.

Gene Vance smiled, pleased with the reaction. He moved the mouse slightly to the right, and the pointer on the screen moved to the right. "We call the pointer that you see a cursor."

"We know what a damned cursor is," Stanley said.

Vance pressed the button on the mouse and the word processor program disappeared. He clicked a second icon and a Tic Tac Toe game unfolded. "The top application is always the active one." He moved the mouse slightly and pushed the button again. An X appeared in the upper right quadrant of the game square which consisted of two horizontal and two vertical parallel lines that crossed and formed nine boxes. Silently, he clicked the box and an O appeared.

"Do it again," Stanley ordered.

Vance moved the cursor to the box on the lower right and clicked the mouse. An X appeared.

"Why in the hell aren't you doing something with this?" Stanley shouted. "It's marvelous."

Stanley leaped up and reached for the mouse.

Vance grabbed Stanley's hand. "I'm sorry Mr. Pitts. We cannot allow you to touch the Alto."

"Shit," Stanley said.

"We can drag the icons around our desktop and arrange them as we wish," Vance said, demonstrating his words. He placed the cursor on the word processor icon and holding the mouse button down dragged it to the lower left corner of the Alto's screen. "If we click it," Vance placed the cursor on the icon and pushed the mouse button. "Our word processor window will open on top of the game." The word processor opened.

"We can now enter text," Vance typed on the keyboard and the words; "Welcome visitors from Astron Computers" appeared on the screen. "If we wish to edit our text, we merely highlight it," Vance dragged the cursor across the word "Computers" and the word on the screen changed colors. "Then if we wish to edit that word, we consult the menu." Vance moved the cursor to the top of the screen, clicked on a button there, and a menu of choices appeared. Among them were words like replace, clear and delete. Vance moved the cursor to delete, clicked the mouse button, and the word "Computers" disappeared leaving the sentence "Welcome visitors from Astron" on the screen.

"Please explain bit mapping," Tom Williamson asked.

"As I am sure you engineers know," Vance replied. "Every pixel on the screen is mapped to a bit in the computer's memory. Alto turns the pixels on and off to obtain dark and light and thus creates the pictures that you see displayed. We call this whizeewig."

"What you see is what you get," Tom said. Whizewig was the phonetic rendition of WYSIWYG. What You See Is What You Get.

"This is a tool, not a toy," Stanley declared. He understood exactly what he was seeing. The Alto symbolized the revolution he liked to talk about. Stanley was so excited he began pacing while the others asked their questions. PARC had made it clear that the center would not open the Alto or divulge how it created its fantastic technology, so Stanley and the others did not ask.

"You did not answer my question," Stanley overrode a question from Dick Ruffin. "Why aren't you doing something with this wonderful technology?"

"I know," Gene Vance answered evenly. "I hoped you wouldn't notice."

"Well, I did," Stanley persisted.

Gene looked about the room as if checking to see that he were the only PARC employee present. "I'm supposed to say that our job at PARC is to think about the future not to produce it."

"But this technology is so revolutionary..." Stanley began.

"We should not place Mr. Vance in a compromising position Stanley," Willard intervened to protect their host from Stanley's impulsiveness.

"That's all right," Vance said. "I'll try to answer the question if you don't quote me."

"Certainly," Stanley stopped pacing and returned to his seat.

"We're a think tank. I honestly believe we may have some fifty of the best scientific minds in the country engaged here. We don't know why our corporate headquarters does not take advantage of the knowledge PARC has advanced."

"You're not dealing with just theory," Stanley interrupted and looked meaningfully at the Alto that was patiently awaiting further instructions.

"I know," Vance admitted. "It's heresy for me to say it, but sometimes we think that corporate has forgotten we are here."

Stanley took a card from his pocket and handed it to Vance. "I won't."

Vance nodded and placed the card in his pocket. "I am here to answer those questions that I may," Vance said.

During the ride back to their Cupertino headquarters, Stanley could not stop talking. "We've just seen the future," he declared at one point. He turned to Tom Williamson. "How long will it take us to do that?"

"Three months," Williamson replied.

Willard said nothing. He was not an engineer, but he knew enough about computers to agree with Stanley. He knew what stood ahead. Stanley would seize control of the Stella Project and do everything in his power to duplicate the Alto.

Chapter 18

"Forty thirty, set point," Terry called from her position at the net. David watched from the base line as his doubles partner bent over to retrieve a ball. He really liked the neat pleated skirt that Terry wore. She had great legs, nicely muscled, probably the best legs he had ever seen, and the short skirt showed them to advantage. Terry bent from the waist, her back to David, and the skirt rose displaying a shapely stretch of white tights. David really stared. Terry said something that David missed and bounced the ball back towards him. She smiled coquettishly, signaling she knew exactly what had preoccupied David, and turned back to the net. David blushed.

"Any time you're ready over there," Jeffrey Madison III, the Director of the IBM Research Center, Boca Rotan said.

Mitchel Webber, Madison's deputy and regular tennis partner, smacked the web of his racquet against the palm of his hand.

"Brace yourself," David called. He threw the ball high and delivered what he liked to consider his cannonball first serve. He hit the ball squarely with a resounding thump but the ball nicked the top of the net and bounced high in the air.

"Out," Jeff announced smugly.

"Get it in," Terry ordered.

Great legs and all, Terry took her tennis seriously.

David prepared to deliver his weak second serve. He always tried to put as much English on the ball as he could, but in six months of regular tennis at Boca Rotan, he had yet to perfect his technique. David positioned his feet, glanced across the net, and saw a smiling Jeff inch forward. The move irritated David. It indicated a lack of respect that was too much for the competitive athlete in David to accept. David changed his mind and decided to chance another cannonball. He tossed the ball high, waited patiently until it reached its apex, then swung his racquet with all his might. Again, he caught the ball just right. This time it cleared the net, landed in the forecourt and whistled past Jeff who did not have a chance to swing.

"Ace, game, set, match," Terry shouted triumphantly.

Terry ran to a smiling David, embraced him lightly with sweaty arms, buzzed him on the cheek and whispered. "Lucky shot ace. Next time don't chance it."

David said nothing. Terry hated to lose.

"Good game guys," Jeff called from the side of the court where he had retrieved his towel and was now wiping his perspiring brow.

David and Terry joined Jeff and Mitchel.

"How's the project going?" Jeff asked David.

David smiled and shrugged as he always did when Jeff asked the same question. At first, David had worried his answer, but after six months in Boca Rotan he had learned that Jeff, despite his position as the center director, did not expect a response. Over time, the secrecy surrounding David's special project had dissipated. All the engineers and scientists at Boca Raton now knew that David was researching a new microprocessor. They respected David's privacy and did not press the subject in his presence because they did not want to embarrass him. They knew to a man that he was wasting his time, not because he lacked the expertise to design a competitive small computer, which he did lack, but because they knew that IBM management would never undermine the main frames. Because the big computers carried IBM on their electronic muscles, a foray into the world of microcomputers could only lead to a dead end; management would not initiate a product that would place them in competition with themselves.

David was aware of the skepticism of his colleagues, but he prided himself as not being like them, denizens of an extraneous company division marking time while they indulged in research on projects IBM would never use. David's real secret was that he was on a mission assigned by James Archibald Crane, himself. David had worked hard during his sojourn in Boca Rotan. He had traveled to the Silicon Valley, to Albuquerque, to Boston, and he had consulted with those engaged in the forefront of the microprocessor advance. David had educated himself. He was familiar with the state of the art and with the dreams for the future. Still, he was not an engineer. He knew what the microprocessors could do, what their problems were, what they could not do, and thus, he hoped, was in a position to avoid the obvious pitfalls.

There were not many components in a microcomputer. David could now take the cover off his Astron 2's box without fear. He could recognize the disk drive, the power supply, the motherboard. He could add a memory card, install software, and connect peripherals, but explaining how these things worked was beyond him. Since his assignment did not require him to create these things, all he had to do was know what parts he needed. Ernie had made it clear. David's task was not to take IBM to the forefront of the existing technology. All he had to do was assemble a reasonably reliable computer with tested components that ranked somewhat behind technology's buggy frontiers.

Jeff and Mitchel departed the courts first. David waited for Terry.

"What we will do for dinner tonight?" Terry asked.

"Seafood?" David responded. It was a safe suggestion. Terry, a Florida coastal native, preferred seafood.

"Great."

David, despite his relationship with Margaret in Armonk, had succumbed. For the first two weeks, he had merely admired Terry from afar. He had phoned Margaret daily, at IBM expense, and ostensibly reported on his activities. Then, he had impulsively invited Terry to a movie. This had led to other things. His calls to Margaret had grown progressively less frequent until now, six months later, he phoned Armonk less than once a week. Six weeks ago, he had moved into Terry's apartment. Now, they shared everything, work and leisure.

"Terry, I've got some news," David decided not to wait until dinner to inform Terry that he had been summoned to Armonk. This was a sensitive subject because Terry knew all about Margaret even though Margaret, David assumed, knew nothing about Terry.

"I know," Terry swatted her leg with her racquet as they crossed the grass to the main building. "I listened."

David nodded. Her blithe words troubled him. He knew that Terry who served as receptionist, secretary, telephone operator and all-purpose administrative assistant to the center frequently listened to telephone calls to pass the time. Somehow, he had expected that his conversations would remain sacrosanct. He thought about protesting the invasion of his privacy but reconsidered. Terry considered herself the queen of the Boca Raton Research Center and did exactly as she chose.

"Margaret sounded a bit put out. Like she's been sucking lemons," Terry said.

"How many of my conversations do you listen to?" He asked contritely, not protesting, trying to sound amused.

"All of them old son."

"Even at the beginning?"

"Particularly then. Margaret was not too happy with you. Are you looking forward to a triumphant return to Armonk?"

"I need to talk with Ernie."

"And Margaret too, I'm sure."

"Margaret and I are history," David said with a certitude he did not feel.

"Does she know it?"

"I haven't talked with her for a week."

"Until today."

"Terry, can you get me reservations for tomorrow?" David asked as they approached the center's main building.

"Already have," Terry smiled as she turned left toward the women's locker room.

David watched her legs as she made her way down the corridor. Just before she entered the swinging door, Terry waved her free hand, signaling she knew David had been watching.

David picked up an economy rental car at the airport and drove directly to the IBM headquarters. To his dismay, he felt slightly intimidated in this his first visit to Armonk in six months. Life in Boca Rotan was great, and David was apprehensive this summons might signify the end of what he turned into a dream assignment. David hated headquarters with its intimidating and irritating conformity to the IBM lifestyle. David had been a halfback, and runners were known for their free style individualism. There was no place in Armonk for independent players. David wondered if he would have the nerve to quit IBM cold if Ernie ordered him back to Armonk.

He parked the rental car in a visitor's space, flashed his identity badge to the guard at the door, and took the elevator to Ernie's office. David had tried to script his opening exchange with cool Margaret and had failed. He was going to have to wing it. David, dressed in IBM blue, nodded to a familiar face in the hallway then marched into Ernie's outer office with as much swagger as he could muster.

Nothing had changed. Margaret, as pretty as ever, sat at her desk. David's old office when he had been Ernie's assistant remained to Margaret's left, the door open, just as the door to Ernie's office behind Margaret, the traffic cop, was open.

"Hi Mag," David said. He tried to put Margaret off balance. She hated being called Mag, Maggie or any other derivation of Margaret.

Margaret frowned and deliberately looked at the clock. "You're five minutes late Mr. Howard."

"How're things?" David whispered as he kissed Margaret on the cheek.

"Do that again and I'll file a complaint against you," Margaret snapped.

"I don't have any reservations," David said, indicating he was available for an invitation to stay in her apartment.

"Yes you do. The Holiday Inn," Margaret smiled thinly. "You know where that is don't you?" Before David could respond, Margaret continued. "You may go right in. Mr. Hendricks is expecting you. And has been for five minutes."

"May I stop in my office first?"

"You don't have an office here," Margaret said.

David looked in the direction of his old office.

"That belongs to Speedy." Margaret clicked the intercom. "Speedy, Howard's here, finally. You can join Mr. Hendricks in his office."

Before David could think of anything to say, a tall red head appeared in the doorway of David's old office. David estimated him at six four and two hundred pounds. David hated looking up to him. David outweighed the redhead, but the height made him feel like a barrel.

"Hi," the redhead said, slipping an arm into his blue suitcoat before offering his hand to David.

David responded and felt the redhead's fingers wrap around his. David prided himself on his size, but the red head's long fingers exceeded David's by a good inch. David decided the redhead was too slender to be a football player.

"Hi, I'm George Spencer. Call me Speedy," the redhead said.

"Are you?"

"Nope. That's why they call me Speedy."

David decided he liked Speedy. "Are you Ernie's assistant?"

"Yep. For now."

David wondered what that meant. Was he just filling in for David?

"We have a lot in common," Speedy said.

"The office?"

"And other things. Speedy smiled at Margaret who smiled back.

David got the message and was relieved.

"When you two old friends are ready," Margaret said. "You may go in."

Speedy hesitated and let David precede him, just as David had when escorting visitors into Ernie's office. The funny thing was, David felt like a visitor. He found Ernie sitting behind his desk. At least that appeared normal. "The prodigal returns," Ernie said, rising. "Nice tan," he appraised David's sun darkened skin. "Playing a little ball?"

"Too busy," David said, taking Ernie's hand. "Just a little tennis."

"Thought so," Ernie said. "A lot, I would say." Ernie closed the outer door and indicated with a wave that David should take the visitor's place on the couch. "How's Terry?"

David sat on the couch and placed his briefcase on the floor beside him.

"She's a good looking girl," Ernie said before David could think of a response. "Likes her tennis. I didn't know you were into the game."

"Didn't have time for organized ball," David said lamely. "So I took up tennis."

"Bet she whips your ass."

"Sometimes."

"You met Speedy?" Ernie glanced at the redhead who had taken a place next to David on the couch.

"Yes," Speedy answered for David.

"He's a pretty nifty right fielder. Batting four hundred. If I recall, you hit three seventy-five last year."

"About that," David said. He had actually hit three seventy-six.

"I want you two to get to know each other."

"Margaret and I plan to have him over," Speedy said.

"Better do it tonight because David's heading home tomorrow," Ernie said. "Enough of the small talk. Tell me where you stand."

David reached for his briefcase.

"The frequency and quality of your reports have fallen off the past couple of months, David. Had the Boca Raton flu?"

"I can explain..."

"That's OK," Ernie said. "It's my fault. I didn't give you any feedback. I wasn't ready for you."

David's interest picked up. Ernie was implying that he now was ready.

"Has Mr. Crane given us the go ahead?" David asked.

"Bring me up to date."

David set his briefcase on the coffee table and opened it. He had carefully arranged his papers for a presentation but was dismayed to find that travel had reordered them.

"Never mind the papers, David, just tell me."

"It's doable," David said. "I know where we can get the components we need. I have the specifications from all the companies I consider potential subcontractors here." David patted his briefcase.

"Reliable technology?"

"As reliable as new technology can be," David evaded a firm commitment.

"Don't cover yourself David," Ernie said. "I understand. Did you visit Astron Computers?"

"No sir."

"Why not?"

"I've been using an Astron 2. It's a good machine and reliable. We don't want to duplicate the Astron 2, we want to improve on it."

"I've heard they're working on a successor?"

"Yes sir, the Stella."

"And?"

"It's not going to make it."

"You're sure?"

"Yes sir."

"Mind telling me why? What do you know that Astron Computers doesn't?"

"Stanley Pitts is heading the development team himself."

"And he gave us Astron 2."

"No sir. Harold Dumbroski is the genius who gave us the Astron 2. Stanley Pitts gave us Astron Computers."

"There's a difference?"

"Yes. Pitts is impulsive. He has a good eye for design. He visited PARC and has not come down to earth since seeing their demonstration."

"Did you visit PARC?"

"Yes sir. Very impressive. Their Alto probably is the machine of the future."

"But?"

"But it's too expensive."

"And the Stella?'

"Pitts won't bring it in for under ten thousand a machine."

"And Pitts knows that?"

"He does but doesn't care. He's shooting for perfection."

"And he'll hit himself in the ass."

"Exactly."

David turned to include Speedy in the conversation. He remembered what it was like to be an assistant and to sit in on the fringes of interesting talk. "We're aiming at the small office, the professionals, and the home market. Price is important. Anything over a thousand, a thousand two hundred takes us beyond the market's reach."

"Speedy's an engineer in addition to being a right fielder. He's got a Ph.D. from MIT in Computer Science," Ernie said.

David was impressed. He assumed he should be the one listening, and Speedy should be the one talking. Suddenly, he remembered Ernie's opening comment. "I want you two to get to know each other."

David proceeded with his briefing. He covered the specifications and cost estimates that he had accumulated and passed the literature to Speedy and Ernie. When David had exhausted his report on the research, Ernie asked another question.

"Do you have everything we'll need for our microprocessor?"

"I still need software and an operating system." Operating systems were software that translated commands from the applications into instructions that the hardware could understand and implement.

"You couldn't find an operating system?" Speedy blurted.

Ernie smiled.

David responded: "That's not the problem. I just haven't focused on software or the operating system."

"Oh," Ernie encouraged David to elaborate.

"I plan to visit Digital Software next week."

"The Seattle outfit?" Speedy asked.

"Yes,' David said.

"Charles Swift runs Digital Systems," Speedy spoke directly to Ernie.

"I've heard of him," Ernie said. "He's the Harvard kid that wrote the BASIC for the Taurus."

"That's right," David said. "He visited me here once."

"Looking for a job?" Ernie asked.

"No, wanting to sell software."

"We write our own," Ernie said.

"That's what I told him, then."

"And now we need an operating system," Ernie smiled.

"Where do we stand with the Management Committee?" David asked.

"With the Management Committee, we don't stand anywhere," Ernie smiled.

David's face registered his disappointment. He had come to believe in his project.

"But we stand very good with Mr. Crane who you may recall pulls a lot of water around here."

David smiled. He still had a chance.

"We've got a meeting at eleven in Mr. Cranes office."

David checked his watch. It was ten forty-five.

"Today?"

"In fifteen minutes."

David hurriedly began to assemble his papers.

"Leave that stuff here," Ernie ordered. "Just give him the oral briefing you gave me."

David instantly realized that Ernie had been rehearsing him for his big act. He had never met Crane. As a trainee, he had seen him from a distance in the auditorium when the chief operating officer had welcomed their class, and David had sat in the back of the room at Management Committee meetings. David began to worry. If he had more time, he might even get nervous.

The briefing went better than David could have hoped. Crane listened intently, asked a few questions, almost exactly those Ernie had asked, and David did not falter. Ernie obviously had a good working relationship with Crane, and this helped David to relax. When David finished, Crane fiddled with his glasses briefly then looked directly at David.

"Good work young man," he said before turning to Ernie. "Ernie, do it."

"What does 'do it' mean?" David eagerly asked Ernie as they returned to his office.

"You're in business?"

"Me?" David was intimidated by the implications.

"Set up a shop in Boca Rotan. Tell me how many people you need, and I'll arrange them."

"Me?" David repeated, now almost in shock.

"Will you come down to take charge?" David assumed he was to take care of the administrative details.

"No. I have a man for that."

"May I ask who?"

"You."

"I don't have the technical training to handle it."

"You're setting up an assembly plant. I'll give you Speedy to help."

David thought about that.

"How many people do you need, administrative and technical?"

"I'll need forty to start," David guessed.

Ernie smiled. "Good boy. You've got them. They'll begin reporting next week."

"Are we still secret?"

"Does the Management Committee know?"

David shook his head, wondering how he could manage such a large enterprise in secret.

"I still need an operating system."

"Find it."

"What about Margaret?" David asked.

"What about her?"

"Will she be coming down?"

"Do you really want her?"

Again, David shook his head negatively.

"She'll miss Speedy but she'll get over it," Ernie said. "She's had experience getting over things."

Speedy accompanied David on his return to Boca Rotan. This time David was not alone in receiving a cool send off from Margaret who pointedly refused to speak to either of them.

Terry, driving David's Volkswagon, met the two of them at the airport.

"Wow," Terry said, admiring Speedy's height. "I'll bet you've got a great serve."

"Terry, Speedy," David said peevishly. He felt like Mutt to Speedy's Jeff when standing next to him.

"Well, do you?" Terry persisted.

"Tennis? It's not my thing," Speedy said. His eyes, however, indicated that Terry might be.

"I've heard that before," Terry said, preening to Speedy's frank scrutiny.

"Down Speedy, she's spoken for," David tried to break up the mutual admiration society.

"Really? I haven't heard any speaking going on around here," Terry smiled brightly.

"We'll drop Speedy off at the Sleepy Haven on the way home," David emphasized the last three words.

"Sounds good," Speedy said, still admiring Terry.

Within three days, the lone ducks began to stagger in. Lone duck was a phrase used by IBMers to describe the odd employee who refused to fly with the others behind the leader. Administrative details deterred David from making his planned Seattle trip to scout Digital Software. The arrival of an administrative officer finally freed him three weeks later, and David phoned Seattle to schedule an appointment.

"May I speak with Mr. Swift please," David asked the efficient sounding secretary who answered the phone.

"May I say who is calling sir?"

"David Howard," David said. "From IBM," David added for emphasis.

"Hey Charles. Pick up the telephone," the secretary called. "It's a guy named Howard from IBM."

:"Charles Swift," the Digital Software president came on the line.

"Mr. Swift," David began, smiling at himself for so formally addressing the kid who had visited him at Armonk. "You may not remember me..." David paused deliberately, giving Swift the time to place him.

"Oh yes I do," Charles responded immediately. "You were so kind to take me to lunch when I visited Armonk."

"Cheeseburgers," David laughed.

"Right. My favorite staple."

"We would like to come out to chat with you," David said.

"Certainly," Charles agreed. The mainframe Goliath ranked at the top of his list of potential clients. As far as Charles was concerned, IBM, the proverbial two-ton guerrilla, could do anything it wanted. Charles leafed through the pages of his daily calendar. "I'm free next week."

"What about day after tomorrow? Friday?"

Charles checked Friday. As he knew, he was booked solid. "Friday's fine with me. What time does your flight arrive? I'll meet you."

"Not necessary," David replied. "We'll gather at your office at ten if that's convenient."

"I look forward to it," Charles agreed.

After David hung up, he turned to Speedy who had been sitting in David's office listening to the conversation. "We're on for Friday."

"I'll call Armonk and alert the others," Speedy said.

David frowned. "That won't be necessary."

Speedy raised his eyebrows, then nodded assent. Speedy had witnessed the bureaucracy's initial attempt to undermine David's independent command. At his first staff meeting that included the new administrative officer, David had casually mentioned that he planned to fly to Seattle to confer with Digital Software.

"Will you be talking specifics, David?" Horace Simple, the administrative officer, asked

"Specifics? Who knows," David said. "I want to explore our software options."

"There's rigid policy on that," Simple insisted.

"Speedy will accompany me," David said. He had planned to travel alone as he had become accustomed to do, but he decided to use Speedy to head off Simple's concerns.

"He's not qualified to manage the secrecy agreements," Simple countered.

"Then I will."

"I'm sorry, sir," Simple said. "The regulation requires that the agreement be negotiated by a company lawyer."

"Hell, I've been meeting with other companies for almost a year now."

"Be that as it may. I'm required by regulation to ensure that company procedures are followed and to report any deviations directly to Armonk."

"Don't do anything until I check with Armonk myself," David ordered.

Immediately following the meeting, David phoned Ernie.

"General Products Division," Margaret answered the phone.

"Hi Mag," David tried to hustle Margaret with camaraderie. "Ernie in?"

"Just a minute Mr. Howard," Margaret said coldly.

"Hi David," Ernie came on the line.

David did not hear a click, so he concluded Margaret was listening.

"I've got a slight problem," David said. "The bureaucracy is trying to close in on me."

"We expected that. You report directly to Crane through me," Ernie said, assuming that solved the problem.

"What about company regulations?"

"Elaborate."

David explained.

"I'll tell you one more time," Ernie said. "You are in charge of a special project under the chief operating officer's direct cognizance. Do whatever you need to do."

"Thank you Ernie." David had just received his carte blanche with Margaret as a witness.

David was confident his mentor would back him to the fullest, but David had learned that in bureaucratic battles the senior on occasion had been forced to sink or discard his protégé with full honors. David had not shared this conversation with Speedy or others. He had, however, decided to protect himself by showing good faith. That meant he would review the regulations himself and have Swift and his people sign the required agreements. David would file them in his safe and deal with the bureaucracy as required.

David and Speedy arrived in Seattle on Thursday evening. David found Speedy a congenial companion, but he still felt like he was traveling with a chaperone. At least, they could discuss sports. They spent the night at an airport hotel and reported to the Digital Software headquarters at exactly ten o'clock. They found Charles waiting for them at the front door.

"That kid is the company president," David spoke softly to Speedy as they approached.

"You've got to be kidding me," Speedy said as he carefully studied the young man in the white shirt, open at the neck, sleeves rolled to the elbows, rumpled khakis and dirty Nikes.

"Mr. Howard, welcome to Seattle," Charles smiled broadly.

David took his hand. "Good to see you again, Charles. This is George Spencer."

"Pleased to meet you sir," Charles said.

"Christ. That's the first time anyone said 'sir' to me. Call me Speedy."

"And I'm David," David said.

"Anyone from IBM is sir to me," Charles said.

As best as David could tell, Charles and Speedy were of the same age, and David was only two years older. "Please, sir," David smiled. "Speedy and David." Two could play the deference game.

Charles laughed. "OK David, Speedy, I'll show you around then we can retreat to my office."

Digital Computers proved to be a larger and more impressive company than David had anticipated. Their offices and working areas were clean, modern and efficient. He counted forty employees. Most of them were young, and their shabby dress marked them as college kids or at most recent graduates. All seemed to be dutifully occupied, indicating a company with a surfeit of orders. They greeted their company president casually, unintimidated by the visitors from IBM in the dark blue suits.

David was glad that he had insisted that he and Speedy look like IBMers when they visited Digital Software. Special project or not, they represented a dignified company with a worldwide grasp.

After the tour, they returned to Charles' office. It was far better appointed than David had expected. "Frankly, Charles, I'm impressed," David said.

Charles smiled. "After one look at me, most people expect Digital Software to resemble a college fraternity."

David almost responded to the candor by noting that most of the employees did look like frat members but caught himself in time. He knew some of the older members of his team had their doubts about him and Speedy. Youth was a problem, but David liked to console himself with the thought that except for IBM and its big mainframes the computer industry belonged to the young.

"I've asked Mike to join us," Charles nodded at the doorway.

David turned and found himself face to face with another young man, about Charles' age, but surprisingly dressed in a dark blue suit, white shirt and maroon power tie.

"Hi, I'm Mike Hamilton," the smiling young man said. "I'm the only one around here that owns a suit. Charles thought you would be more comfortable."

"I'm David Howard, and this is George Spencer." David was impressed with the young man's composure. He did not appear to be the least bit intimidated. Mike was about David's height, porky where David was muscular, and had a prematurely receding hairline.

"Speedy," Speedy said.

"Track?" Mike Hamilton asked.

"A little baseball."

"Pitcher?" Mike examined Speedy's height, long arms, and big hands.

"A little."

"I'll bet."

"Mike's our chief operating officer," Charles said. "He runs things while I dabble with technology and hunt for customers."

"Looks like you're doing pretty good," David said, taking the subject toward the purpose of this visit.

After they were seated, Charles took control. "What can we do for IBM, gentlemen?"

"We..." David began, but Speedy coughed.

David smiled. "As my colleague has so gently reminded me, my bureaucratic leaders require me to do something first." David hoped he appeared apologetic as he opened his briefcase, sorted through the papers, then took out two identical documents and handed one each to Charles and Mike. "I apologize but our bureaucracy requires us to ask you to sign an agreement stipulating that neither of us will disclose to outsiders any proprietary information that should be tabled in this meeting, that either party will be free to discuss the subject matter of this meeting without limitation, and that IBM guarantees no further action as a consequence of our discussions."

"No sweat," Charles said grabbing a pen and signing the agreement without reading it.

Mike read his copy carefully. "Right. We will not disclose any of your proprietary information that you chose to discuss here and the same applies to ours. You promise nothing, and retain your right to discuss later whatever we discuss. Certainly, we expect nothing from you without a contract." Mike took out a pen and signed the document. He collected Charles' copy from him and handed both to David.

"All of this is implicit," David said, "but our superiors require us to handle these stipulations formally." This was not quite true. David assumed Ernie had given him considerable latitude, but David felt constrained to say this in front of a creditable witness, Speedy, who would be available to come to David's defense if necessary.

"We understand," Charles said. "We're not going to show you the family jewels," he laughed, "unless you ask for them, and we want to brag about them." Charles then leaned back and waited for David to begin.

This was the part that always caused David difficulty. Ernie had directed that the fact IBM was building a small computer be kept secret. At the same time, David had to speak frankly with potential suppliers to the point they would be willing to reveal to IBM the capability of their products, in this case software and an operating system. David was no longer in an exploratory mode; he was assembling the staff to build a prototype. Given the numbers of people now involved, he assumed that word would begin leaking. Information about IBM's plans was always sought by the media. The fact that IBM was building a prototype microprocessor would grab Wall Street's attention and shake Silicon Valley.

"Can we talk in confidence?"

"There's just me and my roommate here," Charles said.

David looked at Mike. "We were roommates at school," Charles said. "Mike is my right arm, and we have no secrets from each other."

David recognized that Charles had deflected his question. He took a breath and decided to chance it. "I'm from Boca Rotan," he began.

"IBM's answer to PARC," Mike said.

"Not quite," David said. "We're nothing as grand. To tell the truth most of us are lone ducks." David explained the metaphor.

"Smart guys the company doesn't know what to do with," Charles laughed.

"Need a job, see me," Mike smiled.

"Speedy and I head up a special project unit in Boca Rotan."

Charles could have cracked "So young," but said nothing.

"Probably nothing will come of it. What we're doing is counter to everything IBM stands for." David paused. Neither Charles nor Mike asked a question. "Speedy is our technical guy. We're assembling a prototype, a personal computer."

"Assembling?"

David nodded at Charles, acknowledging that he had picked up on the key word. Charles was really sharp.

"Right." David looked Charles squarely in the eye. All thoughts about him being a kid disappeared, and David was talking to a peer. No more than a peer. It was like talking to Ernie or Crane himself. David felt Charles' hard eyes penetrating his thoughts, divining what he was going to say next. "The microcomputer has IBM over a barrel."

"Concerned that you will lose market share," Mike said.

"More than that. If we don't get in, we will have missed an important opportunity. If we do, we run the risk of hurting our mainframe sales."

"Even if you don't jump in, the microcomputer is going to hurt mainframe sales," Mike said. "I don't see where you have a choice."

"I don't either," David said.

David looked at Charles who waited silently. "Mainframe sales are targeted at big business," David said the obvious. "Belatedly, we recognize there are other computer markets out there, and now we are playing catch up."

"And you do not have five years to implement the usual IBM new product cycle," Charles said.

"Exactly," David agreed. Charles' perception matched his own, and Ernie's, and James Archibald Crane's. "Left to its own devices, IBM would take five years to get where Astron Computers and Digital Software are now today."

"And when you get there, where do you think Astron Computers will be?" Charles said.

"We have to break the mold," David said. "Otherwise, we will never catch up."

David noticed that Speedy was nodding his head in agreement, and David realized that this was probably the first time he had heard an IBMer speaking so frankly with an outsider.

"You've heard of the Alto and the Star?" Charles decided to throw his visitors a bone.

"The Alto, yes," David answered honestly. "The Alto is PARC's prototype with a mouse and graphical user interface, the future, but I haven't heard of the Star."

"It's PARC's attempt to build a commercial version of the Alto," Charles said.

Mike frowned. Charles had revealed a major industry secret that they had acquired only after diligent effort—they had hired a disgruntled PARC software writer at a very high salary.

"That's news," David said. In fact it was shocking news.

"Good thing we have a secrecy agreement," Charles smiled.

"But its not proprietary information," Speedy protested. The Armonk bureaucrats would have to be informed of Xerox's plans.

"We're among friends," David warned his companion off, then returned to the main subject. "You are absolutely right Charles. We're assembling a microprocessor using components from other companies."

"That's hard to believe," Mike played the heavy.

"But it's true," David said.

"State of the art?" Charles asked.

"Something a little less than that. We're using tested and reliable products. Mid state of the art, I would say."

"Good luck," Mike laughed.

"Mike is business management not an engineer," Charles said, warning Mike to back off.

"I'm phys ed," David laughed. "I'm not an engineer."

"I'm nothing," Charles laughed. "I dropped out after two years."

"So much for academics," Speedy said. "I have advanced computer science degrees from MIT, and I'm junior man in the room."

"And what do you need from us?" Charles asked, refocusing the discussion.

"I wanted to see your setup, get to know you better, and determine if we could rely on your company for software and languages," David said.

"If we don't' have the languages or software you need, we will write or acquire them," Charles said confidently. "Tell me about your machine."

David hesitated. "We have yet to present our configuration to the Management Committee," David said, telling more than he had intended.

"Maybe we can advise you," Charles said. "You already have your security commitment on proprietary information. Anything you tell us about your machine will be respected." Charles assumed his guests realized that the secrecy agreement did not cover the most important facet of the meeting: IBM was entering the microprocessor market.

David thought about answering, then decided against it. "We'll have to defer that discussion to another day."

Charles nodded. That was the response he expected.

"May I ask you a question?" David asked.

"Shoot." Charles responded.

"How many customers do you currently have?"

"Over a hundred."

"Are you having difficulty meeting commitments?"

"No."

"If IBM were to contract for languages and software, could you handle it?"

"We would have to expand faster than we planned, but we can handle it. IBM would immediately rank as our number one customer."

"Are you sure?"

"Certainly," Mike joined the discussion.

"Charles?" David asked.

"You have my word on it. Yes," Charles said.

David fixed on Charles' icy gray eyes. Charles did not waver.

"Very good," David said. "I've learned what I came to find out. Speedy, ask your technical questions."

Speedy immediately focused on languages, particularly FORTRAN. After a full half-hour of technical discussion, Speedy declared himself satisfied. Charles offered to host lunch, but David declined. "We have a flight to catch."

Chapter 19

Bill Oldham, after a few months at the "Fairfax Journal," decided he had found a vocation but not a base. He liked his colleagues, even respected some of them, but reporting on community events did not challenge him. Bill had earned his way off the life and death desk and had worked up to covering county government but not even that interested him. One could carry stories of zoning disputes, real estate and personal property tax rate hikes, minor jealousies and traffic congestion only so far. Humphrey Tydings resisted corruption stories and accountings of bureaucratic inefficiencies. The one time Bill had tried to submit a mildly critical report on county government, Tydings had rejected it with a terse comment: "Our job is to entertain and inform the residents. Leave the muckraking to the Washington Post."

Tydings insisted on simple, concise prose, and Bill obeyed. He concentrated on learning his craft and building a resume, but he found it difficult not to fantasize about a day when one of his stories might catch the attention of one of the majors, all of whom were represented across the river.

Bill researched his target audience and decided that the business coverage of the Journal offered him his best opportunity. The leadership of the county government was in the hands of those who believed in development; some might say in the hands of the developers. For many years, Northern Virginia had been used as a bedroom community for federal workers. Starting in the booming post World War II era, this privileged class had moved across the river seeking their personal Valhallas along the highways that stretched out from the Fourteenth, Memorial and Key Bridges. Then, came the beltway, originally designed as a circumferential highway around Washington, D.C. The planners had intended that the beltway serve as a safety valve that moved the heavy north/south traffic around D.C. without contributing to an already unpleasant traffic congestion. They failed to take into account the pent up pressure of a population eager to trade urban realities for a patch of grass and trees.

The beltway evolved into a suburban commmuterway, and Fairfax County with its willing leadership became a haven for developers. Then, the federal government built Dulles Airport to handle international traffic and relieve the stressed facilities of National Airport on the banks of the Potomac. This in turn led an imaginative developer to create a new self-contained city, Reston, where residents would live in a parklike setting and walk to work. Since the federal government was centered in the District of Columbia, the developer had to attract businesses to Reston to provide employment for his lucky new residents. To everyone's surprise, the concept caught on. A few major industries, Mobile for example, moved their headquarters to Fairfax County. No dummies, county leadership and the developers who would build the structures to house and to offer gainful employment to a growing population caught on quickly. Building roads, schools and infrastructure became a county preoccupation. Bill reasoned that with all this money floating about there had to be a number of stories he could write to advantage.

Firstly though, he had to persuade his curmudgeonly editor to let him infringe on the exclusive domain of the middle aged Phil Turner who zealously protected his turf. Predictably, Tydings had counseled: "Be patient. Wait your turn." Then, Tydings had announced that he would take his three weeks of leave in one lump so that he and his family could spend quality time together on the beach at Ocean City. Assuming that business news would be slow during the dog days of a steaming Washington summer, Tydings gave Bill his chance.

"Fill in for Phil, Bill," Tydings had ordered Bill, smiling at his weak alliteration.

"Yes sir," Bill said, eager to seize his chance.

"Don't get carried away kid," Phil had counseled. "Nothing happens in Fairfax County in July."

"Don't go clever on me," Tydings had ordered. "I'll mark the wire service stuff I want you to tone down for me."

"Yes sir," Bill agreed. "Just simple rewrites. Subject, predicate, no adjectives, that kind of stuff."

Tydings smiled. He knew better. The kid would knock himself out trying to come up with a headline grabber, but Phil was right. Nothing ever happened in Fairfax County in summer. All the government workers went on leave, most to the shore, some to the mountains, but all someplace else. When the government workers left town, those who lived off their deep pockets—the realtors, storekeepers, repairmen, and salesmen—went on vacation too. Despite air-conditioning, the tropics had nothing on Washington in July. If most left town, few remained to read the Journal.

For two weeks, Bill patiently rewrote wire service stories and frantically searched the Post and the Times and the newsmagazines and even watched the television news hoping to find a peg for his ambition. As the time available for his breakthrough ebbed, Bill grew desperate. After a year on the Journal, Bill had an opportunity, and it was slipping through his fingers. Saturday morning Bill lingered in his apartment and listened to the roar of beltway traffic. He considered a trip to the Fairfax City branch of the county library. He decided against it. If a story were already in print, it would do him no good. He needed something dramatic, new. He leaned over his littered coffee table and idly leafed through a two-month-old issue of Time magazine. In the business section, one story caught his eye.

Astron Computers had sold 170,000 computers in 1980.

Impressive, Bill thought, but old news.

Astron Computers had gone public with its stock. According to the story, Astron drew in an astounding $82 million dollars from investors and leaped into the Fortune 500 list of companies faster than any other in history. In total value, Astron was now larger than Ford Motor Company. Three principal stockholders had become multimillionaires over night. Stanley Pitts' stock was worth 256 million dollars, Harold Dumbroski's 135 million, Stuart Miller's 239 million. Bill could not imagine what it would be like to have so much money, particularly if you were Pitts and Dumbroski who only three years earlier had nothing but some great ideas and a workshop in a garage.

Bill enviously had to admit that his old Harvard roommate, Charles Swift, and his friend, James Clapper, had been right. At the time, Bill had thought Charles foolish to drop out of school before completing his third year. Bill had not thought of Charles for some time. Bill had spent a summer in Albuquerque working with Charles' startup company, writing manuals and starting a newsletter. It had been interesting, but Charles' company had been small potatoes, a few kids writing code and having a good time. Bill had not bought into Charles' big talk about a technological revolution. During Bill's final year at Harvard, he had maintained sporadic contact with his old roommate. After moving to Fairfax, Bill had become preoccupied with his own career. He had received a note from Charles announcing his return to Seattle.

The Times story did not mention Digital Software. Bill wondered if Charles had shared in the wealth created by the technological revolution. He was not sure he wanted to know. If Charles had hit it big, Bill was sure he would have read about it somewhere. Bill was no technology buff, but he had learned some of the jargon during his summer in Albuquerque. He began to speculate about coming up with some kind of computer story. Recognizing that Charles might be able to help him, Bill decided to call his old friend, pump him a little, and see if he could find some kind of peg that he as acting chief of the Journal business section could exploit.

Bill retrieved his little red phone book. He did not have Charles' new business number, but he was sure Charles' mother could help him. Bill checked his watch. Ten o'clock, too early to call the West Coast. At least, he remembered that. He cleaned his apartment, went to MacDonald's on Route 236 across from the entrance to his apartment complex, then phoned Seattle. Mrs.

Swift answered the phone promptly. Bill identified himself. Mrs. Swift remembered him. She gave Bill Charles' home and office numbers.

"Better try the office," Mrs. Swift said.

"On Saturday morning?" Bill joked.

"Saturdays, Sundays, days and nights," Mrs. Swift said. "I don't know what he needs a house for."

Bill detected a note of pride in her voice that surprised him. Charles' mother had been the one most disappointed by her son's decision to drop out of Harvard.

"Charles has his own house?"

"Just across the lake."

"I always figured Charles for an apartment guy."

"Why would you think that? Charles has always lived in a nice house."

"No, no. I know that," Bill tried to recover. He had offended Mrs. Swift. "I just thought he was so interested in business that he would not bother with the burden of a house."

Mrs. Swift greeted that comment with silence. Bill should have known better. Charles' mother might criticize her son, but she did not allow others to do so in her presence. A sudden thought struck Bill. "Is Charles married?"

"Not that I know of."

Bill thought he detected a wry tone. "A girl friend?"

Charles at Harvard and in Albuquerque had not evinced much interest in girls. To the best of Bill's knowledge, and he thought he would have known, Charles had never had a real girl friend.

"No." Mrs. Swift seemed to hesitate. "He has a live-in maid. A pretty thing, but I don't think she's anymore than that."

Bill decided this was not a topic he wanted to pursue with Charles' mother. "Well, I'll give him a call," Bill said.

"It's been good talking to you again, Bill," Mrs. Swift said before hanging up.

Bill was left with a dead telephone. Charles' mother's abruptness made him wonder if she somehow blamed him for Charles' lack of a marital state. He decided that was impossible. It had been over three years since Charles had left Harvard.

Bill dialed the office number Mrs. Swift had given him. While he waited, he tapped on the notepad with the pencil.

"Digital Software," a female voice said.

Bill was impressed. Charles had a secretary.

"May I speak with Mr. Swift please."

"May I say who's calling?"

"Bill Oldham."

"Who do you represent?"

"Fairfax Journal," Bill said automatically before catching himself. "I'm an old friend."

"Charles," the secretary called loudly. "There's some newsguy on the phone. He says he is an old friend."

"What paper?" Bill heard his old roommate's voice.

"Some Journal."

"Tell him I'll call him back."

"Could I have your number, sir?" The secretary returned her attention to Bill.

"Ma'am, tell him to answer the damned phone. I'm his roommate."

"Oh?"

"From Harvard."

"Oh! Just a minute, sir."

"Charles!"

"What?"

"He says answer the damned phone. He's your roommate."

"Hey, Bill, what's up," Charles came on the line immediately.

"Nothing. Just thought I would check in with the big businessman. Made your first million?"

Charles paused, and Bill wondered if Charles had hit it big.

"Company or me?" Charles asked.

Bill laughed, a little embarrassed by his own stupid question. He had been trying to be funny while establishing their old rapport.

"I'm impressed."

"Need a job?" Charles asked.

"Not really, Charles," Bill said. "I'm a newspaperman."

"What's this journal business?"

"The Fairfax Journal," Bill said. "You probably haven't heard of it?"

"You're right. Where's Fairfax?"

"Fairfax County. A bedroom community for Washington, D.C."

"Good going? When are you moving over to the Washington Post?"

Bill suddenly remembered how competitive in everything Charles was.

"When I get my big story."

"Good luck. You're bigger than some little local newspaper. Haven't they heard of Harvard?"

"Doesn't seem to make any difference here."

"That's what I always told you."

"I don't remember that, Charles."

"I was teaching by example," Charles laughed.

"Playing any poker?"

"My business is a card game."

"Girl friends?" Bill remembered Mrs. Swift's reference to a live-in maid.

"No time."

"Not even a live-in girl friend."

"You've been talking with my mother."

Bill laughed. "She gave me your phone number. What's up with the maid?"

"Not much."

Bill knew better than that. He had once told Mike that Charles would marry the first girl he got close to.

"Mike's here," Charles changed the subject.

"Hamilton? What's he doing there?"

"He's chief operating office of Digital Software."

"Your Digital Software?"

"Is there another?"

"What do you do if he runs it?"

"He handles the business end, manages things. I work on technology and new customers."

"I'm impressed. How many employees?'

"Forty and growing."

"That's great Charles. I'm jealous. I was just reading about Astron Computers going public. Are those figures real?"

"We not in that league, yet," Charles said. "We're still private, but Astron's success helps us all."

"How do you mean?" Bill sensed a story.

"They sell hardware. We sell software. They grow; we grow. Nothing complicated. Sure you don't want to join us? It would be great getting the three of us back together."

"I still don't know anything about computers," Bill equivocated. He was sure he did not want to work for Charles, and he was surprised at Mike's surrender. He and Charles had always been competitors.

"You started our newsletter and wrote our early manuals. You could head up our public relations department."

"Are you that big?"

"Not yet. But we will be."

"Hey Charles, I need help," a distant voice called. "Forget the newsman."

"You're obviously busy."

"No. That's all right. It's just one of our programmers."

"Hang loose," Charles shouted back.

"I just wanted to check in," Bill said.

"We've got to keep in touch," Charles said. "How close is Fairfax to Washington?"

"Just across the river."

"I might need someone I trust to keep me informed on things."

Bill laughed. "We're a suburban community newspaper. We leave the important stuff to the big boys across the river."

"Then get a job on the <u>Post</u>."

"How do I do that?"

"You know better than I do."

Bill hesitated. He could tell Charles why he really called, but he was not sure he wanted to be in debt to Charles. He liked it better when they were equals.

"What do you need?" Charles asked.

Still, Bill hesitated. Charles had always been surprisingly perceptive. When he concentrated, it was almost as if he could read your mind. Bill had once told Mike that Charles was so bright that after he got to know you he could almost put himself in your place and think like you did.

"I could use a good story," Bill said.

Charles laughed. "We all could."

"What's new in the computer world that nobody knows about?"

"That's a way of life here. Everything's new, and every company has its proprietary secrets."

"Tell me some."

"Can't do that, but..." Charles hesitated.

Bill could almost see Charles leaping up and touching the ceiling in his version of a slam-dunk. "Did you just do it?" Bill asked.

"Do what?"

"Touch the ceiling."

"Very perceptive."

Charles did not laugh, but Bill did. Charles was still sensitive about his mannerisms.

"Maybe I can help you," Charles said.

Bill leaned forward and grabbed his pencil. "Shoot."

"This is just between us old friends," Charles said cautiously.

"We newsmen will die before we reveal our sources." It was a line from an old television movie that Bill had watched.

"So I heard. An old Katherine Hepburn movie?"

"Probably. What's the news?"

"You've heard of Big Blue?"

"A football team?"

"No. IBM. International..."

"I've heard of IBM. They make computers, don't they? Are they one of your customers." Bill thought he was making a joke.

"Maybe."

The answer surprised Bill. He decided to stop playing Bob Hope and listen.

"Their business is mainframe computers, the big stuff. IBM dominates the world market in mainframes. So far, they've had a lock on the big business. When Ford or General Motors needs computers they call IBM."

"They don't make little computers?"

"No. IBM has stayed out of the microprocessor market. Too small for them, and they fear that if they started making minicomputers it would eat into their main business, mainframes."

"Why sell a product that competes against yourself in a market you already own."

"Exactly.

"Then what's your news?" Bill dropped his pencil. This sounded too esoteric for the Journal.

"You mentioned Astron Computers."

"Yes." Then, Bill caught Charles' pointed reference. "Astron Computers has IBM worried." Bill wrote "IBM worried" and "Astron" on his notepad.

"Do you have firsthand knowledge of that?"

"Yes."

"You've talked with IBM?"

"Their representatives. Look, let me tell the story and stop acting like a journalist."

"That's what I am."

"OK. Listen. If IBM were to change its position and decide to build and sell microprocessors, it would shake the market."

"Wall Street."

"Wall Street and the real marketplace. IBM has a reputation for quality. Why buy an Astron computer, or Tandy computer, all new technology and buggy, when you can get a reliable IBM product cheaper?"

"You know for a fact that IBM is building a small computer?

"Yes."

"Can I use this?" Bill was starting to get excited.

"You can't reveal your source, not even by implication. It could cost me big time."

"You've got my word. You're selling IBM software?"

"You can't say that."

"If I don't identify Digital Software?"

"I don't know who else they've talked to. Don't even mention software. That's a giveaway.

"What do you mean?"

"Oh Christ, I shouldn't have started this. Look. IBM builds its own hardware and writes its own software. It's a matter of corporate pride, and everybody knows this."

"I didn't."

"Anyone in the business does."

"If they're buying software from you, that means they're changing corporate policy."

"That's right."

"What about hardware?"

"What about it?"

"Are they going to make their own?"

"Bill. I thought I was giving you a teaser. Now, you're getting in deep water."

"I need a big story, not a teaser. You want me to work for the Post don't you?"

Charles thought that over. "Hardware too. I have it on good authority that they are too far behind to wait. Astron Computers caught them by surprise. They're running scared, I think."

"Why is that?" Bill wrote "IBM running scared" on his pad.

"IBM is a big bureaucratic company. It will take them five years to build their own microcomputer, five years just to get where Astron is now. I understand Astron is working on a new graphical interface on the PARC model that will make today's computers look like the Model T compared to a Thunderbird."

"How is that? What is PARC?"

Charles explained graphical interface and PARC's relationship to Xerox. Bill noted "PARC, research Xerox."

"Their new desktop will let the computer illiterate, like you, use the computer. It'll open the market beyond comprehension, worldwide."

"You're talking a few bucks here."

"Don't even joke about it. This is the beginning of the revolution, and IBM is behind. It might even mean their corporate demise."

"If it's so important, how did IBM get so far behind?"

"They got old and bureaucratic. It's a smart, young man's game."

"Your revolution?"

"Exactly."

"Then why sell software to IBM?"

"Because they need it. If they get into the market now, they might dominate it."

"How are they going to do that if they are so far behind?"

"By assembling components they buy from others, putting the IBM brand on the product, and driving the upstarts like Astron Computers into the ground."

"How do they do that?"

"You ask too many questions."

"Tell me, Charles. How do you destroy a company that's bigger than Ford Motor Company?"

"All these computer companies are built on two things: debt and technological leap forwards."

"Explain, please." Bill was now writing quickly.

"They have weak financial bases. They borrow from venture capitalists. They go public. Mr. John Q. Public buys the stock because they read all the financial page hype about the technological future. Thus, their public worth is all relative. Stocks go up and down based on nothing more than public confidence. They have to invest heavily in R & D. Computer speed, memory, peripherals, chips all get better, faster, cheaper. Today's microprocessor is out of date tomorrow."

"Then if IBM comes in, builds a state of the art computer, puts their brand on it, they'll cut into sales by the others and collapse the bubble."

"Maybe. IBM will not build a state of the art computer. They'll build something less, something they can support and guarantee reliable, then update it in time."

"And you'll get rich selling them software."

"We plan to. Sure you don't want to join Mike and me?"

"God that's tempting," Bill said. "I'll think about it, seriously, but journalism is my game. You've given me what I've been looking for."

"Protect your source. I had to sign a secrecy agreement with IBM before they would talk with me."

"So I can put you in jail?" Bill laughed.

"Not quite. I only agreed to not to divulge proprietary information."

"What's that?"

"Information about the specifications of their computer. I didn't promise to conceal the fact of our meeting or the news they're assembling a computer."

"If I write a story, will they trace it back to you?" Bill worried that Charles might reconsider.

"Not if you write it right."

"Besides, who reads the Journal?" Bill tried to joke.

"You might be surprised by this story."

"I hope so. Charles, I have one last question."

"Be my guest."

"What if I need a second source." Bill knew that sensitive stories required confirmation. "Is there someone else I can call? At IBM?"

"Not on your life. The information is reliable. Use it."

"I know. I trust you Charles, but I may need to tell my editor that I confirmed my source with another."

"You cannot tell him who your source was."

"I know. I won't. I just have to be able to say I confirmed it with a second source."

Charles met this comment with silence.

Bill waited, not wanting to plead.

"OK," Charles said. "You can say you have a second source."

"I mean it."

"I do too. Mike sat in on all my meetings. He's not here now, but if you need a second source, he's it."

"Need I call him?"

"Not necessary. I'll brief him. He'll back you up."

As soon as Bill hung up, he abandoned his apartment-cleaning project and headed for the library. There, he researched the more recent news stories on IBM, Astron, PARC, and the microcomputer industry. He checked out two books on IBM. From there, he went to the Journal offices and searched the archives. There, he found little of interest. He returned to his apartment and reviewed his notes. The IBM books appeared useless, both were obviously written by IBM sanctioned authors who had delivered monuments of printed praise for the Watsons. The late night television news was on when Bill finished his first outline of the story. He listened closely to the newsreaders, worried that someone might have scooped him. Nothing came close. The computer industry was not mentioned. Pictures of the morning traffic jam on the Bay Bridge and vacationers playing in the sand dominated.

Bill thought about going to bed, then rising early to start writing when fresh, but he was too excited. He retrieved his old Corona typewriter from the closet and began to write. He finished his first draft by two AM. He sipped coffee and picked up the first page. His prose disappointed him. He had written like a college kid. He knew how Humphrey Tydings would react. Too many complicated sentences. Trying to cover all bases, describing a subject he had not mastered, Bill had wound his sentences into their own knots. Emerson had worshipped the paragraph. Humphrey lived for the simple sentence. Bill grabbed his pencil and began editing. He forgot his coffee. By the time the clock reached four PM, Bill surrendered. He took a quick shower and climbed into bed. He had forgotten to turn off the light on the drop leaf table where he worked. He was exhausted. He thought about getting up and turning it off, then decided against it. He turned on his side and fell into a deep sleep.

The light was burning when Bill awoke, still on his side. He had a terrible taste in his mouth, and his shoulder hurt. He rolled on his back and closed his eyes again. Then, he remembered his story. He visited the bathroom, dropped bread into the toaster, and heated water on the small electric stove. Still in his pajamas, he carried his toast and coffee to the table. He sipped the coffee, nibbled on the toast, then picked up his draft. He had difficulty deciphering the penciled inserts that he had written a few hours earlier. Within minutes, he was typing a fresh draft, the coffee and toast forgotten.

Bill worked through the day. He wrote and re-wrote. He had gone through five drafts before he produced copy that he thought Humphrey might accept. He tried several headlines and finally decided on: "Big Blue Declares War."

Monday morning Bill eagerly arrived early at the deserted <u>Journal</u> office and carefully placed his story on Humphrey's desk. Then, he waited, and waited, and waited. His coworkers finally straggled in and went to work with a minimum of conversation. Finally, the editor made his appearance. Bill discreetly watched as Humphrey read Bill's story. Humphrey, as usual, read with his pencil in hand, an editor's trait, but as best as Bill could tell, he did not apply it to paper. Bill couldn't believe the story was that bad. Humphrey finished reading, returned his attention to the front page, wrote something, then set it aside, not in his out box.

Disappointed, Bill waited for the summons. None came. Finally, he swallowed his pride and approached the editor.

"Morning, Mr. Tydings."

Humphrey looked up, then smiled. "Morning Bill. Guess you don't like it here."

"Sir? It was that bad?"

Bill grabbed his story from Humphrey's desk and noted what the editor had written. Two things. Across the top he had scrawled: "Tuesday's paper, page five lead." He had also written under Bill's headline: "by William Oldham, staff writer."

Bill had his first byline story.

"I don't understand," Bill said.

"Are you sure of your source?" Humphrey asked.

"Yes."

"Did you get confirmation from a second source?"

"Yes sir, but why do you ask?" This was the first time the editor had asked these questions about anything Bill had written.

"Because this is a dynamite story. There'll be reactions."

"We don't have to print it."

"That's not what I meant. We have to print it. Who were your sources?"

"I'm supposed to protect them."

"Not from me. We'll protect them. Who were you sources?"

"Contacts in the computer business."

"From around here?"

"No, the west coast."

"Do they know what they are talking about, have access to IBM?"

"Yes sir.

"Names?"

"Charles Swift. I promised to protect him."

"How does he know what IBM is doing?"

"He's president of a software company. IBM is talking with him about writing software for their new computer."

"Why is he telling you?"

"He was my roommate at Harvard. I worked one summer for his company."

Humphrey smiled. "Not bad." He held out his hand, and Bill passed him his story.

Humphrey started to toss it into his outbox then drew his hand back, still holding the copy. "Who was your second source?"

"My other roommate. He works for the first roommate and participated in the IBM meetings."

"Do you trust them?"

"Yes."

"You've got your scoop. Hang on." Humphrey tossed the story into his out basket.

Bill turned to leave then stopped. "Why did you say I didn't like it here?"

"Wait and see."

Bill's story appeared on page five of the Journal's Tuesday edition. His header "Big Blue Declares War" appeared exactly as he had written. Bill studied his byline with pride, then began to worry. He hoped that Charles was not ragging him. Charles had played practical jokes before, but this was too serious for that. Bill was sure Charles understood. He thought about calling Charles back, but then decided against it. Instead, Bill phoned his mother in Clarksburg. Bill explained about the byline and the story's importance. "I hope you're right son."

"It's OK mom. I doubled checked my source." Bill spoke with more assurance than he felt.

"Good. I'll tell your father as soon as he gets home from the store. He'll be proud of you."

"I'll get a copy of the newspaper for you," Bill promised.

When he hung up, Bill hurried out to buy five more copies from the metal dispenser in front of the building. He stashed them in his car's trunk and returned to the newsroom. His colleagues congratulated him and returned to their desks. It seemed old hat to them. Bill worked on a routine county government story while he waited for something to happen. Nothing did. Humphrey passed Bill's desk on his way out. He patted Bill on the back and said: "Just wait."

Bill feared his fifteen minutes of personal glory had passed. The next morning he made his rounds of the Fairfax police and fire department headquarters before reporting to the office. When he entered, he felt the others staring at him. On his desk, he found a note. It contained a telephone number and instructions in Humphrey's distinct scrawl: "call." Bill assumed it was a story assignment. He hung his jacket over the back of his chair and dialed the number.

"Business Section," a gruff voice responded.

Bill was momentarily flustered. He was the <u>Journal</u>'s business section, at least this week. He looked back at Humphrey who sat at his desk reading copy. Humphrey seemed to be ignoring him.

"Business Section," the gruff voice made no attempt to conceal its irritation.

"This is the <u>Fairfax Journal</u>," Bill said.

"Well good for you. What do you want?"

"My name is William Oldham, and I received a message to call this number. Who am I speaking to, please?"

"Anyone in here call somebody named Oldham?" The gruff voice called.

"Oh," the gruff voice said. "Hang on, Oldham, I'll transfer you."

"You Oldham?" another male came on line.

"Yes. Who am I speaking to please."

"Didn't anyone tell you?"

"No sir."

"Williams. Learn how to answer the damned phone," the voice shouted.

"Answer your own damned phone," the gruff voice responded.

"Are you the guy that wrote this 'Big Blue Declares War' story?"

"Please tell me who I'm talking with," Bill began to lose his patience.

"This is the <u>Post</u> business section. Where did you get this story?"

"None of your damned business," Bill said.

"Have you seen the wire services?"

"No."

"You better be right. They're all carrying you, and the market is jumping around like popcorn in hot grease. You better not have any IBM stock. Do you?"

Bill laughed. "What if I do?"

"Then brace yourself. The SEC will be all over you. If your story is not true, you're dead meat."

"It's true."

"How could it be. IBM is denying everything. If they are to be believed, they don't even know what a computer is."

"You're shitting me," Bill said. "IBM is in the computer business."

"Mainframes. You know what they are?"

"Read my article."

"How come you know this and we don't?"

"That's your problem." Bill hung up, then realized what he had done.

Bill went back to Humphrey's desk. "Did they offer you a job, kid?"

"The <u>Post</u>? I just hung up on them."

"Good for you. They'll call back."

"Why?"

"You scooped them kid, and they don't like that. Our dinky little rag made them look bad. You seen the market ticker?"

"No."

"IBM is up four points and a lot of those Silicon Valley assholes are dropping like shit from a cow's ass."

"I think you mixed your metaphor."

"Don't go big shot on me kid. You're not a prime time reporter, yet."

The" yet" made Bill's day.

Chapter 20

David and at least ten members of his team huddled around the workbench in the basement of the Boca Rotan main building and studied their first try at assembling a prototype. It had an operating system, but Speedy was not satisfied with that. This machine was merely a mockup designed to give them a starting point.

"Doesn't look bad, but I'm not sure I like that white cabinet." David still did not pretend to understand how a computer worked. Like Stanley Pitts, however, David could comment on style.

"We've got to get back to Digital Software and find out what kind of operating systems they have," Speedy said.

"Can you turn this on?" David asked. He did not know if components selected from different manufacturers would work together.

"Ye of little faith," Speedy said, flicking a switch on the front of the box. "This is known as an on/off switch."

"Thanks," David accepted the gentle prod.

A phone on a desk in the front of the work area began to ring. Everyone ignored it as they waited for the monitor to light up and the computer to commence its labor. The phone persisted.

"Terry, will you please get that," David asked.

Terry Mitchel, whose apartment David now called home, had exercised her prerogatives and had abandoned her lobby desk to watch the testing.

"It's not my phone," Terry smiled sweetly. David had insisted that a direct line be installed in the lab after learning that Terry routinely listened in on all calls.

"Terry, please," David pleaded.

"Having trouble with staff, David?" One of the older engineers laughed. They all knew David was courting Terry.

"Boca Rotan," Terry spoke into the receiver just as the screen lighted up.

A picture of a beautiful nude filled the screen. All the engineers cheered. "Is that Terry?" One of them asked.

Terry replied with a finger as she said: "Yes sir. Just a moment please. Mr. Howard is in conference, but I will interrupt him."

"Is that noise coming from the conference?" Ernie Hendricks asked.

"No sir," Terry said. "Some of the children are at recess."

"Please put him on," Ernie said.

Terry covered the receiver and called to David. "Better take it. Armonk, your Lord and Master."

"Hold it down guys. Your paychecks are in jeopardy," David said as he rushed to take the receiver from Terry. He had no idea why Ernie was calling. It had to be bad news. He immediately speculated that the project had been cancelled.

"Hi Ernie," David said. Since his assignment as Boca Rotan Special Project leader, he had begun to treat the Chief of the General Product Division a little more casually than he had as Ernie's special assistant and chief gofer.

"Am I interrupting a tennis match?"

"No sir. We've just started up our first prototype."

'Does it work?"

"We have an image on the screen."

"Good. Pack it up and get your ass up here."

"Ernie. I don't know if it works or not. We've got a lot of work to do before we're ready to demonstrate it, if that's what you have in mind. We need an operating system among other things. I'm scheduled to go to Seattle tomorrow," David improvised.

"Don't bullshit me David. Get your ass up here and bring your machine with you. What do you call it?"

"We haven't researched a name yet. We're just trying to get it to work."

"You said it came on. And shows an image."

"Yes sir."

"That's all we'll need. What do you call it now?"

"The personal computer, that's all, just a way of identifying the thing. We can't just call it "It.""

"OK. Pack up the personal computer and bring it up here. You're scheduled to go on tomorrow at eleven."

"On?"

"On."

"The Management Committee?"

"Yes."

"Oh Christ. I can't Ernie. I'll screw up. Please stall them."

"Your time is up, David. Do what I say."

"Ernie, what happened?"

"Do you read newspapers down there?"

"I read the Boca Rotan <u>Gazette</u> every morning."

"Does Terry let you have the comics first?"

David heard a loud giggle in the background. He assumed it was Margaret.

"I understand that's your preference," Ernie continued.

"No sir, I start with the sports pages."

"David, I'm not talking about the Sandpiper Gazette or whatever it's called in your tropical paradise. Have you seen this morning's Times?"

"The New York Times?"

"Is there another?"

"No sir."

"No sir what? Did you read the Times?"

David looked to his audience for help. All were now carefully monitoring David's end of the conversation, clearly sensing something was drastically wrong. David covered the receiver. "Any of you see the Times this morning?"

No one responded.

"No sir."

"Ever heard the phrase "Big Blue Declares War?"

"No sir."

"Better grab a copy of the Times at the airport. You're famous."

"Me? My name is in the Times?"

"No, but it will be. Look, David, I don't have time for this as much fun as it may be for you."

"Who did Big Blue declare war on?"

"The microcomputer manufacturers. The story of your project has leaked. Did you do it?"

"No sir."

"Then where did this story come from? Somebody leaked. Got any idea?"

"I assure you nobody from here talks with any reporters. We wouldn't recognize one if we saw one."

"Then who leaked?"

"It had to come from Armonk."

"From me? Crane and I are the only ones here who know what you're doing."

"Then one of our subcontractors." David hesitated before continuing. "It was bound to leak eventually. Are we out of business?"

"Just get up here. Bring Speedy to answer any technical questions, and bring your Personal Computer if that is what you have named it."

Ernie hung up before David could say "Yes Sir" again. To his surprise, when he hung up, everybody clapped and some cheered loudly. "Are you guys nuts?" David asked.

Speedy answered: "After you brief the Management Committee, we'll be in business. Everyone here will be division managers."

"I get sales, that's where the money is," one of the younger engineers shouted.

The next morning following a late flight to New York, a drive in a rental car to Armonk, and a few hours sleep in a Holiday Inn, David, Speedy, and their one eyed limited capability Personal Computer drove to IBM headquarters, arriving with a half hour to spare before the Management Committee meeting. David parked their rental Escort in a visitor's slot, hoping to avoid embarrassing the IBM corps who all drove mid-sized cars or bigger, and turned to Speedy. Both were dressed in the proper headquarters uniform all the way down to their shiny black wingtips and garters.

"You get a cart and take it to the conference room. I'll check in with Ernie."

"Right." Speedy had grown more subdued the closer they got to Armonk. Now, he, like David, was worried.

David did not know what he was going to say in this his first appearance before the daunting Management Committee. Three little words—"Close It Down"—could end his career as a manager. IBM did not like publicity it did not generate. Premature leaks had closed more than one project, and these had involved IBM generated projects. David had to defend an assembled hybrid that would be Big Blue in name only.

David entered the General Product Office and was surprised to find a relaxed Ernie lodged on a corner of Margaret's desk chatting.

"Our commanding general arrives," Ernie greeted David.

"We've got to talk," David said.

"No time, son," Ernie checked his watch.

Ernie took a staggered David by the arm and led him out of the room.

"Hi Mag," David mumbled over his shoulder.

"How's the tennis?" Margaret counterpunched.

"What should I say?" David asked as they waited for the elevator.

"Cite your authority and describe your project, tersely," Ernie said.

"Cite my authority?"

"Working on direct instructions from Mr. Crane and that jerk Ernie Fredricks…" Ernie coached.

"Jesus, Ernie."

"He won't help you, believe me. Relax. Keep your report brief, demonstrate your product, reply to any questions with short, direct answers and leave the rest up to Crane and me."

"What about Parsons?" David referred to the Chief of Sales who had taken an early dislike to David following a collision at second base initiated by a naïve David.

"Leave the politics to Crane. Don't let our future chief executive officer provoke you." Ernie referred to the fact that Parsons was Crane's heir apparent. Director of Sales in IBM was the crown prince slot.

They found Speedy in the conference room setting up the prototype. "I hope it still works, Mr. Hendricks," Speedy greeted Ernie.

"Show me," Ernie said.

Speedy turned on the machine. While they waited, Ernie spoke to David: "Call it a home computer. If the mainframe boys think it is no threat to their relationships with the corporate customers, they'll keep quiet as long as Crane supports you."

"Home computer," David repeated.

The monitor jumped to life and displayed the nude picture. Ernie laughed. "Is that all it does?"

"I told you the software is not ready," David said defensively.

"You said operating system."

"Until we decide on the operating system, we can't design software," Speedy came to David's defense.

"Then show them a picture. They won't expect anything more from this little toy. Do you have another picture you can use?"

"Yes sir. We've got a Cape Kennedy blast off."

"Use that."

Speedy typed some instructions and the picture of a rocket replaced the nude.

"OK. Turn it off and wait outside until I call you," Ernie instructed.

"Where do I stand?" David asked.

"Next to your PC, and face the chairman. Talk to him."

David and Speedy waited in the anteroom as the IBM barons filed past and took their places at the V shaped table. Crane was the last to enter. He winked at David, entered the conference room and closed the door behind him.

"Did you see him wink"" Speedy asked.

"What did it mean?"

"Reassurance."

"I hope so."

They did not have long to wait and fret. After a few minutes dragged past, Ernie opened the door and flicked his forefinger. David, followed by Speedy, entered the room. Ernie closed the door. Speedy positioned himself behind the table bearing their prototype, and David reluctantly stepped in front of it, standing in the open end of the V. He glanced at Ernie who smiled and nodded. David cleared his throat and wondered what to do with his hands. He decided to let them hang at his sides.

"Mr. Crane," David addressed the Chairman. "Gentlemen," he let his eyes acknowledge each of the barons aligned on each side of the V, starting with the lowest ranking division chief on his left. Finally, he reached his nemesis, John Parsons, the Director of Sales, who sat on the chairman's immediate left. Parsons smiled thinly and nodded, signaling he remembered David.

"Working on the direct instructions of Mr. Crane and Mr. Hendricks, we in Special Projects, Boca Rotan, have developed a prototype personal computer to present to the board." David paused.

"Point of order Mr. Chairman," John Parsons, Director of Sales, raised one finger.

"John," Crane recognized him.

"Shall we hold our questions to the end of the presentation or may we raise them when appropriate?"

They all knew the answer. Old Thomas Watson had set company policy on that. Watson believed in discipline from the top down once a decision had been made. Until a decision was reached, he expected contention. Watson wanted to hear honest opinions and not to be surrounded by a bunch of yes men. As a consequence, IBM meetings, even those at the Management Committee level, frequently grew quite heated.

"Interrupt when you wish, if Mr. Howard does not object," Crane smiled at David.

"No objection, sir," David said.

"Mr. Howard," John Parsons seized the floor. "Would you kindly tell us why the news media was given information about this special project before it has been presented to the Management Committee for approval. Were you trying to force our hand?"

David flushed. "I do not know how the information leaked, but I am confident it did not come from the Boca Rotan research facility."

"That's nonsense. No one knew of the project but your Special Group. It had to originate with you," Parsons accused.

"Ultimately, yes, that's right, sir, as far as the information is concerned. I reiterate, however, that the leak to the media did not come from my group."

"Then from where?"

"I can answer your question and begin my briefing at the same time if you agree," David said evenly.

Parsons flipped his palm over, indicating David should proceed. He whispered something to the man on his right that David could not hear, but he did not like Parson's sly expression.

"With the concurrence of Mr. Crane and my immediate superior Mr. Hendricks, we broke with the IBM tradition of creating our own product. In the interest of time, we explored the possibility of assembling a competitive microprocessor, one worthy of the IBM standard of quality, by purchasing quality components from the first line microprocessor companies."

"Assembling," Parsons parroted back. "Then the stories are accurate?"

"Yes sir. I assume the story leaked from one of the several companies we surveyed."

"That supports my contention that IBM tradition should not have been violated." This time Parsons challenged Crane.

"I hear you," Crane said softly. "That policy decision was mine. We need not trouble Mr. Howard by debating it further."

Parsons flushed but did not respond to Crane. He turned back to David. "As Director of Sales, I would like to know who will market this hybrid product." His emphasis on "hybrid" clearly emphasized his distaste.

"I don't think Sales need concern itself with that," Ernie intervened.

"And why is that?" Parsons demanded.

"Because the General Products Division will market the Personal Computer just as it does our typewriters, adding machines and other small products."

"Then, if I understand you correctly," Parsons said, "you do not anticipate your hybrid undermining our main frame sales."

"Our market is the home user not big business," Ernie countered.

David suspected this response would not hold true for the long run. All his contacts in the microcomputer world anticipated that in the near future microprocessor power would increase to the point where it would approximate that of yesterday's smaller mainframes. He doubted anyone could guarantee that the microprocessor market penetration would be limited to the home and small businesses, but he kept his views to himself. Debating policy on this level was not his prerogative.

"You are not afraid you will destroy one of your own products?"

David relaxed. The question indicated that Parsons was looking in the wrong direction.

"I assume you refer to the Selectric," Ernie said. "Are you familiar with the latest Astron Computer sales figures?" Ernie did not pause for a response. "The typewriter is already a dead duck and the calculator will soon follow."

David gave that exchange to the good guys.

"Please continue Mr. Howard," Crane said.

David swore that he almost saw a twinkle in Crane's eyes.

"We estimate we will have a product on the market within three months. If we wait to design and build our own microprocessor, we will never catch up."

"And whose fault is it that we are behind?" Parsons blurted.

No one answered, and a faint streak of red appeared in Parsons' cheeks.

"We all know it would take us five years to design and build a new computer," Crane finally broke the silence. He nodded toward David.

"We have selected components that will provide us with a machine that will not besmirch the IBM label," David said.

"Are we talking top of the line equipment?" The Chief of the Production Division asked. His unit produced the mainframes that were on the front edge of technology.

"No sir. We deliberately avoided trying to assemble a state of the art machine. You know better than I that that approach has too many pitfalls."

"They're rushing microprocessors to market without adequate testing."

"Yes sir, too buggy. Our machine will be reliable and as advanced as we can make it and still keep it properly tested and reliable."

"Does this appliance have a name?" Parsons asked irritably.

"Yes sir. The Personal Computer, the PC." David adopted Ernie's shorthand.

Parsons laughed. "That's the best name you people could come up with?"

"It's what we call it," David answered quickly. "That is what it is, a personal computer."

"Must we go into this?" Crane asked.

No one challenged the chief executive.

"We will use open architecture," David said.

"You mean you will announce to the world how your machine works so any body can copy it?" The Chief of the Production Division's question made this policy sound like heresy.

"Yes sir. We don't really have any secrets since we are using components manufactured by others. If we can set standards for the industry and the others follow our lead, the IBM logo with its reputation for quality will dominate the market even if others produce identical machines."

"That's a risky bet," Parsons said.

"Yes sir, but circumstances require taking risk. We are too far behind the technology power curve at this point to do otherwise. Of course, the decision whether to proceed in this manner depends on policy not my opinion." David did not have to say that the members of the Management Committee were the ones who set policy.

"I agree. The issue of standards aside, I don't see how we can keep others from duplicating our personal computer," the Production Division Chief shook his head.

"I believe Mr. Howard has already addressed that issue," Crane said.

"If this committee gives you the go ahead," Parsons said, "how long will it take you to get into production?"

"Three months," David said.

"And how many of these personal computers do you expect to sell in one year?"

"One hundred thousand."

"At what price?"

"We haven't established that precisely," David said. "Somewhere between one thousand two hundred dollars and fifteen hundred."

"Then we're talking small change here," Parsons laughed. "I don't see any problems with this project." Parsons did not have to say that mainframe sales could reach the tens of millions with a single company.

"May I ask the Director of Sales a question?" Ernie intervened. "Are you aware of the Astron Computer sales figures?"

Parsons did not respond.

"Does one billion dollars strike you as small change?"

David waited for someone to respond, but no one did. "With your permission I would like to demonstrate our prototype. I must caution that we do not yet have dedicated software installed."

"They why waste our time?" Parsons asked.

"I would like to see if it works," the Production Division Chief said.

David nodded to Speedy who stepped forward and turned on their prototype. He mentally crossed his fingers. In its first test after arrival from Boca Rotan, the stubborn machine had refused to respond. A frantic check had disclosed a wire had worked loose despite all of Speedy's careful handling. To his and David's great relief, the Personal Computer started up and the picture of the lifting rocket appeared on the screen. David felt like clapping. After Speedy answered several questions about the machine's capabilities, most asked because IBM after all was a computer company and its leadership was expected to demonstrate their technical skills, David and Speedy were dismissed.

"Thank you very much Mr. Howard and Mr. Spencer," Crane said.

As soon as they reached the anteroom, and the door had closed, Speedy turned to David. "Are they always like that?"

David shrugged. "You know as much as I do. I don't think they will ever pay me enough to play that game."

"You just did, champ."

"How do you think we did?"

"That Parsons guy is an asshole, but I think it was a setup. Ernie and Crane have everything under control."

"Then we're going to have to build this damned thing in three months," David said as the consequence of his bland assurances set in.

"We've got to get an operating system and some software first," Speedy said.

Suddenly, the conference room door opened and the chieftains filed out. A couple congratulated David and Speedy on their presentations.

"You lucked out," Parsons said as he walked past.

Ernie and Crane emerged last.

"Very good presentations," Crane smiled.

Ernie signaled a thumbs up and joined Crane as he walked to his office.

Chapter 21

Charles rubbed the back of his neck and stared across the desk at his colleagues and friends, Mike Hamilton and James Clapper. At times like this, Charles turned to these two for solace.

"You don't think they've identified me as Bill's source do you?" Charles was worried. IBM had not called to schedule further discussions about software as David Howard had promised.

"Bill is solid," Mike said with conviction.

"Do you think I should call him?"

"At the Post? No way. That would be a giveaway."

Charles rubbed the stubble on his chin as he pondered the problem. The IBM story he had suggested to Bill had proved bigger than Charles had anticipated. The media had trumpeted it across the country and frightened and inspired investors, causing Astron Computers to drop ten points and IBM to rise fifteen. On the basis of that one bylined story, the Washington Post had offered Bill a job as a staff writer for their business section. Bill had confided that the Post decided that they needed to broaden their reporting on the growing computer industry and had selected Bill to help meet some of their shortcomings. Bill had not told them that he knew little about computers and even less about the burgeoning industry.

"Do you think we forced IBM to abandon their plans?" Charles asked.

"What do you think?" Mike countered.

"No way," Charles said confidently. "IBM is big, cumbersome and slow. Success does that to a company, but they're not dumb. They wouldn't be where they are today if they were."

"We don't need them," Clap said. "We're doing all right."

"That's not good enough, Clap. We're at the point where we either grow, and grow big, or we shrink and die."

"I don't agree," Clap said. He suspected that the business he and his childhood friend had created had outgrown him. Charles was changing, or maybe the real Charles was stepping forth. Clap had not realized how big Charles' ambitions were.

The phone on his private line rang. Charles slapped the desk.

"That's IBM," Charles said, grabbing the receiver before his secretary could.

"Swift," Charles said.

"Charles," David Howard greeted him. "How are things going?"

"Fine," Charles was not interested in small talk. "When will I see you again?"

"How about Monday?"

"Let me check my schedule," Charles said. He knew IBM took precedence over everything else, but he had his pride.

"David wants to meet with us Monday," Charles covered the receiver and raised it triumphantly.

Mike and Clap replied with thumbs up.

"Looks fine to me," Charles said.

"Great. We'll want to get down to specifics, so I'll have a couple of colleagues in tow."

"Whatever you want."

"Did you notice that our plans leaked?" David asked.

"Big Blue Declares War?"

"Yes."

"Who leaked?"

"Who knows, probably one of our potential suppliers."

"You can't trust anybody in this business," Charles winked at Mike and Clap.

"But it worked out for the best," David continued.

"Have money on the market?"

"Don't even joke about that. I don't need the SEC on my back. No, the story caught our Management Committee by surprise, and I got summoned to Armonk for a special meeting."

"Everything go all right?"

"We're in business. It's full speed ahead."

"What's your timetable?" Charles' question was legitimate. If Digital Software was going to commit to provide software to IBM, timing was important. Charles would have to expand his company exponentially.

"Let's save the details until Monday," David said.

"Who are you bringing with you? We'll need to match up."

"Speedy, a lawyer, and a commercial relations specialist."

Charles whistled. "I've got a lawyer, but I don't know what a commercial relations specialist is." Charles looked at Mike who responded with a shrug.

"I know. I apologize," David spoke quickly. "I've got approval to go ahead, but that means I'm no longer a special project. I've got the bureaucracy on my back now."

"Don't let Armonk weigh you down," Charles said.

"Don't worry," David said. "Ernie, my division chief, and Crane both gave me my marching orders. Get this sucker into production. If the bureaucracy gets in the way, I'm supposed to give them a call. They'll handle it."

"A good position to be in."

"In some ways," David said. "Doesn't make you popular with the other big boys."

"Fuck em," Charles said.

The IBM team arrived in Seattle as promised and in force. Charles had asked Mike, Clapper and the company lawyer to join him, hoping to match numbers with numbers. He had even worn a suit, assuming it would make the visitors more comfortable. When the IBM team were escorted into his office, Charles broke out laughing. He and the Digital Software contingent wore suits, and the visitors were attired in slacks, Nikes, and open neck shirts.

"Let's change sides," David said. "You guys represent IBM, and we'll take Digital Software." Not all members of David's team laughed.

After coffee was served and minimal small talk dispensed with—both Charles and David were anxious to get to the details—the IBM lawyer presented a three-page security agreement.

"I assume I don't have to address the necessity for this document," the lawyer began.

"I understand," Charles said and accepted the proffered document. He scanned it quickly. In excruciating legal language the agreement, tailored for this meeting, took three pages to say that IBM intended to divulge confidential company information and the other participants agreed not to reveal said disclosures to third parties. At the same time, the document in obscure language stated that IBM did not wish to receive confidential information from Digital Software and would not consider itself obligated to treat any DS disclosures as privileged.

Charles handed the document to his lawyer who carefully examined it. "This is certainly one sided," the lawyer apparently decided to demonstrate the reason for his presence.

"That's all right," Charles said, holding out his hand for the document to be returned. "I'll sign it." He did.

The Digital Software lawyer said nothing. Charles could bear the responsibility for his decisions.

"Speedy, brief our hosts on the PC's specs," David said.

"PC?" Charles interrupted.

"Personal Computer," Speedy responded.

Speedy described the PC: Intel's sixteen bit 8088 chip, 32K ROM, 16K RAM, six slot open bus, options up to 256 K RAM, printer adapter, choice of color or monochrome, 8-inch disk drives, floating point processor.

"Digital Software is our first choice to supply languages, BASIC, COBOL, FORTRAN, PASCAL," David said.

"What operating system?" Charles asked. Languages were no problem.

"We're looking to you for that," David said. "Providing we can work out an equitable agreement on everything."

The sour look on IBM's commercial relations specialist indicated he felt David was moving too quickly, but he said nothing.

"We're flattered," Charles said, "but we don't have an operating system to offer."

"What about the OS/M you're offering with your SoftCard?" David demonstrated he had done his homework. "Can't you sublicense it to us?"

OS/M stood for Operating System/ Microcomputers. IMSAI had used OS/M in its computer line.

"No," Charles said flatly. "We don't have the rights to sublicense, and our version won't work on the 16 bit 8088 chip."

David looked to Speedy for advice. Speedy shrugged and said nothing, leaving David on his own.

"But I understand that O-Systems is working on a version that will," Charles continued, referring to Jeff Wayne's company in the valley.

"Can you find out for us?" David asked. "We need an operating system yesterday."

Charles answered by picking up the phone. After a little difficulty, he reached Wayne.

"Jeff, I have some important customers sitting in my office right now, and they need an operating system."

"Who are they?" Wayne asked bluntly.

Charles frowned. He and the ex-professor tolerated each other, but personality differences kept them from being friends or amiable colleagues. Wayne, some ten years old than Charles, tended to talk down to the college dropout.

Charles covered the receiver. "He wants to know who?"

"Tell him," David said, ignoring his lawyer's disapproving expression. He needed an operating system fast.

"IBM." Charles assumed that would grab Wayne's attention.

"Tell them to come on down."

Charles covered the receiver again. "When?" He asked David.

"Tomorrow," David said.

"How about tomorrow?" Charles asked Wayne.

"We'll work them in," Wayne said, then hung up.

"Tomorrow's fine," Charles said to David. "Don't be put off by Jeff's style. He's a little pedantic. Thinks he's still a professor."

Speedy and David then discussed the details of IBM's computer language requirements. Charles assured his visitors that DS could meet their every requirement on a timely basis and suggested they could work out the details of cost.

"Frankly we want your business," Charles said. "We'll tailor our products to your need and work out an equitable licensing arrangement. If we get your business, we'll worry about taking your money later."

Assured of a source to meet their language and software needs, David and his party departed that afternoon for California. At exactly nine o'clock sharp, David and his colleagues, all dressed in their standard IBM uniforms, took a taxi to the offices of O Systems, a handsome Victorian house. David rang the doorbell and waited. Nobody responded. David raised his hand to knock, assuming the doorbell was inoperative. It was then that he saw the small card. "Enter."

Cautiously, David turned the knob on the big oak door. The door opened, but David hesitated. He peered inside.

"It looks like someone's home," he said. "Are we sure we have the correct address?"

Speedy checked the address Charles had scribbled for them. "This is what he wrote."

David entered. The hallway had a huge mirror on the right and a fine Persian carpet on the floor. An umbrella stand stood on his left. "Doesn't look like any computer company I've ever seen," David said."

Speedy nudged him forward. The floor squeaked. David came to an open doorway on his right. Inside, he saw a young lady sitting behind a desk.

"May I help you gentlemen?" She asked.

David, who had been expecting to be confronted by an irate housewife ready to charge him for breaking and entering, sighed.

"We're from IBM," David said. "We have an appointment with Mr. Wayne."

"Sorry. You're out of luck. Jeff's out flying."

"Flying?"

"Yes. He's got a new toy," the secretary said.

"Can someone else help us?" David asked.

"His wife."

Irritated, David blustered: "We're not interested in coffee. We want to discuss operating systems."

"Then you've come to the right place," a voice on David's right said.

He turned to find a woman in a business suit standing in a doorway to what appeared to be the former owner's dining room. The woman was not bad looking, just austere. She appeared to be in her mid-thirties.

"You're the contingent from IBM," she said.

"Yes ma'am," David said. "I think we have an appointment. Charles Swift talked with Mr. Wayne yesterday."

"I know. Jeff asked me to handle it. I'm Cynthia Wayne. Please come in."

She stepped to the side, and David, followed by his team, filed past. The room was impressive, paneled, carpeted and huge. A large table flanked by six chairs on each side stood in the center of the room. Imposing, expensively framed oil paintings hung on the walls.

"This is our conference room," Mrs. Wayne said.

"A very unique headquarters," David said, meaning it.

"Nothing like Big Blue in Armonk," Mrs. Wayne laughed.

The comment disconcerted David. It was embarrassing to admit, but the IBM headquarters was renown for being a poured concrete block, functional but not pretty.

"Jeff has his own style." Mrs. Wayne said. "We didn't live like this at Stanford. Please be seated gentlemen and tell me how I can be of service."

David and Speedy selected chairs on one side of the highly polished table, and the lawyer and the commercial relations man sat on the other. David was still not quite sure what his function was, but Ernie had assured him that his presence would be routine. He had yet to say a word. Mrs. Wayne sat herself at the head of the table. She waited for David to begin. He nodded to the IBM lawyer.

"Mrs. Wayne, I am a legal advisor to IBM." He carefully placed his black leather briefcase on the table on opened it.

"I'm a lawyer too," Mrs. Wayne said. "Why are you here?"

"We have certain formalities that IBM regulations dictate," the lawyer said. He took a multipage document from his briefcase. "It's a simple nondisclosure agreement. As a lawyer, I am sure you understand our position."

"I sure do," Mrs. Wayne held out her hand for the document.

David did not relax. He did not like the woman's demeanor; he detected a latent hostility in her manner.

She read the document carefully. After reaching the end, she placed it on the table and pointed at the second paragraph. "This is an interesting clause."

Which one?" The IBM lawyer asked. Legal negotiations were his forte.

"This one where it says our discussions 'will not impair the right of either party to make, procure and market products or services now or in the future which may be competitive with those offered by the other.'"

"Does that trouble you?"

"What ever happened to proprietary information? I'm not about to reveal anything that you can then do on your own without reference to us. I won't sign this."

"Mrs. Wayne, if you don't sign this, company regulations prohibit us from proceeding further," the IBM lawyer warned.

"This is bull shit," Mrs. Wayne said.

"Mrs. Wayne," David interceded. "I agree with you. This agreement is one of the penalties one incurs when employed by a huge, bureaucratic organization. Such things are very irritating. However, I promise you the rewards can be substantial."

"For you, maybe, but not for me. I'm not signing this."

"Possibly we could talk with your husband," David ventured onto treacherous ground.

"Certainly you can talk with him, if you can find him. He might be back this afternoon, and then again he might not. Financial independence has made a very difficult man impossible."

David did not know how to respond to that. "We've come a long way," David began again.

"We didn't invite you. I assume that Big Blue declared war," she smiled, "without first mobilizing its resources. You need an operating system."

"Yes ma'am, and you have one."

"Then I guess you'll have to wait and talk with Jeff because I won't sign that."

"Are you authorized to negotiate O-Systems business?" David asked.

Mrs. Wayne smiled. "What do you think?"

"What do you recommend we do?" David asked.

"Go to lunch, then call back to see if Jeff has returned."

"Mrs. Wayne," David lost his patience. "We represent IBM ..."

"Good for you. Now if you excuse me, I have important business to discuss."

Mrs. Wayne rose from her chair and left the room.

David and Speedy exchanged glances, the lawyer closed his briefcase, and David led the way into the outer room. "Would you please call us a taxi?" David asked the secretary.

After a dispirited lunch, David returned to his motel room. He called O-Systems and asked for Mr. Wayne.

"Who may I say is calling please," the secretary asked politely.

"David Howard. I was in this morning."

"Just a minute Mr. Howard."

David heard the secretary call out: "They're on the phone, Doctor Wayne."

"Hello, this is Jeff Wayne."

"Good afternoon, Dr. Wayne, this is David Howard." David paused and waited for Wayne to apologize for missing their appointment.

"What can I do for you?" Wayne asked.

"We met with Mrs. Wayne this morning," David said.

"So I heard."

David waited. Wayne said nothing.

Again, David began to lose his patience. "Did she tell you what we wanted?"

"No. She said she refused to sign your ridiculous secrecy agreement."

David decided to avoid that subject for a while. "We're interested in your operating system."

"That doesn't surprise me. We have only one product, OS/M."

"You do have different versions?"

Silence.

"Mr. Wayne. IBM is interested in licensing your product if it meets our needs. We must speak frankly with you in order to determine if it is appropriate. This could be a most lucrative sale for O-Systems."

"I don't like the way IBM does business," Wayne replied. "I am not interested in having O-Systems devoured by your corporate sharks."

"That's not at all what we have in mind."

"Mr. Howard, I'm not interested in debating with you. On my wife's advice, I will not sign your insulting agreement."

"Then I'm afraid we have nothing to discuss," David said.

Wayne apparently concurred because he hung up.

David immediately called Charles in Seattle and described his frustrating conversations with Wayne and his wife. Charles commiserated and offered to phone Wayne and to try and sort out the problem. David, desperate, agreed. He paced the floor and waited for Charles to call back. Finally, he did.

"David, Charles, I talked with Jeff."

"What in the world is the problem?"

"Jeff is a difficult person. I warned you."

"I can live with that. What does he say?"

"He doesn't like your security agreement."

"I learned that much. What else?"

"He wouldn't talk straight with me either. As best as I can tell, they're afraid that IBM will swallow them up, take their product, change it, rewrite it, and use it as its own, all for a one time licensing fee."

"That's unprofessional," David protested.

"I understand, but reverse engineering and pirating software are not unknown in the computer world."

"But not by IBM."

"I know."

"You've got to help me," David said. "Charles, where else can I get an operating system that will work on my machine?"

"As I said, your sixteen bit decision causes problems."

"But it makes a more powerful and more efficient system."

"I know."

"I'm not an engineer," David said. "These technical matters are beyond me."

"Let me look around, and I'll get back to you," Charles said.

"I'm finished with the Waynes," David said. "I'm returning to Boca Raton. Please find me a system comparable to OS/M, and I'll deal with you for software and an operating system." David decided that by lumping the operating system with the software he could pressure Charles to come up with a solution.

"We'll be in touch, David."

In Seattle Charles summoned Clap and Mike to discuss IBM's problem.

"If we can come up with an operating system, the return will be huge," Charles said.

After they had discussed all the more obvious solutions and discounted them, Clap came up with an idea.

"Remember Tom Peters?" He asked Charles.

"Yeah, Lake Computers," Charles snapped his fingers. "He's the guy with the sixteen bit computer that needs our BASIC."

"He's been trying to get OS/M from O Systems," Clap said.

"And because they didn't have it ready and wouldn't tell him when and if they would, Peters cobbled together his own operating system. What does he call it?"

"I don't know."

"Will he sell it to us?"

"Let me talk with him," Mike said.

Mike and Clap talked with Tom Peters the next day and reported back to Charles.

"It's doable," Mike said.

"What do they have?" Charles started to pace. "What do they have?" He repeated himself.

"Peters decided to build his own clone of OS/M," Clap said. "He also claims to have made some improvements. I don't know, but he showed us a Lake Computers machine that seemed to work."

"Then get it for us," Charles ordered. "I'll negotiate with IBM."

"Wait until we get it for sure," Mike cautioned.

"You just worry about your end," Charles said. "I'll handle IBM."

Mike and Clap did not move.

"Do it now," Charles said. "IBM is running scared and someone else may scarf them up."

"I'm on my way," Mike said.

Charles winked at Clap as he picked up the phone.

"Hi, this is Charles Swift. May I speak with Mr. Howard."

"I'm sorry sir, but Mr. Howard is at lunch."

Charles checked his watch. Nine thirty, Seattle time."

"You mean he's on the tennis court," Charles laughed.

This provoked a nervous giggle on the other end.

"This is important," Charles said. "Can you reach him?"

"There's an extension out there. I can't guarantee that anyone will answer it."

"Please try," Charles said, "and I'll buy you a big bouquet the next time I'm in Boca Raton."

"That's not necessary sir. Please hang on and I'll try to transfer the call."

Charles was glad he had not promised dinner. The voice on the other end sounded terribly prissy.

The phone clicked, then to Charles surprise an extension began to ring. Charles waited, and the extension rang.

In Boca Raton, Terry was poised to serve. David waited at the baseline on the right. "Try to get it in this time, baby," David called over his shoulder.

On the other side of the court, Speedy waited for the serve. Speedy had proved to be a varsity class tennis player. David and Terry really had to work to beat Speedy and Jeff Madison, the Boca Raton center director. The telephone extension on the table to the right of the net began to ring. David was closest. Terry paused. "Do you want to answer that?" Terry asked.

"Go ahead and serve," David said.

"Forty thirty," Terry called.

She tossed the ball high in the air and swatted it hard, right into the net.

The phone continued to ring.

"Answer that," Terry ordered. "It's getting on my nerves."

David looked at the phone, back at Terry then at the phone again. "OK. Time," David called.

David jogged to the table, retrieved a towel to mop his face, and picked up the receiver. "OK, what is it?" David demanded. He assumed someone was merely relaying a message.

"David? This is Charles Swift," Charles said.

"Hi Charles."

"Who's ahead?"

"Don't ask."

"Come on David, let's get this over with," Terry called.

David held up one finger. "What's up Charles?"

"I've got good news for you. We've got you an operating system."

"You got O-Systems to come through?"

"No. I've got one for Digital Software which we can lease to you."

"Is it as good as OS/M?"

"Better," Charles said. Clap covered his ears with his hands. Charles was trying to sell a system he did not have and claiming it was better without ever having seen it.

"When can I see it?"

"What about day after tomorrow? I can bring it to Boca Raton."

"Come on David," Terry called.

"Play or surrender," Speedy called.

"Do it," David said. "See you Friday."

David tossed the towel on the table and jogged back on the court. "Charles has an operating system for us," Charles called to Speedy.

Speedy, who was waiting for Terry's serve, lowered his racquet. "No kidding," Speedy said.

Terry served the ball that whistled past Speedy.

"Game," Terry called.

"I wasn't ready," Speedy protested.

"Tough shit," Terry said. "Your serve," she bounced the second ball across the net.

In Seattle, Clap spoke: "Christ, Charles, we don't have it yet."

"Rely on Mike," Charles smiled.

Digital Software paid Lake Computers $75,000 for their LDOS operating system. Without informing Lake Computers that they had a customer for LDOS—they let Tom Peters assume they wanted to integrate it with DS BASIC—DS obtained full rights.

In Boca Raton on Friday morning, David and Speedy chatted in David's office while they waited for the Digital Software team to arrive for their final negotiations for the software and operating system that IBM needed.

"We're going to have to sign today, or at least during this visit, if we hope to meet our timetable," Speedy worried.

"Don't remind me," David said. He leaned back and lifted his feet to his desk. "Sometimes I wonder what we're doing here."

"Building a computer."

"Assembling a computer with someone else's components, but that is not what I meant. Here we are, working for a huge company, worrying ourselves to death for a measly thirty thousand a year."

"The pay's not bad," Speedy said. "Nobody is forcing you to do it."

"Look at these guys in Silicon Valley. Christ, look at Stanley Pitts and Dumbroski. They went from a garage to two hundred and fifty million each in three years. I went from twenty thousand to thirty, and I worked harder than they did."

"They earned the same salaries as we do until..."

"Until they went public and opened their personal gold mines."

"Do you want to go west and give it a try? I could play Dumbroski and you could be Pitts." Speedy smiled at the thought, particularly square David playing the flamboyant Pitts.

"Funny. If only IBM wasn't such a damned bureaucracy. We might as well be working for the government."

"Our pay is better and the retirement just as good."

"You got to take a chance. When you're young," David persisted. "Shit, we've got four lawyers sitting out there waiting to negotiate. I bet Charles shows up alone."

"Only two are lawyers."

"And we don't even know what the other two do."

"We'll steamroller Digital Software. They'll look like pancakes when we're finished."

"Don't kid yourself," David laughed. "Charles knows we're desperate for an operating system."

"You're the one who told him we had a deadline. What difference does it make, he's just a kid."

"Yeah," David lapsed into silence. Finally, he spoke again: "Terry's right. Enjoy life while you can."

"There's more to life than tennis."

At the Miami airport Charles and Mike debarked, retrieved their luggage, then headed for the men's room to change into dark blue suits for their meeting with IBM.

"Do suits really make a difference?" Charles asked as he buttoned his white shirt.

"Shit no," Mike said. "But IBM thinks they do."

"Then we ought to dress normally," Charles referred to the rumpled khakis and worn shirt he stuffed into his bag. "It might disarm them and give us leverage."

"We don't need leverage," Mike said. "The suits need us."

"This is IBM, man. We need them."

"I don't know why you're so impressed with Big Blue," Mike used the name disparagingly. "These guys are just like us. I know from Procter and Gamble. They're all bureaucrats."

"They're not like us," Charles said seriously. "We're smarter." Charles buttoned his collar and turned towards Mike. "How do I look? Like an IBMer?"

"Where's your tie?"

"Shit," Charles leaned over and searched in his open suitcase.

"Will you guys shit or get off the pot," a large man with red rimmed eyes demanded.

Charles stepped back and let the man have the washbasin. "I didn't bring a tie," Charles said. "Loan me one of yours."

"I only have one, and I'm wearing it," Mike said.

"You look fine to me kid," the smiling traveler pulled paper towels from the dispenser.

Charles checked his watch. "We're running late."

"I told you we should take an earlier flight," Mike complained.

"We'll grab a car and stop on the way," Charles said.

"You can probably get one in the airport," the traveler said helpfully.

"It'll cost double here," Charles said.

"I drive," Mike said. "I won't ride with you if you think we're late.

During the drive north, Mike continued to complain. "This is taking longer than you said it would."

"I've never been here. Why did you listen to me?"

"You're the company president."

"Partnerships don't have presidents, so I'm told."

"Then why do you call yourself president?"

"I like the sound of it," Charles said. "This president needs a tie."

"You look, I'll drive."

"We sound like my mother and father."

"Why don't you get a wife?" Mike, who recently married, asked.

"Don't have time. Besides, girls don't like me."

"You have a housekeeper."

"That's all she is."

"She's pretty."

"So am I. Stop, there's a shopping center," Charles pointed to his left.

Mike turned into the shopping center and stopped in front of a clothing store. "This do Mr. President?"

Charles got out and approached the door. "Oh Christ," he called to Mike who waited in the car. "It doesn't open until ten."

"We've got five minutes to wait. Don't sweat it," Mike said.

Charles began to pace back and forth in front of the door. "Our appointment's at ten."

"They can wait."

Finally the store opened, Charles rushed inside and grabbed the first tie he saw. He emerged pulling it around his raised shirt collar.

"Dig out," Charles said.

He turned the mirror and knotted the tie. "How's it look?"

"Great Mr. President. Orange is your color."

Charles looked down. He was wearing a bright orange tie with large purple dots. "I can always go into the clown business if we don't get this contract," Charles smiled.

"Now this is my kind of place," Mike said as they pulled into the Boca Raton Research Center parking lot.

Charles approved of the palm trees and the ocean that roared a short distance across the highway and narrow sand beach. "Wonder where the tennis courts are?"

"I'm looking forward to meeting this Terry," Mike said. "She sounds great on the phone.

"You're married, remember? Besides, she and David are sharing an apartment."

"Maybe David is like you and she's just his housekeeper," Mike needled his friend.

"Maybe," Charles said wryly.

Mike parked in a visitor's slot and led the way into the building.

"They don't look terribly busy," Charles said.

"They're all off thinking. This is a research center, remember."

"Good morning gentlemen. You're Digital Software," Terry said. She turned to Charles: "And you're Charles. You match your description despite your disguise."

Charles looked down at his orange tie and blue suit and laughed. "And you're Terry. Where are the tennis courts?"

"You play?" Terry's bright blue eyes sparkled.

"No. He does."

"Hi. I'm Mike," Mike said. "I play, but I'm not in your league." His eyes frankly examined the pretty brunette in the bright yellow cotton dress.

"Relax, Buster," Terry smiled. "You're late," she turned back to Charles. "You should have taken an earlier flight."

"It's my fault," Charles admitted. He did not say that they had needed every minute waiting for the late changes to their proposal to be typed.

"Please follow me gentlemen," Terry turned and started down the hall.

"With pleasure," Mike said staring at Terry's exaggerated hip swing and long muscular legs.

Charles elbowed Mike then followed Terry, staring at the same sight.

"She's a secret IBM weapon," Mike whispered. "Dispatched to divert us."

"She's succeeding," Charles said.

The meeting proved to be satisfactory for both sides, primarily because they both needed it to succeed. In very short order, Charles overwhelmed the skepticism caused by his unusual appearance and soon had the IBM engineers listening as equals. David was surprised when Charles refused to sell the operating system outright. The other component manufacturers, overwhelmed by the opportunity to sell their products to the mighty IBM, had reacted like retailers who allowed the customer set the price. The volume potential was so high, that most accepted IBM's initial offer.

Charles refused to accept the flat fee price quoted by the IBM customer relations expert, one of the company's experienced negotiators. After much haggling, IBM gave Digital Software an advance of $100,000 against a royalty of $50 per copy for its language compilers PASCAL, COBOL, FORTRAN and BASIC. The operating system required more negotiation. Finally, IBM agreed to an advance of $400,000 for LDOS and interpreted BASIC against a royalty of ten dollars per copy. Additionally, IBM agreed to pay Digital Software a flat fee of $200,000 for adapting all its software so it would work on the IBM Personal Computer. IBM was expressly prohibited from licensing Digital's programs to third parties but Digital could. To offset the concessions to Digital, IBM imposed a tight schedule complete with penalties.

The negotiations finally concluded in time for Charles and Mike to catch their six o'clock flight from Miami.

As they were driving back to Miami, Mike summarized their day: "Seven hundred thousand. Not bad for a few hours work."

Charles laughed. "The seven hundred thousand is petty cash. Wait for the royalties to start."

"You think they'll be big?"

"We're on the way," Charles shouted and slapped the dashboard. "Remember IBM has decided on an open source policy. Others are free to clone the PC. And they will all come to us for their operating systems and software."

Mike had been concentrating exclusively on the monetary figures and the lawyer nit-picking. Charles, the poker player, had not revealed his thoughts on the royalty arrangement, not even to Mike.

"Now we have to work on making our stuff and the PC industry standards," Charles said. "Look out Astron Computers," he shouted, raising his fist in a victory salute.

"Astron Computers is a customer," Mike cautioned.

"They are, aren't they?" Charles smiled. "Isn't power great?"

Chapter 22

When Stanley returned from the PARC demonstration, he adopted the Stella as his project much to the dismay of Tom Williamson who held the title of Project Director. The Stella had been his baby, and neither Stanley nor Willard the company president, nor the other division chiefs had paid much attention. Their focus had been on the Astron 3. Harold Dumbroski, Astron's resident genius who had designed Astron 1 and Astron 2, surprised them all with his disinterest in Astron 3. Willard Temple, with the Board's approval in the form of a go ahead from Stuart Miller, assigned the project to Astron engineers. Dum, still a primary stockholder, insisted on only one thing, that the Astron 3 be backward compatible. This made sense to Miller and Temple because the Astron 2 continued to be the company's primary money machine. It was still selling 78,000 units a year.

Stanley, who from birth had always indulged himself, had developed a severe case of tunnel vision. He would focus exclusively on the one project that interested him and dismiss everything else as pure, unadulterated shit. This proved to be a mixed blessing for the employees of Astron Computers. Those whose work was thusly dismissed, worked in relative peace, and this included the Astron 3 engineers. The Astron Management Committee decreed the new machine would be backward compatible, that is be able to run Astron 2 software. They also wanted a more powerful operating system with additional memory and enhanced graphics. The Committee decided the Astron 3 must appeal to the professional business class. The marketing people pressed for an early release and it was demonstrated, prematurely in the view of the engineers, at a National Computer Conference in Anaheim.

Stanley by this time was so involved with the Stella that he did not attend the grand introduction. The industry greeted 3 with acclaim and the first shipments virtually leaped off the shelves of some 3,000 distributors. Willard, sensing victory, doubled their advertising budget, then disaster struck. Astron 3 had inexplicable bugs and the operating system crashed. The engineering staff had difficulty trouble shooting them, and irate customers, vociferously denouncing inadequate support, returned 3s in large numbers. Panic set in and a smiling

Stanley did not help matters. The engineers identified the problem quite by accident. A trouble-shooting engineer, facing a frozen operating system, did what some of the dissatisfied users might have. He lifted the obdurate machine one-inch in the air and dropped it. To his surprise, 3 restarted. Thus warned, he identified a faulty connector. Amid protestations of inept testing, 3 re-entered the market.

To assist Tom Williamson, Stanley hired Gene Vance to assist with Stella. Gene had briefed the Astron contingent during their visit to PARC. Vance added the mouse to supplement Stella's keyboard, and Tom struggled with the graphical user interface and its ubiquitous icons. Stanley, leaping ahead of his engineers, decided that Stella would target large businesses, promising that Stella would increase productivity. They gave Stella two floppy drives, an internal 5MB hard drive, a 68000 Processor, and an attached twelve-inch monitor. Every addition created its own technical challenges, but the graphical user interface posed the most of all. Stubbornly, Stanley insisted on progress, refusing to listen as his engineers tallied their obstacles. Determined to build the best microprocessor ever, Stanley insisted that every advance that he read about, heard about, or imagined, be incorporated.

Costs for the Stella rose. The engineers complained to Willard Temple, the company president.

"You all know Stanley. Live with him. He's one of our founders," was Temple's stock response.

"But he's not in our chain of command," Tom Williamson protested. "He thinks he's project manager. One day he insists on a change and the next day denies he ever said such a thing. My project engineers are ready to rebel."

"Stanley is our major stockholder," Willard countered.

"Give him something to do. Keep him out of our shop," Williamson insisted.

Willard equivocated. He was tired of his shouting matches with Stanley. Stanley refused to discuss anything. There was only one way, Stanley's way.

While Willard wrestled with the Stanley problem, real disaster struck. Harold Dumbroski, now enjoying the beneficial fruits of stock appreciation engendered by the public offering, crashed his single engine airplane while practicing touch and go landings in the company of his fiancee. Both survived, but Dum emerged with face cuts and a faulty memory. He could not remember flying the airplane or the crash. Details of his precautionary stay in the hospital were hazy. It took a month for Dum to recover, and when he did, he decided he had no more interest in Astron Computers.

Although Dum had been only marginally involved in the new projects, the loss of his engineering genius was worrisome. When asked to persuade Dum to return, Stanley declined. "We don't need him," Stanley declared. "Dum is outdated. He doesn't understand gui." Graphical user interface, pronounced gooey, was Stanley's current preoccupation.

The Stella team continued to complain. Willard's hand was finally forced when three top flight engineers submitted their resignation. Willard summoned Stanley to his office.

"Stanley, it's got to stop," Willard said.

"What has to stop?" Stanley played innocent.

"Stuart agrees," Willard said.

"Stuart agrees what?"

"That you are to terminate all contact with the Stella project."

"Bull shit."

"No Bull shit. Not this time Stanley. The entire project team has threatened to quit because of you."

"Let them. They're all incompetents."

"No they're not, Stanley. They're the best we have. You've driven the estimated sales price for the Stella above $10,000," Willard let his voice rise.

"So what? It's worth it," Stanley shouted back.

"It may be," Willard controlled his temper. "But nobody will buy it."

"Sure they will. Big business has lots of money."

"Not for microprocessors," Willard insisted.

"Bull shit," Stanley shouted.

Willard saw Stanley glance through the open door to make sure he had witnesses to his defiance. Willard got up and closed the door.

"Stanley, forget the Stella. You're out of there. If you don't find something else to occupy your time, I'll have security throw you out and keep you out."

"Bull shit," Stanley shouted.

"If you even approach the Stella shop, you're out of here."

"I'm a major stockholder," Stanley said.

"Then get the Board to fire me," Willard called his bluff. Stanley knew he did not have the votes. When Astron went public, Stanley and Dumbroski had gotten rich overnight, but they had lost their leverage on the Board. "Dum won't support you. I've already talked with him. Between Dum, Stuart and the public members, your stock doesn't mean a thing."

"This sucks," Stanley said.

"Suck it may but keep away from the Stella project. Take a vacation. Spend some of your millions. Travel around the world."

"I'm only twenty-six," Stanley said. "I'm not ready to retire."

Willard did not respond. He was tempted to point out that Stanley had never held a position since Astron Computers had incorporated, but he did not. He had won, and he knew it. There was no need to rub Stanley's nose in the dirt.

Stanley stamped from the room, angrily slamming the door behind him. He marched through the outer office, turned, and shouted: "The Stella's a piece of shit anyway."

Stanley submitted to Willard's challenge. He avoided the Stella workshops for a week. Then, he began to stroll through once a day, pause and pretend to study an engineer's work, mutter "what shit" then proceed on to the Asteroid project which he had adopted in the Stella's place.

Having temporarily bested Stanley for once, Willard decided to address other company problems. Willard believed that success had made Astron Computers complacent; stock options gave too many of the older employees financial independence and this impacted on their work attitudes. After giving the matter much thought, Willard concluded he needed to reassert discipline. With Stuart Miller's tacit concurrence, he opted to shake things up by jettisoning some of the company's deadweight. A short time after Dum's untoward accident, really before the staff had time to acclimate, Temple terminated forty employees. He blamed the engineers for 3's shortcomings and fired the project manager and two engineers, including the hapless fellow who had discovered the faulty connector design.

Morale plummeted. Worried about who might be next, key employees organized protests that demanded that those who had been let go be rehired. Willard Temple shouldered the deserved blame. Neither Stanley nor Stuart Miller came to his defense. A dispirited Temple decided the battle was no longer worth fighting and submitted a bitter resignation letter in which he denounced "hypocrisy, yes-men, foolish plans, a cover your ass attitude and empire builders." Few shed tears over Temple's departure, and many agreed that his parting remarks were aimed aptly at Stanley. Willard Temple took his stock millions and departed for Europe with one parting shot: "A company's quality of life cannot be set by committee."

Despite Temple's many contributions to Astron's success, his presence had given the company a divided leadership, and most had long previously concluded that either Temple or Stanley would have to go. Since Stanley was now the major stockholder, few doubted that he would emerge victorious. Stanley did not step forward, however, to assume the position of company president. Stuart Miller, the retired venture capitalist, took charge as acting president. Harold Dumbroski refused to involve himself in "business matters." Stanley, demonstrating his support for this move, consented to become Chairman of the Board of Astron Computers at the ripe age of twenty-six.

Stanley's first act as Chairman of the Board was to suggest that the Stella Project be renamed the Mars Project. Everybody knew why. The majority of the Board, prompted by the Stella Project team, resisted, and Stanley graciously relented, apparently not prepared to make a name change his first major issue. Despite his lofty new position, Stanley continued to manage the Asteroid project.

Stanley moved the Asteroid Group into its own building and promptly hoisted a pirate flag as a symbolic warning to the other divisions. He put a piano in the lobby, freshly squeezed orange juice in conveniently located coolers, and provided first class tickets for flights of longer than two hours. Stanley dubbed his Asteroid group the elite and described the other divisions as bozos.

When IBM formally announced its intention to enter the personal computer market, Astron Computers responded with a full-page advertisement in The Wall Street Journal. Stanley personally drafted the odd banner headline: "Welcome IBM, Seriously." While others in the industry trembled at the threat posed by IBM, Stanley with Astron Computers eighty percent share of the market believed himself invulnerable. The accompanying text read: "Welcome to the most exciting and important marketplace since the computer revolution began thirty-five years ago. We look forward to responsible competition in the massive effort to distribute this American technology to the world."

Acting President Stuart Miller visited Stanley in his pirate lair the morning the Journal challenge appeared in print.

"Do you think this was a good idea Stanley?" Stuart asked.

"Obviously. I wrote it."

"IBM's a sleeping giant. Why torment it?"

"Why not? Asteroid is light years ahead of the PC. Look at their operating system. It's from the dark ages."

"It's better than the Astron 2."

"The Astron 2 is a piece of shit."

"I wish you would stop saying that Stanley. Without the Astron 2, we would all be bankrupt, and you would not be driving that Rolls of yours."

"It's a Harley," Stanley said.

"Do you have to park it in the lobby?"

"Why not?"

"Just remember. The Astron 2 is paying the freight."

"It won't be for long. The Asteroid will." Stanley smiled.

"You said that about Stella."

"I was wrong. Stella is a piece of shit. If you're so big on the Astron 2 why don't you get our resident genius to bring it into the modern world."

"You know Dum had an accident."

"He's recovered."

"He's still suffering head injury problems."

"Wrong. He's at the university working on his degree."

Stanley interpreted Stuart's abrupt departure as capitulation. Later, under prompting by the media, Stuart tried to downplay the fact of IBM's entry to the market. The press described Stuart's reaction as condescending.

Internal dissension had been one of the prime factors leading to Willard Temple's departure. Stuart took on the burden of company leadership in order to bring Astron Computers back together. Temple had introduced the division management system to establish clear lines of command. He made the mistake of leaving Stanley out of the hierarchy, and Stanley took full advantage of his position. As a reaction to Stanley's interference, the division chiefs converted their commands into baronies that were soon feuding with all other divisions. Stuart tried to channel Stanley's disruptive tendencies by giving him responsibility, the Asteroid project, and status, the Chairmanship of the Board. Stuart, like Temple failed, and once again the cause was Stanley. A division flying the skull and crossbones led by a chief who denounced all others as bozos forced the others to react in kind.

Stuart decided the only solution was to hire an experienced, well-tempered and strong willed chief executive. He paid sixty thousand dollars to an executive search firm. The result was heartburn. The prime candidates had all heard of Stanley Pitt's maverick reputation. Not one was willing to take on a company whose Chairman of the Board and principal stockholder insisted on leading a favored division of prima donnas and undermining the command structure.

Bureaucratic lethargy fell on Astron Computers, and Stanley loved it. He had his money, and now he was having his way. Stuart Miller, absorbed by his quest for a new president, allowed the corporation to manage itself. The feuding division chiefs, freed from their former chief's arbitrary management, indulged themselves. Never were the barons more sovereign, and Stanley was the most royal of them all.

Stanley was playing with a mockup of Asteroid's graphical user interface when the grapevine rattled.

"Why didn't you warn us Stanley?" Dick Ruffin, Stanley's ostensible number two, asked from the doorway to Stanley's office, commonly known as "The Playpen."

"Why should I?" Stanley challenged, having no idea what Ruffin was talking about.

"When is he coming?" Ruffin ignored Stanley's sally.

Stanley concluded that Ruffin could only be referring to a new president. "We'll tell you when you need to know. Don't let that hit you in the ass on the way out," Stanley glanced at the door.

"Thanks a lot," Ruffin complained and departed.

Stanley immediately closed the door and picked up the phone. "Give me Stuart," Stanley gruffly ordered the front office secretary who answered the phone.

"Yes Stanley," Stuart came on line.

"What the hell's up?" Stanley demanded.

"I don't know what you're talking about," Stuart said.

"Everybody wants to know when he gets here. You can't hire anybody without my permission," Stanley shouted.

"Nobody's hiring anybody," Stuart tried to calm Stanley down.

"Then why does everyone think you have?" Stanley shouted.

"Some of us think we have a good candidate," Stuart admitted.

"What us?"

"Some of our board members."

"Those damn outsiders," Stanley said.

"Experienced businessmen who know business," Stuart said. He was growing tired of Stanley's adolescent outbursts.

"Trying to mount a coup against me?" Stanley asked. "I'll quit first."

"Calm down, Stanley, I was just getting ready to call you. We want you to go to New York and meet the candidate."

"I'm too busy."

"All right. I'll schedule a Board meeting, and we can vote on the candidate."

"Remember what a mess your last selection created." Stanley assumed Stuart was bluffing, but one never knew. Since Dum had turned his back on Stanley, he had lost the ability to threaten to override Board decisions.

"We want everybody on board this time," Stuart said. "That's why we want you to go to New York."

"Who are we talking about?" Stanley retreated slightly.

"Lawrence Starkey."

"Who the hell is Lawrence Starkey?"

"President of Pepsi Cola."

"What does he know about computers?" Stanley had never heard of Starkey.

"He's got a MBA from Wharton." As soon as he had spoken, Stuart knew he had made a mistake. Stanley was sensitive about his lack of a degree and always tried to cover it with bluster.

"Another educated idiot."

"He hasn't done badly. Pepsi is big time." Stuart did not tell Stanley that Starkey's former father-in-law was chairman of the Pepsi board.

"We don't need another bureaucrat."

"Stanley," Stuart decided to quit humoring his problem. "Go to New York and talk with him."

"Why New York?"

"Neutral ground."

"When?"

"He'll meet you at the Taft for lunch on Saturday."

"I'm busy," Stanley said. Stanley had planned on taking the Harley for a spin in the country.

"Make time or I'll schedule a board meeting for Monday."

"OK, but don't expect too much," Stanley capitulated. Stuart could be stubborn when pressed.

Having decided that he was going to enjoy this round, Stanley caught the red eye to New York on Friday evening. There was no way he was going to accept a cola manufacturer as president of Astron Computers. It would be fun, however, to show the over educated business major how little he knew about the real world. Stanley, eschewing the Taft, a businessman's hotel, checked into the Astoria. The grand old lady of hoteldom had recently been remodeled, and Stanley thought she represented style. He shaved, showered and instructed the operator to wake him at noon. Stanley hated airplanes, never slept in flight, and wanted to nap before the meeting in order to be at his sharpest.

Stanley took a cab to the Taft. He pushed through the crowded lobby and went to the businessman's grill, arriving twenty minutes late.

"I'm looking for a guy named Starkey," Stanley smiled at the hostess.

She took in Stanley's informal dress—blue blazer, open neck dress shirt, tan slacks and garish green striped Nikes. The other patrons to a man wore austere, dark business suits, maroon power ties, white shirts and three hundred-dollar shoes. That is, almost to a man.

"Yes Mr. Pitts. Mr. Starkey is waiting," the hostess said. Starkey, the exception to the "almost to a man" observation, had slipped her fifty dollars to watch out for his guest.

"Lead the way baby," Stanley said. The use of his name had surprised him, but he was determined not to show it.

The hostess, carrying a large menu, led Stanley past tables occupied by over dressed males who Stanley assumed represented New York's power elite. He wondered what they would say if they knew he probably had more money than all of them put together. In fact, the businessmen conferring at the Taft were middle level and together they could approximate Stanley's millions. The hostess led Stanley to what had to be the choicest table in the large room. It faced the New York skyline.

"Welcome Mr. Pitts," Stanley's slender host rose to greet him.

Stanley smiled, silently approving Starkey's attire—tieless, expensive Harris sportscoat, slacks. Stanley could not see his shoes. Stanley liked the fact that Starkey had obviously dressed to impress Stanley. Starkey appeared to be about thirty-six, ten years older than Stanley.

"Mr. Starkey," Stanley nodded.

Starkey waited politely for Stanley to be seated before sitting himself.

"The waitress will be with you shortly." The hostess handed Stanley the menu. She did not linger despite the large advance tip.

"Snippy bitch," Stanley said.

"Agreeable flight?" Starkey ignored the comment.

"Red eye. Hate flying," Stanley replied, deliberately keeping his reply short, putting the burden of conversation on Starkey.

Starkey nodded. He could have said that he had flown up from Atlanta on his private corporate jet, but he did not. Stanley was already living up to his reputation as a rude son of a bitch. Stuart Miller had warned him. "Don't let Stanley put you off. If we have to, we'll fire him." Miller had already committed Astron Computers: "The job is yours if you want it." Starkey had opted to meet Stanley Pitts first, wanting to see if he lived up to his advance billing. Starkey would prefer to start his sojourn at Astron Computers without a bloodletting if he could. He had suggested this meeting to Miller under the guise of Pitts interviewing him.

The waitress appeared as soon as they were seated. "Drinks gentlemen?"

Stanley, assuming the role of host, looked at Starkey.

"Martini," Starkey said.

"Tomato juice," Stanley said. "And bring the Tobasco Sauce."

"The Italian kind," Starkey elaborated with a smile. "No vodka or gin."

"Gottcha," the waitress said.

"Gottcha," Stanley parroted as the waitress departed.

Starkey assumed he was denigrating the waitress' lack of style. "New Yorkers," he agreed.

"Very well, Mr. Starkey, tell me about yourself," Stanley directed.

"I agree, Mr. Pitts, no small talk, but please call me Lawrence."

"Not Larry?"

"No." Starkey decided to show a little steel. "I assume you've read my resume."

"No."

"Or the <u>Time</u> piece." Starkey smiled at Stanley's surprised reaction. The newsmagazine had featured Starkey on the cover of a recent issue.

"Of course," Stanley bluffed. Stuart had not mentioned <u>Time</u> magazine. Stanley consulted the media only rarely and then only when he himself was featured.

"Do you have any technical background?"

"I'm sorry to say, no."

"What makes you think you can manage Astron Computers?"

"I don't think. I can."

"Oh?"

"I know I can. I'm an MBA."

Stanley had wondered how long it would take Starkey to bring that up.

"And I have managed a big business. I know marketing. I don't have to build computers. I can hire engineers to do that."

"How will you know what they're talking about?"

"I'm a fast learner."

Stanley despite his predilections liked Starkey's confidence.

"I understand you are not an engineer yourself."

"I started the business."

"You and Harold Dumbroski."

"Yes. Harold built our first two machines, and I did the rest."

"So I understand. And Mr. Dumbroski is no longer actively involved?"

"He had an airplane accident." Stanley suddenly felt like he was the one being interviewed. "So you can make sugar water," Stanley countered. "How does that equip you to make computers?"

"Would you gentlemen like to order?" The waitress returned and waited, pad at the ready.

"Stanley?" This time Starkey seized the roll of host.

"A salad for me."

"Any preference?"

"Nicoise." Stanley decided to show his sophistication. He would have preferred Caesar.

"Very good choice, sir." The waitress turned to Starkey.

"Caesar."

"Will that be all?" The waitress addressed both men.

"Yes," Stanley said.

"I think I'll have a small cup of French Onion," Starkey said. "Are you sure you'll have nothing first Stanley?"

"Bring me soup too," Stanley ordered. "I'm a vegetarian."

"Anything to drink?" The hostess persisted.

"So I heard," Starkey said, unimpressed. "Bring me a glass of white wine. House."

"Stanley?"

"Why don't we get a bottle of good wine?" Stanley said.

"Not for me. My palette can't tell the difference."

Stanley shrugged. He was indifferent to wine. "Bring me iced tea."

The waitress turned away.

"Unsweetened," Stanley called after her. "And brew it. None of that powdered stuff."

"Where were we?" Starkey asked.

"I was asking why you thought you could run Astron Computers?" Stanley said, not quite accurately.

"Yes, well, I know I can manage a big company. I think I have proven that, and marketing is kind of a specialty of mine."

"The Pepsi Generation."

"Yes. That went over very well."

"Do you think slogans will sell computers?"

"Why should computers be any different from everything else. Yes, good marketing will expand your reach."

"We already have eighty percent of the market."

"For now."

"You're afraid of IBM?" Stanley let a little sneer creep into his tone.

"It would be unwise to underestimate a proven adversary."

"We're not making mainframes. We're talking microprocessors."

"So I understand. Are you interested in selling to the business community?"

"Of course," Stanley said.

"I can address that subject with personal knowledge. We're dependent on IBM for our computer support."

"As I said, for mainframes."

"Will your software be compatible with IBM's Personal Computer?"

"Certainly not. I decided that Astron Computers will make and sell its own software. But that's not important. When business finds out that our new computer, the Asteroid, will do everything the big mainframes can do at a fraction of the cost, they will buy microprocessors."

"Companies like mine are familiar with IBM software. Retraining our employees would be a major cost and problem. People don't like to change you know."

"What are you implying?"

"That we will buy from the company we have learned to rely on for something important like computer support. Particularly when it deals with a complicated but vital subject like computing that we do not understand."

"Our Asteroid will be years ahead of the Personal Computer."

"And how long do you think it will take for IBM engineers to overtake you, technologically speaking. They are a computer company."

Stanley paused. "If they reverse engineer the Asteroid, they could do it in a year, but it will take them longer to go into production. By then, we will have moved on to a new product. Computing power doubles every year."

"It gives us something to think about, Stanley. If you hope to move into the business market, it will not be easy. Astron Computers could benefit from someone who knows marketing."

"And you do?"

"Yes."

Stanley did not say it, but he liked the direction the conversation had taken. If Starkey were president and concentrated on marketing and business management, Stanley could do what he wanted on the technical side.

"What are your views on company organization?"

"I understand from Stuart that your division organization has been a disaster with each division fighting the other."

Stanley had not expected so direct an answer or challenge.

305

"I suppose you've also heard that some think I am a problem."

"Leadership is more important than management," Starkey evaded Stanley's comment. He was not prepared to address that issue. He had not decided whether to keep Stanley on or not. "I believe in decentralization. Delegate authority, give people the resources they need, money and manpower, then monitor the hell out of them to make sure they produce."

Stanley liked the part about delegation but was not sure about the rest. Stuart was always telling him that he had yet to learn how to manage people. Stanley was willing to learn, at least he always said he was. Maybe he could learn from Starkey then toss him out on his ass like Willard Temple.

The waitress served their soup.

Stanley tasted his first. "This is not bad."

"A specialty here," Starkey said.

"Tell me, Lawrence," Stanley said as both men spooned their soup. "What do you think about philosophy?"

"I'm all for it," Starkey laughed. "Man individually and in groups needs a philosophy to guide and regulate their actions."

"To manage their behavior? Like religion, you mean?"

"Religion is just one philosophy, a code of behavior."

"Religion is much more than that. It tells us where we come from, where we're going, as well as how to regulate our behavior."

"Are you religious Stanley?"

"I've studied eastern religions, but I don't go to church if that's what you mean."

"You don't let a religious code dictate how you behave?"

"Hell no. If anything, I believe in Emerson and Thoreau's philosophy of the individual." Stanley decided to throw in Emerson and Thoreau to show that he wasn't illiterate despite his lack of academic credentials.

"That explains a lot," Starkey laughed.

"What do you mean?"

"You have a reputation for highly individualist behavior."

"You mean I'm a pain in the ass."

"To put it bluntly, that's what people say."

"Do I live up to my reputation?"

"Yes."

Stanley smiled. "Ever spent any time in a commune?" Stanley asked.

"No," Starkey replied. "Following graduation, I attended Wharton, then went to work. I was married and didn't have time for communes."

"Married people live in communes," Stanley said defensively. "After a while, you joined Pepsi. Did your father-in-law insist?" Stanley revealed that he knew of Starkey's relationship to the then president of Pepsi. By doing so, he intended to denigrate Starkey's success by hinting at nepotism.

"My ex-father-in-law. I was divorced by then."

"What was your first job?"

"They sent me to Brazil."

"Started at the bottom?" Stanley laughed. He had started in his father's garage.

"I sold potato chips."

"That's how you learned marketing."

"That's where I started. You lived in a commune?"

"Yes."

"Enjoy it?"

"The sixties thing," Stanley smiled. He was beginning to like Starkey. He and Willard Temple had never had this kind of conversation.

The waitress served their salads.

"I thought you were a vegetarian," Starkey said.

"I am."

"Then what about the tuna?" Starkey pointed his fork at Stanley's salad.

"Never thought about it," Stanley said honestly.

"True vegetarians don't eat fish do they?"

"Some do, some don't. It's not a religious thing. I guess I can eat what I want."

"Your choice."

"I just don't eat much meat."

"Poultry and fish?"

"If I have a choice."

"I have friends who only eat white meat: pork, chicken, fish that sort of thing."

"Guess they're not vegetarians either. We all don't have to be consistent."

"That's really Emerson," Starkey said.

"What do you mean?"

"The foolish consistencies are the hobgoblins of little minds thing."

Stanley who had never read Emerson or Thoreau made a mental note to try and remember that quote.

As soon as Stanley and Starkey had parted, leaving New York to the natives, they returned to their respective home towns where each, unbeknownst to the other, telephoned Stuart Miller.

Stanley: "I can work with Starkey, hire him."

Starkey: "I can work with Stanley, don't fire him."

Chapter 23

David Howard initialed "OK" on a request for authority to spend another million dollars of Big Blue's green cash. An intimidating act, approving the expenditure of company funds, had become routine. For the past three months, David had been inundated with paper. He had given up his daily tennis, and his apartment mate, Terry, had enlisted the tennis partnership of a series of newcomers, all more than happy to accommodate the vivacious receptionist. David thought about complaining but decided that it would gain him nothing. He now had two hundred IBM employees under his authority, and more streamed in at the rate of five a day. David even managed the unthinkable: he hired locally without reference to Armonk administrative control. The Management Committee in its supreme wisdom had selected October 1st as the day they would officially launch the PC.

David protested. "There's no way we can be ready."

"Of course you can," Armonk replied. "You are merely assembling."

David argued that assembly was only a small part of the problem.

"What do you need? More manpower?" Armonk asked.

David tried to explain that he could not absorb the employees he already had, but the bureaucrats in Armonk, driven by the media and their superiors, did not listen. David thought of the good old days of the secret project when he had been protected from the bureaucracy by Ernie. Hierarchically, the Personal Computer Division rested under the General Products Division, but now that it officially existed, it was subject to IBM regulatory controls.

David studied the stacks of paper on his right. "I should finish one more," David told himself. He listened to the building. It was silent; he had sent his secretary home an hour previously. He checked his watch. Eight PM. David rubbed his burning eyes, then reached for another column of paper, deliberately selecting the smaller one. He picked up his ballpoint pen with his right hand and the top document with the left. He started to read but stopped. The words appeared to jump around on the page.

"Fuck it," David swore and threw both the document and pen to the cluttered desktop.

He grabbed his jacket and headed for the door where he paused to turn out the lights. Until the past two months, he had never encountered this chore. He had always been the first to leave the office, and his secretary took care of the lights.

David flipped the switch, and the telephone rang.

"Shit." David debated whether to ignore it, but guilt won. Terry might have an urgent housekeeping errand that only he could perform—something like buying a cantaloupe for breakfast. David had developed a preference for fresh pineapple and green cantaloupe with toast for breakfast. Terry cheerfully accused him of being pregnant, a charge he vigorously denied.

David reached across the desk and pushed the button on his speakerphone. "PC," he announced noncommittally.

"David," Ernie said.

"Ernie," David said. "I was just leaving."

"Working half days?"

"Funny."

"Are you standing or sitting?" Ernie asked.

'Oh shit," David said, dropping into his chair. "Sitting, now."

"Good. I've got news."

"Good or bad?"

"You're sure you are sitting?"

"Yes sir."

"Crane announced his retirement today at a special meeting of the Management Committee."

"That's bad," David said. Crane had been their secret weapon. His admonition to David that he should immediately inform him if the bureaucracy impeded the development of the PC had been a great solace. David had never exercised the option, but at times he had been sorely tempted. Just knowing it was there helped.

"Name his successor," Ernie said.

Ernie seemed to be in a good mood, so David made his first wish his question. "You?"

"Don't be ridiculous," Ernie laughed.

David acknowledged this was too much to hope for. Ernie was an outstanding senior director. He had, however, a detachment that David was sure his peers resented. Ernie did not take senior politics seriously. Then, there was his paternal attitude towards his subordinates. Ernie was a senior manager who believed that loyalty ran two ways, down as well as up.

"Try again."

David did not have a clue. He had been away too long from the Armonk gossip vines to even hazard a guess.

"Someone from outside?"

"That's not IBM's way."

"If not you, then it has to be me," David hazarded a joke.

"Get serious. You are involved."

David knew that any Armonk decisions stretching down to his modest level had to be bad news. He was not surprised. He had expected the worst as soon as word of the PC broke. David was too junior for the position he now held. No one cared who led a one-man special project office in the backwater that was Boca Raton, but now the Personal Computer Division was a main line unit and there were, in David's uninformed estimation, probably ten thousand IBM managers senior to David who would kill for the job.

"When do I leave?" David asked. He had personal matters to clear up, his relationship with Terry for example. David liked Terry too much to abandon her as he had Margaret. Terry had too many admiring tennis partners.

"We'll get to that," Ernie said. "Important matters first."

"Who is replacing Crane?" David realized he had to play the game Ernie's way.

"Good man. Now you're asking the right question. Have you heard of a gentleman of the name John Parsons?"

"Oh no," David almost moaned. Parsons, the Director of the Sales Division, had taken a personal dislike to David from the moment they had met at second base when David had crushed him.

"That's right. Your old softball buddy."

"He's fired me," David said.

"No but he does remember that you promised to handle all sales of the Personal Computer from the General Products Division."

"I didn't say that. I think you did."

"Are you sure?"

"No." David with honesty could not remember what he had promised. "Parsons didn't want to have the Sales Division to handle something so petty as a personal computer. I remember that."

"And?"

"And someone said the General Products Division would deal with PC sales."

"Right."

"So?"

"So, the situation has changed now that Parsons is Chief Executive Officer."

"When does he take over?"

"He has. Crane doesn't believe in overlaps at the senior level. He said something momentous like: 'IBM can have only one president at a time.'"

"And?" David knew the bad news hammer was about to fall.

"Parsons, in his first act as CEO, declared that it was intolerable that the General Products Division direct the marketing of a major IBM product."

"He called the PC a major IBM product?"

"That ocean air hasn't affected your hearing."

"I can live with the Sales Division handling our marketing," David said. "I'm overloaded as it is."

"I'm sure Mr. Parsons will be glad to hear that you concur."

"He is going to fire me."

"No, but he has reassigned you."

"I'm sure I'll love New Guinea."

"You have to learn to ask the right questions."

"Ernie. Who is the new Director of Sales?"

"You're talking to him."

"That's great news." As Parsons had demonstrated, the Director of Sales position was a stepping stone to the presidency. "Congratulations."

David suddenly realized he was now at the mercy of the new Director of the General Products Divisions. "Who's taking over odds and ends?"

"The Honorable John A. Holden the third."

"Who's he?"

"Currently deputy chief of IBM Europe."

"That means he is one of the old liners."

"In spades."

"You said I'm being transferred. Am I going to Asia, really?"

"No. You are now Chief of PC Sales."

"Armonk!"

"Welcome home."

David did not know what to say. Like other trainees, the Sales Division had been his goal. That was where the big boys played, but his physical education background had relegated him to the playpen, the producer of the toys like typewriters and adding machines. Then, David had learned to despise Armonk with its regimentation and corporate rules for everything from dress and behavior to where employees lived, rank by rank. David had thrived on his freedom in Boca Raton, and there was Terry to consider. Terry traveled under a false flag. Most, misled by her tennis style, dismissed her as nothing more than a mental lightweight in tights and a short skirt. David knew better. Sharing life with her in the apartment, David had gradually peeled back the defensive layers and had discovered a sensitive, intelligent, ambitious young lady. David had grown closer to Terry than he had ever been with Margaret or the other girls who had filled his Cornell and high school years. There had been a string of cheerleaders and majorettes, all of whom had been enchanted to share the attention of the star running back. David did not want to leave Terry behind.

"I 'm flattered," David said. "Can I home port in Boca Raton?"

"No."

Again, David hesitated as he pondered his options.

"I understand the company needs an experienced production man to set up the PC Division," David tried again.

"Exactly. This move is no reflection on you. John Holden will select your replacement."

"I could stay on here and serve as whomever's deputy."

"You could, but you won't. Pack up David, and take your medicine like a man. I'll see you Monday morning." Ernie's tone made it clear that David had no choice.

After David hung up the phone, he leaned back in his chair. He thought briefly about screening his many contacts in the microprocessor industry for a job. The money was enticing. He frankly admitted that his loyalty to IBM rested on sand. David could leave Big Blue in a second, knowing the company with its thousands of career fledglings would not miss him, but he could not do that. He owed Ernie big time. For some reason, Hendricks had taken David under his protection, and David could not abandon him now with so much riding on the project they had started together. From his new position in Sales, he could help guarantee its success.

David picked up the phone and dialed the apartment.

"I'm waiting," Terry answered brightly.

"I'm on the way," David said.

"About time."

"Need anything?"

"Only you. The sauce is ready."

"I'm flying."

David listened to the click as Terry hung up, then he put down his own receiver. Friday night was spaghetti, one of David's favorites. Terry's mother was Italian, and genes talked. Terry had mastered the spaghetti art under the tutelage of a master.

Terry, who had been watching the parking lot for the arrival of David's battered Volkswagon, met him at the door with a glass of Chianti in each hand.

"You're learning," David said as he threw his jacket towards the couch, took a glass from her right hand, and kissed her lightly on the lips.

"Yes, master," Terry said. She raised her glass. "Congratulations." Terry touched the glass to her lips then studied David with her cool green eyes.

"Let's sit down and talk about it," David said. Somehow, Terry had learned the news.

"The water's boiling. I've got to cook the spaghetti," Terry said. "You sit down and talk."

"I'm serious."

"So am I. Sit down."

David did as he was told. He watched from a chair at their drop leaf table as Terry broke two handfuls of pasta into the water.

"There. Now talk," Terry turned to face him.

David drank half his wine in one gulp. "What about us?"

"What about us?"

"Tennis in Armonk is great in the summer, but winter is something else," David tried to make light of his problem.

"You'll survive. Play basketball."

"I'm not talking about me."

"I'll find another doubles partner. You're not my first."

David thought about challenging the ambiguity of her statement, then decided that discretion mandated that he ignore it. "Terry. I don't have any choice."

"I know."

"How did you learn about it so fast? Ernie just called me."

"We witches use telepathy."

David knew that Terry had many friends at Armonk, visitors to Boca Raton who the outgoing Terry had befriended.

"You'll like Armonk, then. Those fellow witches and all." David made his decision.

"Me? You're the one going to Armonk."

"And so are you."

"Who says?"

"Our marriage would never work with me in Armonk and you in Boca Raton. The temptations might prove too much for me."

"Margaret and the like," Terry laughed.

"Particularly the like," David smiled.

Terry lifted a piece of spaghetti from the boiling pot, tasted it, then turned off the burner. Without a word, she lifted the pot and used a strainer to drain it into the sink.

"Well, what do you say?" David asked.

"It's ready."

"I don't mean the spaghetti," David said as he encircled Terry with his arms from behind as she stood at the sink.

"Don't make me drop it," Terry said. "Get the plates."

David did as Terry ordered. David held the plates as Terry heaped spaghetti on them, then the thick rich sauce loaded with mushrooms and meat as David liked it.

"Get the wine," Terry said as she took the plates and led the way to the table.

When they were both seated, Terry spoke first: "Try it."

David sprinkled the red pepper and grated Parmesan on his spaghetti. He carefully rolled a forkful then lifted it to his mouth. Perfect. "Not bad," he said.

"Will you marry me?" David asked.

"You just like my pasta," Terry smiled, green eyes sparkling.

"I have to be there Monday," David said.

"OK," Terry said. "You go…"

"What about my question?"

"…first," Terry said, completing her sentence. "Tell me again about the tennis in Armonk."

"It snows in the winter."

Terry put down her fork, went into the bedroom and returned carrying her favorite racquet. As David watched his unpredictable friend, she crossed the room and placed the racquet behind David's game ball that David displayed on a small table against the far wall. David had been awarded the ball after the Harvard game his senior year.

"Now we're both ex jocks," Terry said.

"I've always wondered about that," David said. "Why do they call girls jocks? Do they…"

"Don't ask."

"About my question."

"You go first. I'll stay here, sell the Volkswagon, pack out, then join you."

"I…"

"We'll get married in Armonk. Ernie can be best man. It won't hurt to have your boss involved. Margaret can be maid of honor."

"I don't think…"

"Just kidding. Forget the maid stuff."

David studied his spaghetti.

"I mean that about Margaret. Forget it."

****

In Seattle, Charles slumped at his desk, writing code, a demanding chore that he used for distraction when he wanted to ignore a problem. The phone rang.

"Yes," Charles picked up the receiver. He was all alone in the office. Charles had ordered his staff to leave and to enjoy the weekend; he had not admitted he had no plans.

"Charles," his mother greeted him. "Come for Sunday dinner."

"Mother, I can't."

"Sunday is your birthday."

"I know," Charles lied. He had forgotten all about it.

"Your father and I have something to discuss with you. Be here at six."

Before he could protest, his mother hung up. Age has not made her any less imperious, Charles thought. He could visualize his parents' home, the one he had grown up in, across the lake from his new house. Charles seldom enjoyed the view, because he was rarely home. Weekdays, those rare days when he was not traveling, he spent his time at the office, napping on his leather couch when needed. Charles still liked to work around the clock. It seemed to inspire the others to emulate him. The IBM contract had Digital Software under pressure. The thought of IBM raised a problem that had nagged Charles since his last visit to Boca Raton. Digital Software was now much more than a simple software company. Mike had suggested that it was time to change the name. Charles resisted. He had grown comfortable with their logo with the letters DS emblazoned on an unfurled red, blue and green flag.

"There he comes," Charles' father announced when he heard the throaty roar of his son's new Porsche approach.

"I wish he hadn't bought that car. He drives too fast," his mother said.

"He's a big boy now."

"Then it's time for him to grow up and take on a man's responsibilities."

"I hope you aren't going into that tonight," the elder Swift said. He tended to be more forgiving of what his wife considered their son's shortcomings. "Charles will do what he wants, when he wants," he said for the thousandth time.

"He was a very difficult boy."

"But he's quite a success," Mr. Swift said proudly.

"Hi guys," Charles burst through the front door.

"Wipe your feet Charles," Mrs. Swift admonished.

"Yes mother." Charles rubbed his Nikes lightly on the expensive Persian carpet.

"Happy birthday, son," Charles' father said.

"Happy twenty-sixth," Charles' mother said.

Charles smiled. He could tell, however, that his mother had something she wanted to discuss. He had learned as a very, very young boy that the only way to handle her at such times was to patiently listen, agree, promise to do what she wanted, then to ignore his promise.

"Shall we have some wine?" Charles' father asked.

Mrs. Swift frowned.

"Sure," Charles agreed. "Want me to get it."

"I will," Charles Sr. said. "I've got some champagne in the fridge."

"Charles, I have something I want to discuss," his mother said as soon as his father left the room.

"I didn't do it. I promise to do it, whatever," Charles attempted humor to deflect her.

"When are you getting married?"

"I don't have any prospects. Besides, I'm not ready."

"Charles, tell me honestly. Are you gay?"

"Oh shit, mom."

"Don't use profanity in this house. Answer my question. Are you gay? If you are, it's all right. I understand. It's genes."

"I'm not gay, mother. I like girls."

"Your father's family has had its problems. We never told you."

"It's not genes, Mom. I'm not gay?"

"You've never had girl friends."

"Yes I have."

"Not in school."

"The girls didn't like me."

"Did they think you were gay then?"

"We didn't know about such things."

"Do you have a girl now?"

"I have a housekeeper."

"Are you sleeping with her?"

"No," Charles lied.

His mother glared at him in disbelief. "It would be very inappropriate if you were. Why don't you get a more mature housekeeper, then people wouldn't talk."

"Is that what this is about? Somebody at the Club?"

"Don't be ridiculous," his mother lied. "You are twenty-six. It's time you settled down."

"Here we are," Charles' father entered the room, carrying a silver tray, balancing a frosty bottle of champagne and three fluted glasses.

In Cupertino, Stanley Pitts, dressed in a five hundred dollar beige sportscoat, two hundred dollar slacks, three hundred dollar tassel loafers, and a gray turtleneck jersey shirt he had picked up on sale at Penny's for fifteen bucks, guided his new Mercedes into the handicapped space at the local Hilton motel. Stanley ignored the honk of an elderly couple in a ten year old Cadillac who angrily protested Stanley's disregard of the handicap sign that threatened to tow away transgressors at their expense. Stanley paused in front of the polished glass door and admired himself. He looked like a million bucks, not a paltry thousand.

Stanley opened the door, stepped back to allow a young married couple to pass, turned to study the young wife's legs, decided she had fat ankles, then entered. He headed for the reception desk to ask for his date but stopped abruptly when the familiar voice called.

"Stanley, I'm over here."

Stanley pivoted, smiled, then frowned. Ann Page, dressed in a tight black dress, waved. She looked great, but Stanley did not like the handsome visage of the man in the dark black suit standing with her.

"I hope I'm not interrupting something," Stanley let his tone convey his displeasure.

"Stanley, this is Jim Slater, my boss."

"I didn't know we were meeting for business," Stanley said coldly. He ignored Slater's offered hand.

"We're not. Jim has other plans," Ann said.

"Good."

"I've been looking forward to meeting you," Slater said.

"Are you ready?" Stanley looked at Ann. "We're late."

"Fuck you Pitts," Slater said to Stanley.

"Fuck you," Stanley said, taking Ann by the arm.

"Are you sure you want to go with this guy Ann?" Slater asked.

"It's all right Jim. Stanley's rude but an old friend."

As they walked toward the door, Ann pulled her arm free of Stanley's grasp. "That was rude."

"He was rude."

Ann stopped. "I changed my mind. I'm staying here."

"The food's terrible," Stanley said.

"But the company's better."

"You don't know what you're missing."

"Yes I do."

Ann turned on her slim high heels and ignored the stares as she walked back to join Slater.

"He's worse than his reputation," Slater glanced at Stanley who watched them from about twenty feet away.

"I won't argue with that. Mind if I join you and Helen?" Helen was Slater's wife.

"I guess we won't plan on doing much business with Astron Computers," Slater, the Director of Marketing for INTEL said.

"They're in trouble anyway," Ann said.

She turned and looked in Stanley's direction. He raised his middle finger at her, smiled, then turned around and left.

Chapter 24

A Japanese delegation occupied Charles when a Xerox engineer named Alec Krzinski appeared at Digital Software's main offices. The secretary asked Mike Hamilton if he would fill in for Charles.

"Christ Kathy. I don't have the time," Mike demurred.

"He's from Xerox."

"Charles would kill me if I refused to talk to Xerox," Mike changed his mind. Xerox, the parent of PARC, was a magical name in the Valley.

Mike was not impressed when he first viewed the visitor. A short dumpy man in an ill-fitting suit, Krzinski looked like somebody's black sheep uncle.

"Good morning," Krzinski said, speaking with a heavy accent.

Mike wondered if the secretary had made a mistake. If so, he planned to lecture her severely on the need to screen visitors. With its elevated profile, Digital Software was attracting all kinds of creeps.

"Good morning Mr. Krzinski. I hope I pronounce your name correctly."

"No worry. Polish names very difficult. We even have difficulties."

"What can I do for you?"

"I am engineer. Formerly I work at PARC. Now in System Development Division."

Krzinski immediately had Mike's attention. Everybody in the Valley had heard rumors about the Alto.

"I have business with Boeing, and I impulsively visit this wonderful software company I hear so much about. I write word processor for Alto," Krzinski paused.

Mike leaped to his feet and shut the door. "Please have a seat, Mr. Krzinski. I'm very sorry but our president Mr. Swift is not available. My name is Hamilton."

"I know you Mr. Hamilton. You were schoolmate Mr. Swift."

"At university," Mike corrected.

"I also know DS write software for the IBM."

Mike wondered where his visitor had heard that. IBM still insisted that their contractual relationship be kept secret.

Krzinski accurately read Mike's reaction. "We all know each other business, no?"

"I hope not," Mike said. "Tell me about the Alto." Mike did not expect a frank answer, but he wanted to test his visitor.

"Such beautiful machine with its funny mouse and gooey."

"You actually worked on the Alto?"

"Software, yes."

"Congratulations. I am not a technical person myself, but I understand, correct me if I err, that software is what makes the Alto do such marvelous things."

"Yes."

"And you had business at Boeing?"

"Yes sir. They testing Altos and had problem someone thought I fix."

"Did you?"

"Yes. Xerox so dumb. Not even rich American corporations pay $15,000 for workstation, not while they buy big computers too from IBM. Would you like see my software?" Krzinski opened his fat briefcase.

"Please Mr. Krzinski, just a minute." Mike picked up the intercom phone. "Kathy get me Charles."

"He's with the Japanese visitors, Mike."

"I know that. Interrupt and ask Charles to join me immediately."

"Charles is giving some foreign visitors a tour of our facilities," Mike said. "He's our software genius."

"I hear that," Krzinski said. "That why I here. Please call me Alec."

"I'm Mike."

"Coffee?"

"No thank you."

Mike and his visitor waited in silence. Suddenly, the door burst open and Charles entered. He glanced at the visitor, dismissed him, and looked at Mike. "What's up?" Charles asked.

"I think you ought to hear this," Mike said.

"I Krzinski," the visitor rose and offered his hand.

"Charles," Charles said, a look of puzzlement on his face as he took Krzinski's hand.

"Alec is from Xerox. Their Systems Development Division."

This caught Charles' attention big time. "You work on copiers?" Charles knew better but tried to prime the visitor.

"No Alto."

Charles sat down. "Mind if I join you?" He smiled.

"Alec is a software engineer. He works at PARC where he wrote their word processor."

"You're just the man I need to talk with," Charles said. "We're writing a word processor for a client ..."

"IBM," Krzinski said.

Charles looked sharply at Mike. Their work for IBM was a big secret. IBM lawyers had even insisted on an escape clause that would let them void the contract if Digital Software revealed that they were working on software for IBM.

Mike shrugged.

Krzinski smiled. "IBM like secrets, no, but everybody knows. We at Xerox laugh because the IBM needs someone to build their little machines."

"Mr. Krzinski," Charles said. "This is a subject we cannot discuss."

"Big bucks." Krzinski said cryptically.

"Tell us about your work at PARC," Charles said. "I'll bet that's interesting."

Krzinski described the Alto in technical terms that flew over Mike's head, but Charles quickly demonstrated his understanding. Soon, he and Krzinski were talking in engineering shorthand. Charles quickly focused on the mouse, bitmapping, and particularly the use of icons and menus in the graphical user interface system.

"I would really like to see that," Charles said.

"Can arrange demonstration," Krzinski said.

"I understand others have seen the Alto," Charles began.

"Yes, Mr. Stanley Pitts. That stupid. Now, Astron build Stella."

"And Asteroid."

"And if succeed...poor PARC."

"Mr. Krzinski, may I ask why are you here?"

"Why not? It your privilege. Xerox senior management too stupid to know what have at PARC. Many engineers leave."

"Yes. Astron hired Mr. Vance," Charles said.

"Poor Gene, working with Stanley."

Charles smiled. "Stanley can be difficult. I can't say anything. He is one of our customers."

"You have many customers. Make much money."

Charles recognized their visitor was hinting that money was important to him.

"Where are you from, Mr. Krzinski?"

"Poor computer engineer from Poland."

"Working for Xerox, you can't be too poor," Mike laughed.

"Stock options in Xerox only for presidents not engineers."

"We are a young company that cannot pay our employees top scale salaries," Charles cautioned.

"You have that reputation."

"But we are generous with stock options," Mike joined the discussion. Hiring new employees was Mike's responsibility.

"That too know. Must go. Catch airplane," Krzinski said, closing his briefcase.

"Do I have time to look at some of your work?" Charles asked.

Krzinski opened his briefcase and took out a stack of computer printouts. "This my word processor."

Mike glanced at it. The lines of code meant nothing to him. Charles, though, leafed quickly through the paper. "I'm impressed," Charles said.

"Can you arrange a demonstration for us at PARC?" Mike asked while Charles read.

"But certainly. Will call before next visit."

Charles handed the papers back to the visitor. "I would like to study your work in more detail," Charles said.

"When return," Krzinski took the paper. "Would like know about stock options. Poor engineer not know such things."

"Mr. Krzinski, may I drive you to the airport?" Charles asked.

"Of course," Krzinski said.

"Mike, please take care of our Japanese visitors," Charles called as he followed the engineer out the door.

One month later Charles and Clapper flew from Seattle to the Valley to join Alec Krzinski for the promised tour of PARC. For Charles, the trip served as a vacation. Charles did not take vacations to sit in the sun; for him, they were a time to contemplate the future free from the demands of daily business.

Alec met them at the front door and escorted them to the demonstration room.

"We give pony and dog show," Alec said.

"Hold nothing back," Charles laughed.

"No, nothing," Alec's chuckle made a lie of his words.

"I wouldn't hold demonstrations if Alto were mine," Charles said. "Not until we hit the market."

Alec smiled.

Charles and Clapper took seats facing the modest table on which an uncovered Alto waited. The demonstrator, a bored PARC engineer, introduced himself.

"We have only a few ground rules," he began. "You can see but not touch. Ask any questions you like. We reserve the right to not respond to technical questions. Since you represent a software company, we will demonstrate our unique software but not discuss specifics. Any questions?"

James deferred silently to Charles who smiled and turned his head from side to side indicating he had none.

The demonstration lasted one half-hour. Neither Charles nor Clapper asked a single question. Both paid careful attention, watching the demonstrator's actions as well as listening to his words.

"Any questions?" He asked when finished.

"None that you will answer," Charles said.

The demonstrator checked his watch: "We set a record today. Usually, we go on for at least an hour. Are you sure you have no questions? Comments?"

Charles rose to his feet. "We thank you very much," Charles said. "Please convey our appreciation to your Center's Director."

"Thank Krzinski. He's the one who made this demonstration possible."

"Thank you Alec," Charles turned to their host.

Krzinski escorted the silent Charles and James to their rental car.

"I sorry this was not good," Alec said.

"To the contrary," Charles said. "I was spellbound. Your engineers and scientists are the smartest in the country."

"PARC think so."

"With good grounds. You just gave us a look at the future."

"Beg pardon, Mr. Charles, but you hide good. Visitors always say dumb things. 'So that mouse.'" Krzinski laughed.

"Alec," Charles said. "Correct me if I am out of line. Would you consider joining Digital Software?"

"Of course. We both know why I visit you. My bosses so dumb. They think Alto like toy."

"An expensive toy," Clapper said, referring to the Alto's fifteen thousand-dollar basic price.

"Mike Hamilton will be in touch," Charles said. "I'm sure we can reach an agreeable accord."

"Thank you, Mr. Charles," Krzinski said. "May I ask what position we discuss?"

"Director of Advanced Product Development."

"You like our little toy," Alec smiled broadly.

"The world will."

"And many stock options?"

"Of course."

Alec joined Digital Software one month after the PARC visit, just in time to help celebrate DS's incorporation. Almost from his first day at DS, Mike Hamilton had argued that Charles and James Clapper's company had outgrown its partnership status. Initially reluctant to change, Charles had finally succumbed to Mike's persistence. In the new corporation, Charles with fifty-one percent of the shares became the majority stockholder. Clapper accepted thirty one percent, and Mike eight percent. As a consequence of shrewd bargaining, Alec Krzinski, DS' newest employee received one percent. The remaining nine percent went into a new stock option plan for employees. DS remained, however, Charles' baby. Charles, Clapper his schoolboy friend and first partner, and Mike Hamilton, his Harvard roommate, controlled a dominating ninety-percent of the company.

Shortly after Digital Software incorporated, God in the form of Stanley Pitts descended to bless Charles and his company. For Stanley, the visit took the form of virtual slumming. Over the past year, Astron Computers had amassed a total sales of $335 million with a net profit of $40 million. DS by comparison had a sales of $15 million and a net profit of $1.5 million. Stanley, whose face had appeared on the cover of <u>Newsweek</u>, needed software for his personal project, the Asteroid. He deigned to visit Seattle to be honored as well as to discuss software.

Charles personally escorted Pitts, Dick Ruffin, the engineer who had first conceived Asteroid, and Gene Vance, a former top engineer at PARC, into Digital Software's small but traditionally furnished conference room. Mike Hamilton, James Clapper and Alec Krzinski joined Charles for DS.

The elegantly clad Stanley, wearing his one thousand dollar sports outfit and Penny's turtle necked jersey shirt, disdainfully inspected the room. He noted the lack of expensive oil paintings on the walls and assumed the furniture had been purchased at Office Depot. He smiled thinly until he noticed that before each place at the table rested a copy of the recent <u>Newsweek</u> featuring Stanley's grinning visage.

"Please sit here, Stanley," Charles indicated his customary position at the head of the table.

Charles waited until the regal Stanley had seated himself, then took the chair on Stanley's right. Charles, Clapper and Mike had discussed this tactical move. Both Clap and Mike had argued that Charles should take his customary command chair, but Charles had overruled them. "If we can buy him with a chair, we'll do it," Charles had declared.

After Mike, Alec and Clap had joined Charles on the right side of the table, the rest of the Astron Computer contingent aligned themselves on the left, each facing their relative counterparts. Only the chair opposite Charles, the one on Stanley's left, remained vacant. Mike, Alec and Clap each noticed this discrepancy and shared a common thought: "Charles does not have an equal."

"Tell me Stanley," Charles decided to speak first and ease the formality that had followed them on the regal procession from the front door. "Is it true you dated Joan Baez?" Charles patted the newsmagazine cover.

Stanley smiled. He picked up his copy of <u>Newsweek</u>, admired himself, and then tossed it aside. "We don't have time for small talk, Charles. Tell me what Digital Software can do for me." Stanley's tone indicated that he doubted that Digital Software could do anything for him.

"Our market research tells us that consumers today worry about applications, and the applications they need most are word processors and spread sheets."

"You are talking about the business market."

"For spread sheets, of course. Everybody will need word processors."

"We already have a word processor. Tell me about your spreadsheets."

Charles had anticipated this question. He knew that Astron Computers already had a spreadsheet program, one that Stanley was dissatisfied with because of cost. Astron was paying a seventy-five dollars royalty for every copy installed on the Astron II. DS had a spreadsheet application devised by a DS programmer and improved by Alec Krzinski. He had added a database capability. Mike estimated they could sell it to Astron for twenty dollars a copy and still make money.

"Let us show you," Charles responded. He nodded at Alec who silently rose, walked to the curtain behind Stanley's chair, pushed a button, and the curtain rolled back to expose an Astron 2 on a table with the Astron logo already beaming on the monitor.

Stanley pushed his chair to the side and took the position opposite Charles where he could comfortably view the demonstration.

"Mr. Krzinski joined DS a few months ago from PARC," Charles said, deliberately playing on Stanley's known affinity for things PARC.

Stanley nodded.

Alec demonstrated their spreadsheet with its database capability.

"This works on old technology," Stanley referred to the Astron 2 when Alec finished. "Will it work on the Alto?"

Charles fought a smile. Although he readily disparaged his own company's hardware as old technology, Stanley was not prepared to discuss the capabilities of the Asteroid. Charles assumed Stanley referred to the Alto, the PARC product, because it used the same graphical interface that Stanley had adapted for Asteroid.

"Yes sir. DS can adapt," Alec said.

"Have you given any thought to licensing costs?" Stanley asked Charles.

"Speaking in round numbers," Charles said, "we can halve your spreadsheet license costs."

Stanley smiled but made no commitment. "Tell me about your other products."

Alex returned to his chair.

"I have a confession to make," Charles said. "We're a young company, still expanding. Our sales of our languages and operating systems have put us under considerable pressure. Only recently have we been able to turn our attention to the application market. The responsibility is mine. I say responsibility, not oversight or fault. We at DS are focused on quality."

This made the entire Astron Computer delegation smile. They knew that Charles had a reputation for turning out product before its time, before testing had removed all bugs.

"I suggest you also take a look at our word processor and games." Charles ignored the smiles. He knew what was behind them but decided that now was not the time for contentious pride.

"You're committed to the PC," Stanley referred to IBM's entry into the microprocessor market.

"I'm sorry, I cannot discuss another customer's product," Stanley smiled. In the room behind Stanley's chair were two IBM Personal Computers that DS used to test dedicated applications.

"Not necessary," Stanley dismissed the thought. "You realize, of course, that Astron Computers has an eighty percent share of the microcomputer market."

"Yes sir," Charles said. He did not observe that he doubted this lopsided situation would continue for very long after IBM unleashed its monster sales division.

"Also, IBM with its open system will soon have a multitude of copycat competitors."

"And we hope to sell software to them too," Charles said.

"Our Asteroid will have a closed system, just like our other models," Stanley continued, ignoring his gaffe.

"Yes sir," Charles said. "We understand software written for Astron Computers will work only on your product.

"Are you willing to make this commitment?"

"Yes sir."

Both Mike and Clap wished silently that Charles would stop fawning. The arrogant Stanley was the same age as Charles. They both knew that when it came to technology Charles was a genius and Stanley a self-promoter.

That evening Charles hosted a dinner for the visiting Astron dignitaries at the Seattle Tennis Club. Stanley dominated the conversation with his plans for the future. He described a futuristic factory where sand would come in at one end and emerge at the other as an Astron information appliance. Everyone at the table knew that sand was a basic element of the computer chip, but all felt Stanley's allegory a bit of a stretch. Stanley projected that the day would come when his information appliances would be as common as toasters. Charles and the DS contingent were relieved when the long evening dominated by Stanley's tour de force came to an end. All came away convinced that Stanley, an arrogant bully with highly imaginative visions for the future, was both boring and a spellbinder.

Charles followed up the session with a visit to Cupertino where Stanley demonstrated his beloved Asteroid. Charles was impressed. Clearly, Stanley and his engineers were on the threshold of a technological breakthrough of startling significant proportions. PARC with its Alto had shown them all the future, and now Stanley was creating it. Charles gave Stanley full credit. Despite his preening, his arrogance, and his lack of technical skills, Stanley was driving his engineering team into the future. They had a graphical user interface that would change computing. They had a ways to go before they could consider production, but the project had changed from a dream to reality.

Stanley and Charles reached agreement. Stanley would loan DS four Asteroid prototypes which DS would use to produce three applications: a spreadsheet, a database, and business graphics. Stanley declined Charles' persistent efforts to sell him a word processor. As to financing, Charles gave Stanley options: bundle the applications with Asteroid and pay DS five dollars per program, or sell the programs separately and pay DS ten dollars a copy. Stanley agreed to put up a $50,000 advance but insisted that DS commit to not selling, leasing or distributing the software written for Astron to any one else.

The arrival of the four Asteroid prototypes created a carnival type excitement that lasted for several days as everyone at DS, including Charles, insisted on trying the new technology. Charles was particularly fascinated by the mouse and its ability to open menus and drag windows about the desktop. He summoned Alec, the ex-PARC engineer, and insisted that he explain in detail every function. After a full two hours, Charles turned to his Director of New Projects and issued an order: "This is your priority number one."

"What is?" Alec asked. He had quickly learned that the only way to cope with his new boss was to stand up to him.

"Make LDOS do this."

"You want the PC to do what Asteroid does?"

"Yes."

"But that's a major project. Astron has been working on this for three years."

"It's based on PARC's Alto."

"That's right."

"Astron did it."

"They reverse engineer much, and Gene Vance help."

"We have you, Alec."

"You want whole gui?"

"You got it."

Chapter 25

David arrived in Armonk as ordered and immediately found himself swamped far beyond anything he had experienced in Boca Raton. The entire company was in a state of uproar. On his first day as chairman, Parsons announced his plans to reorganize the company from top to bottom. To make matters worse, Ernie was reshaping the Sales Division. Ernie met briefly with David on arrival and authorized him to set up his newly created Personal Computer Sales Group any way that he wished. To assist him, Ernie assigned ten junior sales technicians as a starter.

"Can I at least have a couple experienced sales reps?" David asked.

"You don't want them," Ernie replied.

"What's my target date for kickoff?" David asked.

"1 August."

"Where?"

"Decide."

"Agenda?"

"Prepare one. Look David, I've got my hands full. The PC thing is your baby. Set it up."

"Do I get help from the Sales Division pros?"

Ernie laughed. "These guys are all main frame specialists. Your commitment to market the PC through Sears and Montgomery Wards stands. The Sales Division knows nothing about retailing computers. They sell directly to the major customers."

David's mouth dropped open. He was tempted to remind Ernie that Ernie had come up with the idea to market the PC through retailers when they encountered resistance from Crane and the Sales Division.

"You've got a week to develop your marketing plan and your proposal for the introduction in New York. Now get busy and let me worry about the Sales Division. You're not our major product."

"Yes Ernie," a chastised David retreated.

In the outer office he approached an unsmiling Margaret who Ernie had moved over to Sales with him.

"Margaret, I need help."

She smiled but said nothing.

"Can you tell me where to find my office?" David asked.

"I hear you're getting married," she said.

Her frosty words caught David by surprise. He had counted on having time to figure out how to break the news.

"I hope you enjoy tennis," Margaret turned, scribbled a room number on her notepad, then ripped the page off and handed it to David.

"Thanks," David said, thinking he had escaped.

"Poor girl," Margaret called after him as he hurried out the door.

David had to ask for assistance twice before he located the room number Margaret had given him. A sign on the door gave him pause. "Storeroom A." David hesitated, wondering if Margaret had sent him on a wild goose chase. Finally, he opened the door and found himself in the midst of chaos. Painters were daubing a sickly green on the walls; three phone technicians were stringing wire across the floor to accommodate five rows of five desks each. Two groups of chattering men in dark blue suits stood in the back of the huge room, arguing and gesturing, several with angry red faces. No one acknowledged David's entry, and he did not recognize any of the youthful combatants. It did not take long for David to learn that his work force came exclusively from the latest IBM trainee class, all of whom had been assigned to the PC Marketing Team on an interim basis.

"Listen up," David shouted.

Nobody responded. They continued to debate who would receive which desk.

"Listen up," David shouted again. "I'm David Howard."

This time several recognized his name and quieted the others.

"Come into my office, and we'll get ourselves organized."

This was greeted by a loud joint laugh.

Determined not to be daunted by a bunch of kids, David went to the doorway at the back of the room that he assumed led to his office. Inside, he found his desk, a large executive model which IBM regulation authorized for his position. It filled the entire room that measured no more than six feet by nine, leaving space for one straight backed chair for visitors. David returned to the bullpen, a name that just occurred to him, and shouted again: "Follow me."

David led the chattering group down the hallway and out the door to the rear parking lot. There, David gathered his team around him, feeling much like a coach on a practice field. All he needed was a whistle on a string hanging from his neck.

"This is the Personal Computer Marketing Team," he announced. "Obviously, it is not much," David reverted to a coach's practice field demeanor, disparaging his team, "but it is all we have." Recalling his own experience that trainees were an arrogant and at the same time an insecure lot, he continued: "What we do here during the coming six weeks will impact on your future careers. This may be an interim assignment, but it is not recess time. Any questions?"

"What are we supposed to do?" An eager young man in the front rank asked.

"Good question. We're charting virgin waters. You know what that means don't you?"

"Yeah. There's a first time for everything," A wise guy in the back shouted.

"As you've learned from your lectures, the Sales Division is responsible for marketing all IBM products. Do you know how we do that?"

"We lease mainframes to big business," the lean young man in front responded.

"That's right. Then, we spend the rest of our careers catering to our customers. Leasing means we rake in our commissions, and we keep the computers running."

These words brought a smattering of applause, disconcerting David. After a pause, he continued: "But we are not the Sales Division. We are the Personal Computer team. We've got a new product and a new way of doing business. We sell microprocessors. We do not lease them."

"What about the commissions?" The wise guy shouted.

"What commissions? We work on salary."

"I like the old way best," the wiseguy challenged.

"During your interim assignment, guys, you will like our way best."

"How do we sell microprocessors? Go house to house?"

This provoked weak laughter.

David smiled. "You tell me how we sell them."

In the silence that followed, David pointed to three young men selected at random. "You are the team leaders. Choose two others to join your team then come up with your recommendations."

"What do you mean recommendations?"

"You are my sales force, such as you are," David smiled. "You will identify the most promising retail chains—Sears, Monkey Wards, major electronic chains—use your imagination and knowledge. Tomorrow at eight, I will expect each team to submit its recommendations. We'll discuss them and decide how to proceed from there."

"How do we meet in that mess?" The wise guy who David did not select as a team leader asked.

"That's your problem. You are our administrative maven. Have the bull pen ready for action tomorrow morning."

"What if it takes all night."

"Good."

"Can I authorize overtime."

"Whatever it takes."

The wise guy smiled, apparently pleased at being given individual administrative responsibility.

"What's your name?"

"Smith, William."

"All right Smith William, get busy."

David turned to the eager young man in the front row. "What's your name?"

"Barrington Seldon."

David was not sure if the young man were Barrington Seldon or Seldon Barrington. "OK, BS you've got a special assignment."

The others laughed at the BS designation.

"I want you to confer with the legal and financial people in the Sales Division and come up with a proposal we can use when we contact the retailers."

"Where do I find them?"

"Wherever you can."

"That's it. Get to work." David did not expect much from their initial efforts, but he knew trainees were eager. They might accumulate enough information to give him a place to start.

"Sir," BS raised his hand.

"Yes?"

"I prefer SB."

After dismissing his young staff, David retreated to his office where behind a closed door that only partially shut out the bullpen turmoil, David spent the rest of the day making lists of things to be done, people to be called, assignments to be given. About once every five minutes, somebody knocked and entered to ask a question or to introduce themselves. The trainees appeared, one by one, each intent on establishing a relationship with their new boss. To a man—there were no female trainees in this group.—.they were bright, well educated, most from Ivy League schools, and eager, particularly eager. David remembered himself as a trainee. He had not been a part of the ardent crowd. He had felt like an outsider, the only ex jock in the group. He still wondered why he had been hired.

Late in the day, a gentle tap disturbed him. "Yes?" David called.

"Sir, I apologize for interrupting," a middle aged female with traces of salt in her hair entered.

David smiled and waited for her to reveal her mission.

"I am Mavis Brown. Admin assigned me as your assistant. Margaret told me where to find you."

"Our secretary?"

"Whatever."

"Mavis, how long have you worked for IBM?"

"Thirty years."

"Where?"

"Mostly in the Sales Division. Before that, in the front office."

"The front office?"

"Yes, for Mr. Watson Jr."

"Himself?"

"I was one of three secretaries, the most junior."

"Thank the Lord," David said. "Mavis, you are the only person in this group that knows what she is doing."

Mavis smiled.

"Please tell that young man out there..." David hesitated. "Smith, William, what to do to get that mess cleared up."

"Yes sir."

"I want all the workmen out of here by opening of business tomorrow."

"Yes sir."

David worked through the dinner hour all the way to nine-thirty, then, overcome by fatigue, decided to quit. When he paused, he was suddenly conscious of the silence. He opened the door and found a much more orderly scene. The painters and the phone men had vanished, the desks aligned, and Smith William and Mavis busy polishing the desktops.

"Looks great," David said and meant it.

Mavis and Smith smiled.

"You two should call it a day."

"We will shortly," Mavis said.

David realized she was waiting for him to leave. "See you both tomorrow at eight," David said and left.

He retrieved his rental car and headed for the Holiday Inn. He stopped at a 7 Eleven and indulged himself by purchasing a six pack of Budweiser. He parked directly in front of his first floor room, entered, flicked on the light, pulled a can of Bud from its plastic strap and sat on the bed. He lifted the metal tag, took a deep drink, then turned toward the phone. Before he could reach the receiver, it rang.

"Yes?" David said.

"David?"

"Terry. I was just getting ready to call."

"I'll bet. How's Margaret?"

"I don't know. She's not speaking to me."

"Good."

"How are things coming?" David expected a report on Terry's progress in preparing to join him. They had estimated it would take a week.

"I've got news. I'm not sure you'll like it."

"You lost," David joked, referring to Terry's daily tennis.

"I had a long talk with Mom and Dad." Terry ignored David's sally. "I spent Sunday with them."

Terry's parents lived in Miami Beach in a modest home a few blocks off the main highway not far from a large golf course where Terry's father worked as pro and operated a caddie shack. David's relationship with Terry's parents was an uneasy one. He had expected to get along well with Terry's father, but he had encountered a cool reception despite their respective athletic backgrounds. Terry's explanation that her father treated all her boy friends that way did not make matters any easier. Terry's mother was a different matter. An older version of Terry with bottle blonde hair, she lived for her church.

Because of the father's aloofness and the mother's intense piety, David avoided the older Mitchels whenever he could.

"How did they take the news?" David was not sure he wanted to hear.

"I spent Sunday afternoon with them," Terry said. "I told them about your transfer and our plans."

Neither Terry nor David wanted a fancy wedding. David had suggested a quick visit to a Justice of the Peace in Armonk, and Terry had agreed.

"Dad grunted and mother was excited. She likes you a lot."

"I hope so," David said.

"But she has one small problem."

"And that is?"

"She wants a church wedding."

"Oh shit. Terry, I don't have the time. I'm in over my head. I've got six weeks to get ready for the PC's launch."

"Well, if you don't want to get married..."

"I didn't say that," David interrupted. "I want married. I just don't want a church wedding."

"I know."

"Neither of us are churchgoers. We agreed on that. It would be hypocritical to stand there and let some strange preacher mouth words over us as if they meant something."

"I know."

"That's all preachers do. They sprinkle water, they marry, they bury and every Sunday they spout nonsense."

"I know."

"Then?"

"Mom wants a church wedding, and I agreed."

David, who had a running back's ability to make quick decisions when he had to, capitulated: "When? Make it a weekend so I can fly down. I can't get off during the week. No way."

"I know. Mom says August first."

"August first? Impossible."

"OK."

"The Management Committee set August first as the day for the PC's debut."

"Tell them to change it."

"I can't. What about this Sunday?"

"No way. We've got to get invitations written and sent out, arrange for the church, flowers, bridesmaids, a best man and grooms. There's too much to do."

"Pick another day. What about September?"

"What about September?"

"There's a holiday then," an idea occurred to David.

"Yes. Labor Day."

"It's a Monday. I could fly down on Saturday or Sunday, we get married on Monday morning, then fly back here."

"What about a honeymoon?"

"No time. We'll have one later."

"I'll bet."

"I promise. Christmas. We'll fly to the Bahamas for Christmas."

"Big deal."

"We'll take a cruise. It's not like we haven't been living together forever."

"Longer than that, David."

"OK. It's Labor Day and a cruise at Christmas."

"You're so romantic David."

"Ahhh, Terry, I'm so busy."

"That raises another question. What am I supposed to do in Armonk while you work all the time?"

"Play tennis."

"I retired my tennis racquet, remember?"

"You can unretire it."

"What do I do in the winter? It snows in Armonk. You said so yourself."

"I know." David sipped his Bud while he tried to come up with something to divert Terry from this problem. "What about skiing?"

"Too cold. I'm a hot weather girl."

"We'll find an indoor court somewhere."

"I'll get a job. I'm an IBM employee too, remember?"

"You don't want to work." David liked the idea of coming home to be greeted by Terry with dinner prepared.

"That gives me two months to work something out."

"You'll hate Armonk."

"I already do."

"Terry, I'm tired and haven't had dinner yet. We'll talk about this tomorrow."

"Ten o'clock and you haven't had dinner?"

"I just got to my room."

"Order pizza."

"I was just going to."

"Honey, get used to pizza because there is no way I am going to fix dinner at ten o'clock at night."

"I know," David said.

"We're set for September fourth?"

"Yes."

"No shit?"

"Yes."

"OK. I'll make arrangements here. You find us a house, lover, a nice one, and furnish it. I won't have time if I have to report to a new job September 5th."

"Terry…"

"Be talking to you lover," Terry said and hung up before David could reply.

David wearily studied the receiver before hanging up. He loved Terry, found her independence and impulsiveness enchanting, but he wondered what they would mean to their married life after the ceremony.

The pressures in the office almost overwhelmed David. Ernie, preoccupied with his reorganization of the Sales Department, had little time for David. He expected David to attend the daily Sales Department senior staff meetings, where his reports on his progress met with indifference on the part of the older Sales executives, and to prepare for the launch of the PC while establishing his retail network. Gradually, David's little empire expanded. His staff now numbered forty people, the new additions coming largely from the main line divisions. Most were the usual types that office managers took great pleasure in dumping on new groups—the lame, the weary and the indifferent; retreads, worn out executives marking time to retirement, garrulous personality problems that created more difficulties than they solved. David welcomed them all. Now, he needed warm bodies. Later, he could worry about upgrading his personnel. Now, he needed spear-carriers. Unfortunately, trainees and problem employees require close supervision, and this ate into David's time. Before long, he too was rushing about with a harassed look on his face.

On weekends David searched for a house. His choices were limited. As a trainee and junior employee living in an apartment, David had heard comments about residential stratification, but he had paid them no attention. Now, he was forced to. The friendly realtor explained the facts of IBM life to him.

"The members of the Management Committee reside on 'King's Row.' Senior and upper middle level managers live in 'Seniorville.' Lower middle and junior executives have 'Juniorville,' and everyone else resides in town."

David took one look at the small ramblers that were crowded on Juniorville's quarter acre lots and decided that with Terry working they could stretch and handle Seniorville. He was after all a group chief even though his salary had yet to equal his title. The realtor had only a handful of available homes to show him. David, weary of the process, selected a handsome colonial on an acre of land with a view of a distant artificial lake when the leaves were not on the trees, or so the realtor alleged. David put down his deposit, cadged a sales photograph of the house from the realtor, and mailed the papers to Terry to sign before a Notary Public. Terry confided that her father had grumbled this was no way to buy a house, but Terry had signed the papers and returned them to David who was anxious to escape from the tedium of his motel room.

Once this hurdle was surmounted, David devoted a Sunday afternoon to furniture shopping. He contracted for a bed, a dresser, a recliner, and a television set to be delivered on the afternoon of August fifteenth, two hours after he was scheduled to close on the house. David decided that discretion required him to wait for Terry's arrival before hazarding a choice of the rest of their furniture. He reassured himself with the thought that the bed and dresser could be moved from the master bedroom to a guest bedroom if Terry refused to pass on it. The recliner would go to the family room, and who could object to a television set.

As time passed, the pressures at the office did not ease. The media picked up on Big Blue's plans, and stories began to leak. Many described the PC in detail, generating even more interest. Retail agreements were signed with Sears and several chains. Fortunately, David was able to defer to Boca Raton for details on shipping dates. Unlike the main frame salesmen, the employees of the PC Marketing group were not expected to handhold retailers or customers. Boca Raton established a support unit to respond to purchaser technical problems.

David with Ernie's concurrence selected a mid sized Manhattan hotel as the site for the PC introduction. The mainframe salesmen made their displeasure quite clear—IBM was expected to go first class. David would have gladly accommodated them, but all of New York's four star hotels were booked a year in advance. Ernie counseled David not to worry, so David concentrated on trying to organize his first computer media event.

"What will happen if we have a party and nobody comes," David silently worried.

A technical/legal problem developed two weeks before David's gala. Jeff Wayne, the owner of O-Systems, the erratic engineer who had gone off flying when David had tried to negotiate permission to use his operating system for the PC, had decided that Digital Software had pirated his copyrighted OS/M. Since David's replacement had yet to make an appearance, Speedy called from Boca Raton in a panic. "Without an operating system, we have no computer."

"Have you talked with Digital Software?" David asked, trying to remain calm.

"No. Charles is your buddy. You call."

Although David no longer had line responsibility for the production of the PC, hardware or software, he phoned Charles in Seattle.

"Jeff Wayne's making waves," David said.

"No surprise. That's his personality," Charles said. "What's the problem?"

"He claimed you ripped off his operating system."

"No way. We bought LDOS from Tom Peters. Wayne has no claim on IBM or Digital Software."

"But if Lake Computers copied it from O-Systems, the sale is invalid."

"Tom Peters is a reputable engineer," Charles replied calmly. "He may have reverse engineered some of OS/M, but LDOS is his product. I'm confident we have no problem. Tell Wayne to piss off."

"We'll send a couple lawyers out to talk with him and find out what he wants," David said.

"Keep in touch," Charles said and hung up.

The dispatch of IBM lawyers proved to be a fortuitous decision. They quickly learned that Wayne and O-Systems belatedly wanted to offer his next version of OS/M to IBM, and the threat of a lawsuit was his way of getting their attention. David decided to oblige Wayne. He agreed the next version of the PC would offer customers a choice of operating systems, O-Systems' OS/M and Digital Software's LDOS. Wayne priced his optional operating system at $250, and Charles set LDOS at $40. David doubted that many PC customers would opt to pay $210 extra.

Ernie hosted a dinner at the Manhattan hotel the night before the August first unveiling. David, several of his team, representatives from Boca Raton including Speedy Spencer, several public relations specialists from Boca Raton, and, of course, three ever present company lawyers attended. David, who had worried that nobody might show up, decided that the IBM contingent alone was sufficient to ward off embarrassment. As inevitably happened, technical problems appeared when the prototypes that had been so carefully escorted from Boca Raton were unpacked. David, who had anxiously watched the unpacking, approached the first machine to be hooked up. He fondly patted his baby and received a sharp shock.

The engineers gathered and quickly uncovered a short circuit caused by dust. Each of the demonstration models was opened and a piece of cardboard inserted between the case and the offending wiring.

The next morning a modest crowd numbering one hundred and two attendees appeared. Ernie presided. He gave a brief history of the PC then turned the demonstration over to Speedy who demonstrated IBM's first microprocessor. The press contingent included representatives from the Times, the Post, several West Coast papers, and the wire services. There was no television coverage. David circulated through the small crowd and listened for reaction. He was surprised when one of the reporters singled him out.

"You're David Howard aren't you?"

David turned. "Do I know you?"

"My name's Oldham. I'm from the Post."

"Thank you for coming," David played host.

"It's my job. I'm a staff writer for the business section, and they think I know something about computers."

"Do you?"

"Not really."

"Neither do I," David confessed. "I'm not an engineer."

"We got a lot in common then," Oldham smiled. "I'm an English major pretending to be a staff reporter."

"I'm a jock," David laughed. He instinctively liked Oldham. "Call me David."

"I'm Bill."

"Oldham? That name's familiar," David said.

"I was afraid you might recognize it."

"You're the guy who broke the story that IBM was building the PC."

Oldham nodded. "I didn't have the name PC."

"You caused me a lot of heartburn. Armonk thought I leaked the story. Who did?"

Oldham smiled.

"Ahh it doesn't matter. Think we can get any free publicity today?"

"You don't have a bad turnout. Most of the important media are here."

"How come no TV?"

"It's not as if you're unveiling a state of the art machine."

"I know."

"Don't worry about the turnout. IBM producing a microprocessor should rattle the market."

"Thank you," David said. "That your story line?"

Oldham winked. "I'd like to visit with you when you have more time."

"Me? I'm not in Boca Raton now. I work in the PC Marketing Group."

"That's what I hear. I'd like to do a piece on IBM's marketing strategy for the PC."

David considered the request for a few seconds. "Why not? But on one condition. You don't use my name."

"Agreed."

They exchanged cards. David offered his with the expectation that he would not hear from the young reporter again.

The next morning, back in his Armonk office, David scanned a full-page advertisement in the Wall Street Journal. Under the headline: "WELCOME IBM. SERIOUSLY" the Astron Computer advertisement read: "Welcome IBM. Seriously. Welcome to the most exciting and important marketplace since the computer revolution began 35 years ago. We look forward to responsible competition in the massive effort to distribute this American technology to the world."

"They don't take us seriously," the trainee who had brought the advertisement to David's attention said.

"Don't worry about it," David said. He found the challenge amusing. "I assume Stanley Pitts is behind this."

Before he could comment further, his intercom buzzed.

"There's a William Oldham on the line. He says he's a reporter for the Washington Post," Mavis Brown said.

"I'll take it," David said, picking up the phone. "Hi Bill, how are you?"

"I know you didn't expect to hear from me so soon," Oldham said. "After we parted, I decided to overnight in New York and take you up on your offer to let me interview today. Doing it now will save me another trip."

"Sure," David said. "When can you get here?"

"I'm here now."

"In the lobby?"

"Yes."

"Stand still. I'll send someone for you."

David dispatched Mavis to fetch the young journalist. He wondered why he really had stayed. Within minutes Mavis appeared with Oldham in tow. After they were seated, Oldham retrieved a notepad from his jacket pocket and looked meaningfully at David.

"Did you see the Journal advertisement?" Bill asked.

David smiled.

"Any comment?"

"I think Mr. Stanley Pitts is in for a surprise."

"Did the advertisement irritate IBM?"

"I can only speak for myself. I find it amusing."

"Do you consider it a challenge?"

"It reads like someone speaking with tongue in cheek. We welcome the media attention it's obviously generating." David referred to Oldham's presence in his office.

"Oh," Bill said. "I didn't see it until I arrived here. Your receptionist had a copy of the <u>Journal.</u> I just thought I would take advantage of your presence to see if I could get a useful reaction."

"Did you?"

"Oh I might get a couple of lines out of it. No big deal."

"I must confess, Bill, that I haven't had much experience dealing with the media."

"I'm harmless."

"You've already proved that," David meant the opposite of what he said, referring to Oldham's initial story about IBM's plans. "You're good. You sound sincere."

"I am. I won't print anything you say that you don't want me to, and I won't deliberately misinterpret any comments you might make."

"Good."

"May I ask another question?"

"Please."

"Obviously, the PC is not state of the art."

"That's true."

"How do you expect to compete with Astron Computers' Asteroid?"

"We have IBM's proven track record for providing reliable equipment which we support. State of the art microprocessors come with their problems."

"Are you saying that Astron Computers sells faulty computers?"

"Certainly not. It's a given in the computer industry that state of the art equipment will come with bugs. Computers are highly complex machines, sometimes it seems they have wills of their own, and only real time use will surface some of these evasive problems. That's why we have been so selective in choosing the components we incorporated in the PC. We're confident the marketplace will support us."

The interview lasted for a full forty-five minutes before David called a halt to it. At no point did Oldham stray into sensitive areas, and David bid him goodbye with a comfortable feeling.

Chapter 26

Stanley did not know what Lawrence Starkey was thinking the first day he reported in and took over the president's office, but Stanley was determined not to let the new executive become another Willard Temple who actually tried to manage Stanley's company. Stanley, without being invited, joined Stuart Miller and Lawrence Starkey on the latter's initial tour. Stanley insisted that they start with the Asteroid Division where quite naturally Stanley played host and introduced his staff one by one with a succinct description of their duties. When they moved on to the Stella Division, Stanley elbowed Tom Williamson out of the way and assumed charge of that briefing also. Neither Starkey nor Stuart Miller objected, so Stanley led them throughout Astron's seven buildings. Starkey was determined to put a good face on his relationship with Stanley, but he learned first hand the problem he was facing.

The first day Starkey smiled a lot and said little. He had already set his private goals. Firstly, he had to stifle internal discord and bring the various divisions together to achieve common goals. Secondly, he had to rebuild the bridges to the industry at large that had deteriorated over the years. Starkey knew he had to rein in Astron's Chairman of the Board. The flamboyant Stanley personified Astron Computers to the world at large. It was Stanley who grabbed the headlines and set the tone. Unfortunately, Astron Computers' reputation had grown to resemble Stanley's own—that of an arrogant, self-centered company willing to trash any competitor.

Starkey hoped to avoid self-defeating conflict with Stanley, thus from the very first day he followed Stanley's lead in creating an in-house Stanley/Lawrence show. Their motives were far different. Stanley intended to dominate the new president who lacked a technical background, while Starkey simply wanted to harness Stanley to a positive, unified company plan.

Two months later, after Stanley had relaxed his guard without changing his style, Starkey acted. He reorganized Astron Computers. He abolished six of Astron Computers' nine feuding divisions. In their place, he substituted three super divisions, engineering, marketing and production, all three under his direct command. Stanley retained his operational control of Asteroid, which was folded into the engineering division.

To Stanley's delight, his new boss focused on what he knew, marketing and advertising. To Stanley's dismay, Starkey decided to change an emphasis that Stanley had labored hard to create. Stanley took pride in Astron's image, which portrayed the company as the industry's innovators. Starkey insisted that they had to emphasize the proven quality of their microcomputers if they were going to best IBM. This of course downgraded Stella and Asteroid with their revolutionary graphical interfaces and acknowledged the Astron 2 as the company's tried and proven money earner.

Stanley immediately denounced this policy as "marketing bull shit." He continued to characterize the Astron 2 as outdated crap and to herald the Asteroid, conveniently ignoring Stella whose five thousand-dollar price tag placed it out of the reach of their target markets. Everyone knew that Stanley's insistence that the engineers load Stella down with every fancy new application was responsible for the price inflation, but few said so in his volatile presence. All could see that the Stanley/Lawrence act was losing its novelty, but none dared test its solidarity until the principals reacted.

In the privacy of his Asteroid group, which still defiantly flew the skull and crossbones despite many hints to do otherwise from the new president, Stanley began to castigate Starkey: "He's just another Temple, a businessman who knows nothing about computers."

Then, the market began to erode Stanley's base. The Personal Computer caught on, buoyed by the sterling IBM reputation. Astron Computers profits fell for the first time in history. Market share dropped two percent. Starkey, the businessman, fretted, while Stanley ignored the trend. Despite its marvelous new technology enhanced with every available flag and whistle, the Stella flopped. Starkey suggested a merger of the Stella and Asteroid groups, but Stanley, now convinced he was riding a company saver, resisted. Stanley went so far as to order his engineers to configure Asteroid so that it would not use either Astron 2 or Stella software.

Stanley proposed a hundred million-dollar advertising budget to launch the Asteroid. Recognizing that all their money was riding on Stanley's horse, Starkey concurred. Stanley decided to aim high: to challenge IBM in the corporate marketplace. The media loved it. Thanks to ample advertising dollars, television covered the launch.

David in Armonk envied the hoopla.

For three months, Astron blanketed the country with media advertisements including a twenty-page spread that was carried by several magazines. Stanley presented an Asteroid to Mick Jagger and personally oversaw a dramatic sixty second Super Bowl advertisement that ran once to popular acclaim. Astron even tried a test drive program—take an Asteroid home, drive it for a day, then bring it back if you dare not to like it. In his enthusiasm, Stanley forgot his target, the corporate market. The massive hype produced one major corporate sale, 4,500 Asteroids. At $1,500 per unit, this sale failed to offset the cost of the hundred million-dollar campaign.

IBM's mainframes owned the corporate market. When big business decided to give the microprocessor a test, they turned to the company they knew and trusted. Thanks to Stanley's insistence that the engineers design Asteroid's operating system to be Astron software specific, Asteroid was not compatible with IBM software. Companies that depended on IBM software were not inclined to risk their bottom line to experiment with an untested state of the art product.

As Asteroid sales stagnated, Astron stock slid downwards. Starkey directed that Stanley's engineers devise a new operating system that would run PC programs. Stanley stalled. Starkey developed a secret plan that would eliminate 1,200 jobs and deprive Stanley of his operating command. Harold Dumbroski, who had returned to the company his inventions had founded, retired in disgust and sold all of his Astron stock.

During the ensuing months, two constants prevailed: Stanley continued his old ways, sowing dissension and creating discord; and, ominously, the PC eroded Astron's market share. The Board of Directors, the august body that over the years had indulged Stanley's idyiosysncracies, turned on him. Concern over declining profits replaced the willingness to let Stanley have time to outgrow his extended corporate adolescence. Individually, Board members, particularly the seasoned businessmen, encouraged Lawrence Starkey to take over the company, to rein in the undisciplined, selfish Stanley, removing him from operational command if he resisted.

Starkey discarded the old premise that neither he nor Stanley had ever believed, though they mouthed it frequently: Starkey would learn technology from Stanley, and Stanley would learn management from Starkey. "Stanley will never change," Starkey began to confide to his supporters, some ninety percent of the company. Recognizing that the day of confrontation approached, Starkey secretly identified Stanley's replacement as head of the Asteroid Division: Pierre Dupris.

Starkey discussed the transfer with Dupris, the head of Astron France, during a European tour of overseas divisions.

"We'll move you in gradually," Starkey said.

Dupris responded with a skeptical look. He was no novice to corporate politics. He had earned his spurs at Hewlett-Packard and Intel before signing on with Astron.

"Stanley does not strike me as the kind of man to back down gracefully," Dupris said.

"Stanley is destroying Astron Computers," Starkey said. He felt he had to be candid with Dupris, the one man he felt had the toughness to take on Stanley.

"To be honest, it's Stanley's company to destroy. He's our major stockholder," Dupris said.

"Only thirty percent."

"Will the Board support you?"

"The majority of the Board feels more strongly about Stanley than I do."

"That Stanley must go?"

"Yes."

"Then why are we talking about moving gradually? You're the president. Fire him."

"We have to think of the company's image," Starkey said, maneuvering to place Dupris in the position of arguing Starkey's own views.

"Have you looked at that image lately. Stanley has isolated us in the backwater of the industry."

Starkey smiled.

At the very next Board meeting with Stanley presiding as Chairman, Starkey threw down the gauntlet. Unbeknownst to Stanley, Starkey had divulged his plans to sandbag Stanley to a majority of the Board and had obtained their prior commitments of support.

After reporting the latest deteriorating sales figures, Starkey presented a dismal picture of internal anarchy. Stanley listened silently with a deepening frown on his face. This only encouraged Starkey to table more detail. After describing the problems, Starkey, like a good chief operating officer must, turned to the solutions.

"We must act decisively, now, if we are to save the company," Starkey said.

"Aren't we over-reacting Lawrence?" Stanley spoke for the first time since Starkey had begun his presentation.

"No, Stanley, if anything I understate our problems."

"Bull shit," Stanley reacted predictably.

"Stanley, let Lawrence complete his presentation before we enter the discussion period," Stuart Miller interceded. It had been Stuart's money that had given Stanley and Harold the opportunity to escape from the confines of the Pitts garage. He was the only member of the Board who Stanley allowed to counsel him.

"I reserve the right of reply," Stanley, the Chairman, said.

"I plan a drastic reorganization and a personnel reduction," Starkey said.

"Not another silly reorganization," Stanley challenged. "When are you going to learn that it is technology that matters not bureaucratic structure. Shuffling blocks on a paper chart might work in the sugar water industry, but not here."

"Stanley," Stuart warned.

"I might as well say it right now," Starkey, irritated by Stanley's outburst, decided to react to the challenge. "I plan to bring Pierre Dupris in from France."

"To do what?" Stanley asked.

"To head up the Asteroid Division," Starkey said.

"And why would you want to disrupt the only division in this staggering company that has a product that can save it?" Stanley demanded.

"I'm prepared to bet our future on the Asteroid," Starkey said.

Stanley smiled and turned his palms upwards asking a question. "Why interfere then?"

"But only if Stanley devotes his full attention to his proper position as Chairman of this Board."

"Isn't that what I'm doing?"

"And refrain from interfering with the management of this company by professional managers."

"I thought you were going to teach me management?"

"Stanley, I have tried and failed. Your participation as a line manager is no longer an option."

"We are obviously in total disagreement. Shall we poll the Board?" Stanley asked.

"Please do, Stanley," Stuart Miller said.

Stanley's pained expression indicated that he understood Stuart's warning.

"Let's take an informal poll," Stanley said. "Not for the record," he turned to the secretary who was taking notes. She nodded understanding and placed her pencil on her steno pad.

"All those who support Starkey's re-organization raise your right hand."

Every member of the Board raised their right hand.

"That clarifies the matter," Stanley said. "As the major stockholder I must express my disappointment at this lack of confidence."

"Stanley, there is a big difference between being a major stockholder and being the majority stockholder. You may be the major stockholder, but the members of this Board represent the majority of the stockholders. Need I say more?" Stuart Miller asked.

Stanley glared at his old benefactor. "Please continue," he ordered Starkey.

"I request the Board's concurrence to the appointment of Pierre Dupris as chief of the Asteroid Division," Starkey said, deciding to take advantage of the moment.

"That's not necessary," Stanley said, not wanting to have the Board on record as approving the transfer.

"Please put it to a vote, Stanley," Stuart Miller said.

"All in favor signify by saying aye," Stanley relented.

All answered aye.

"Motion carried," Stanley said. He turned, looked at the stenographer and spoke clearly and slowly. "The Chairman notes that no timetable has been established for this transfer. Timing will be discussed at a subsequent meeting. Meeting adjourned."

"Stanley ..." Starkey started to protest, but stopped when Stuart Miller shook his head negatively.

After the meeting, Starkey phoned Dupris in Paris to inform him of the Board's action.

"You can pack your bags," Starkey said after briefing Dupris.

"Lawrence, I trust you but not Stanley. He still has influence and will do everything in his power to impede my appointment. This is not the first time that someone has challenged him."

"It is the first time that someone has challenged him and won. Stanley is on his way out of the company and into the boardroom."

"Could you please send me a letter and confirm the date I will take over?" Dupris asked.

"That is not necessary," Starkey said, making no attempt to conceal his irritation at the implied lack of trust. "You have my word that you are the new Director of the Asteroid Division."

"I appreciate that. I trust your word. You are an honorable man," Dupris tried to placate his chief executive officer, "but I do not trust Stanley Pitts."

"Stanley has no choice. The Board unanimously supports me. We have the votes."

"I still must ask for a date in writing," Dupris said.

"Very well," Starkey said. "I'll get back to you." He conveyed his displeasure by abruptly terminating the conversation.

Stanley immediately began his counterattack. He summoned his branch chiefs to an Asteroid Division management meeting.

"I've just returned from a Board meeting at which the asshole has tabled his plans for another drastic reorganization of this company. He did not say how many employees will be summarily fired but the number could reach as high as a thousand." Stanley knew that his words would reverberate throughout the company in a matter of minutes. The already disgruntled employees would immediately react against management.

"Should we be talking like this, Stanley?" Gene Vance asked. He had taken a big chance when he had left Xerox and moved to Astron Computers, and he did not like the prospects if what Stanley said were true. Of course, he had underestimated Stanley's capacity for creating discord when he had succumbed to Stanley's blandishments.

"We are in a war," Stanley replied. "I'm going to fight Starkey. The master business professional thinks the way to build a company is to tear it apart first. I'm going to ask each of you if you support me. You have to decide between Starkey and me. Do you want an Astron sugar water company or an Astron computers with the best damned microprocessor ever built?"

Nobody replied.

"Do you support me or not?" Stanley demanded.

Stanley polled the table, and all responded with a faint "Yes, Stanley."

It was clear to all that Stanley had intimidated them. They each knew a negative vote would result in immediate dismissal because Stanley did not tolerate challenge to his leadership. Each, however, also knew that intimidation did not guarantee ironclad support.

"You will hear rumors alleging that Pierre Dupris will be moving from Paris to Cupertino and that he will take over this Division. You can authoritatively quote me as saying this will not happen."

Two days later Stanley received a phone call from a Washington Post reporter.

"Mr. Pitts, this is William Oldham. I am a Post staff reporter."

"I've heard of you," Stanley said. "You're the guy who shills for IBM. I've wondered when you would get around to us."

"I'm sorry you think that, Mr. Pitts. I assure you that everything I've written about the PC has been accurate. My editors would fire me immediately if they thought I was shilling for anybody."

"What can I do for you?" Stanley asked, pleased that he had Oldham on the defensive.

"I've been hearing rumors."

"About what?"

"About a Board fight and a struggle for control between you and Lawrence Starkey."

"Do you believe every rumor you hear?"

"That is why I am calling you Mr. Pitts. I always check my facts."

"I'm glad to hear that Oldham," Stanley tried to put the reporter in his place. "Where did you hear the rumors?"

"Are they true?"

"You know better than to ask that," Stanley condescended. "I never discuss such sensitive matters on the telephone."

"Will you discuss them in person?"

"I have no plans to visit Washington in the near future."

"I can be there tomorrow," Bill Oldham said.

"I'll be here. Give me a call."

"One thing, Mr. Pitts. I will report your interview accurately, but I will also try to meet with Mr. Starkey and others."

"A rarity. An objective reporter?" Stanley smiled to himself. He knew that Starkey would be traveling for the next three days, visiting Astron plants in a desperate effort to calm employees who were in a state of near revolt over his pending reorganization and downsizing.

After terminating his conversation with the Post reporter, Stanley resumed his informal poll of the Board. Stanley was exploiting his position as Board Chairman to elicit support for his continued tenure as Asteroid Division Chief. He began each conversation with the same refrain:

"You know the Asteroid is my baby. If I hadn't visited PARC and seen the future, Astron Computers would still be preoccupied with Dumbroski's shitty Astron 2. Starkey may be a big deal businessman, but he doesn't know squat about computers. Who do you want to run this company? A sugar water salesman or the man who crated Astron Computers."

When his target demurred, Stanley would continue:

"I grant you Starkey has let IBM impact on our market with its assembled PC. You know their components are ancient history. As soon as we shake down the Asteroid, they're dead. Now are you going to support me or not?"

To Stanley's dismay, the Board members refused to bend to his individual intimidation, a tactic that worked effectively with Astron employees. Failing in his efforts to sway the Board, Stanley turned his attention to his old mentor Stuart Miller. Impulsively, Stanley hosted an impromptu pizza party at his home and issued command performance invitations to his Division Management team and asked Miller to drop by. Stanley hoped to sandbag Miller by demonstrating the "spontaneous" depth of employee support for Stanley over Starkey. Stanley knew the Board had two choices, himself or Starkey. If he could entice Stuart Miller to return to his side, Stanley could survive Starkey just as he had Willard Temple.

The early chatter concerned Astron's future; Stanley lieutenants dispiritedly followed Stanley's lead and talked about technological marvels of the future. Stanley listened then made his move. He grabbed the bowl of cherries that Stuart had dominated and took the cherries and his former mentor to one side. There, Stanley made his pitch.

"Starkey doesn't know computers. He thinks we're selling sugar water. We've been a very special company, and we will not be managed by an outsider."

Stanley repeated the whole pitch that Stuart knew by heart. He listened with a sympathetic expression on his face and munched cherries. Stanley talked for a full hour, growing more enthusiastic with each devoured minute, while his stern faced lieutenants watched from across the room, only half interested in their talk about the Forty-Niners' upcoming season.

"What do you think?" Stanley concluded his pitch.

Miller ate the last cherry, smiled then devastated Stanley: "The Board will support Starkey's reorganization."

"Including Dupris?"

"Stanley, you're a young man. You've had your chance. You have lots of money. Get a new life."

After his guests departed, Stanley worried his way through the night. After reconsidering all of his options, he decided that he had to take his campaign public. He would use popular opinion to influence stockholder views and force the Board to retain him. The <u>Post</u> reporter offered an exploitable opportunity.

"Welcome to Cupertino, Mr. Oldham," Stanley greeted his surprisingly youthful interviewer. "Is this your first visit to the Valley?"

"No sir, but it is my first visit to Astron Computers," Bill Oldham replied.

"I'm very busy," Stanley said. "I hope you don't object if we get right to your questions.

"No sir," Bill responded. Stanley's demeanor belied his reputation. Bill anticipated that some of his questions would provoke the real Stanley. At least he hoped so, because he was here to probe the Astron Computers reactions to rumors that alleged the Astron Board was on the verge of dethroning the wunderkind.

"May we start with your decision to keep the Asteroid's architecture closed?"

The question surprised Stanley who was prepared to respond dramatically to a query about the rumors. "Certainly," Stanley smiled, working for rapport. "What is your question?"

"Obviously closed architecture denies other companies the opportunity to clone the Asteroid."

"Obviously. Would you prefer that we emulate IBM and encourage our competitors to copy the Asteroid?" As soon as the words were out of his mouth, Stanley regretted their barbed tone.

"I don't have a preference, sir," Bill did not back down. "I have heard others more knowledgeable than myself opine that IBM's open architecture policy promotes the sale of computers with a LDOS operating system and that this in turn encourages developers to write software applications."

"And the increase in the number of software applications promotes the sale of LDOS based computers for everybody including IBM," Stanley said. "This of course is fallacious reasoning. Only Astron Computers manufactures the Asteroid. We have no competitors, and we subcontract for our Asteroid exclusive software. Our computer is ten years ahead of the PC. Informed consumers do not hesitate to choose Asteroid."

"But...."

"Please let me elaborate," Stanley interrupted. "We are not opposed to cloning. We encourage developers to write applications for the Asteroid. It is only the hardware that we retain exclusive."

Bill wrote a few words on his pad, stalling while he framed his next question. He decided to shift his focus. "The Asteroid has been highly praised by the industry for its style." Bill hoped that Stanley who prided himself on his sense of design might be distracted by the not so subtle flattery.

"That's correct. A pleasing design remains one of our priorities," Stanley responded.

"There has been considerable customer criticism of Asteroid's limited memory and single disk drive. I understand that graphical user interface is a real memory hog."

"Memory is expensive. We will shortly offer our users options to upgrade their system's power if they wish to spend the money. As far as the disk drive goes, I personally made that decision. If we had included a hard drive, we would have had to include a clunky fan. Who wants that distraction and besides we would have had to redesign our case which you earlier praised. It's easy to criticize but all these decisions are interrelated."

Recognizing that Stanley was sensitive to criticism, Bill pushed harder. "Your critics claim that Stanley Pitts never saw a new program that he did not want to shove into his machine. They say your frequent additions added years and millions of dollars to the Asteroid development program."

"That's bull shit."

"They also complain because there are no cursor keys on the Asteroid's keyboard."

"The mouse moves the cursor. I deliberately decided not to include the keys, not because of cost, but because I wanted to force the user to learn to use the mouse. It's not a toy."

"Did you include too many frivolous bells and whistles on the Asteroid, driving the price above a competitive level?"

"$2, 495 for a basic machine of this quality is competitive," Stanley slapped the desk for emphasis. If users want to purchase an inferior machine for $1,500 let them. All they will get is a box of disparate components assembled by IBM and its pack of cloners, most of whom don't even know what a microprocessor is. Why are you wasting my time with this nonsense?"

Assuming he now had Stanley in the proper frame of mind, Bill went for the jugular. "Is it true that Mr. Starkey, your president, has issued an ultimatum to the board?"

"What ultimatum?"

"Either Stanley Pitts goes or he does?"

"That's bullshit. Those bozos know that I founded this company. I made it what it is. If I were to go, Astron Computers would be finished."

"You did not answer my question," Bill persisted.

"If those bozos tried to remove me, it would mean war."

"Who do you refer to when you say bozos?"

"Starkey and his sycophants."

"Have you declared war on the bozos?"

"Yes."

Bill knew he had his headline: "Stanley Pitts Declares War on Bozos."

"Have you counted votes?"

"I've met with my senior executives and to a man they support me."

"You mean in the Asteroid Division?"

"Certainly where else? The rest are bozos."

"When I asked about votes I referred to the Astron Computers Board of Directors. How would you characterize your support there?"

"I am the company's major shareholder."

"Am I to infer that means you have the support of the majority of the stockholders?"

"How do I know?"

"If you have declared war on the bozos, I assume that means you are prepared to fight?"

"Certainly. The bozos are in for a surprise."

"Mr. Pitts do you have an engineering background?"

"If you are asking do I have an academic degree in computer science or engineering, the answer is no. I do have a practical experience degree."

"I don't understand."

"I'm not surprised. How did you get your job? Did you create your newspaper? I created my company."

Bill scribbled Stanley's words on his pad.

"How much do you earn? $30,000? I have two hundred and fifty million dollars, and I started with nothing. What do you say about that?"

"Mr. Pitts, I apologize if I offended you," Bill said. In reality, he was not sorry at all. He had deliberately probed to find out if the rumors about Stanley's disruptive personality were true.

"Very well," Stanley said. "If you want to write a story about Astron Computers management, go ahead. Just make sure you say that Stanley Pitts has declared war and is confident of victory. As soon as our stockholders learn that the bureaucrats are attempting a coup d'etat, their voices will be heard."

"Mr. Pitts ..." Bill began another question.

"This interview is over," Stanley declared. "I have a company to run."

The next day Stanley awoke with a sense of great anticipation. This was going to be his day to fight back. If he had gauged his meeting with the young Post reporter correctly, today Stanley would make headlines. Stanley shaved, dressed, grabbed a cup of tea and hurried to his white Mercedes. Stanley mouthed the words as he listened to his favorite tape while he plotted his pending conversation with Starkey. Stanley decided he would be a gracious winner. He parked in his reserved space directly in front of the door that led to the Asteroid Team Building. Overhead, Stanley's flag, the skull and crossbones, flapped in the light breeze.

"Morning Mr. Pitts," the guard at the door greeted him.

"Morning John, great day," Stanley surprised him. Usually Stanley simply ignored the guard, treating him as just another fixture.

Stanley crossed the lobby. He smiled and nodded but noticed that those he passed greeted him with curious blank stares. Stanley assumed they had read his challenge and did not know how to react. Unlike Stanley, they had no way of knowing if his words came from weakness or strength. Stanley knew.

"Morning Jane," Stanley literally bounced into his office. He was eager to get started. He wondered how Starkey felt.

"Good morning, Mr. Pitts. You have had several calls."

The news pleased Stanley. "What does Starkey want?" Stanley knew but wanted to hear his secretary say it.

"Not Mr. Starkey. The media," Jane said. "Have you seen the papers?"

Stanley shook his head. He did not need to read the morning news. Stanley had made it.

"You might want to start with the paper on your desk," Jane said. "Before you return any calls."

Stanley detected a note of concern in her voice. "We're on top Jane, don't worry," Stanley said.

"I hope so," Jane said.

Stanley entered his office and closed the door. He hung his jacket on the coat tree next to the door and approached his desk. Along the right side, Jane had arranged a series of phone messages. In the center, a copy of the Washington Post waited. Stanley sat down and paused a few seconds before picking up the paper, savoring the moment. Stanley so liked to win. He wondered what Starkey was thinking. Certainly, he would realize that Stanley was appealing to the stockholders over his and his parrot board's heads.

Stanley was disappointed to see he had not made page one. He didn't bother to read the news stories. Stanley had no interest in the Beirut crisis or the European reaction. Stanley leafed through the pages until he found the Business Section.

"STANLEY PITTS DECLARES WAR ON BOZOS"

Stanley's story topped the Business Section, but that headline was not what he had in mind. The entire focus was on the bozos. Stanley was quoted as saying that the Board, the Company President, all the division chiefs and employees of every division except the Asteroid Division were bozos intent on driving Stanley out and destroying his company. The reporter described Stanley's pirate flag and quoted Astron employees who to a man depicted Stanley as a disruptive influence. Then, the reporter baldly disclosed Stanley's plan. He was appealing to the stockholders to reject the Board and senior management and install Stanley in their place.

Stanley angrily threw the paper to the desktop. Before he could consider how to react, the intercom rang.

"Yes," Stanley barked. All of his good cheer had vanished.

"Mr. Starkey asks that you join him in his office now."

Stanley slammed the intercom receiver down. He thought about phoning Starkey and telling him to stuff his meeting but decided against it. That would be admitting defeat. Instead, he decided to meet Starkey head on and brazen it out. Stanley vowed to show him he still had fight left in him.

"He's expecting you, Mr. Pitts," Starkey's secretary waved him through.

Stanley noticed she averted her eyes as he passed.

"Shut the door and have a seat," Starkey greeted him gruffly.

A copy of the Post open to the Business Section lay on Starkey's desk.

"Did you talk with this reporter?" Starkey demanded.

"Oldham?"

"Yes, Oldham."

"Yes. What of it?"

"You called employees of this company bozos?"

"In a matter of speaking."

"And the Board?"

"Indirectly."

"And me personally? Am I a bozo?"

"Yes, definitely you."

"And you declared war on the bozos?"

"A turn of phrase."

"Is this some kind of coup attempt?"

"I don't understand what you mean?"

"Pitts. Stop playing games with me. It won't work. You don't have the votes. After this story, you don't stand a chance with the stockholders."

"We'll see Lawrence."

"Don't call me Lawrence. I am going to do what I should have done when I first came here. As of this moment, Gene Vance is acting Chief of the Asteroid Division."

Starkey's words stunned Stanley.

"Take leave, whatever, but stay away from Astron Computers until we decide what to do with you."

"I'm the major stockholder," Stanley said.

"Then vote it at stockholders meetings. You will never hold another operating position in this company."

"But I founded Astron Computers. I named it. I built it."

"Thanks. Talk with Harold Dumbroski about that."

"We'll see what the Board has to say."

"Yes we will," Starkey glared at Stanley. "I've called an extraordinary Board meeting for Friday."

"But I'm Chairman," Stanley stammered.

"We'll see about that too."

The two men sat in silence. Starkey was so angry that he had difficulty controlling his temper.

"That'll be all. You're dismissed," Starkey said.

"I'll see you at the Board meeting," Stanley's voice cracked despite his effort to sound menacing.

"And don't talk with any more reporters."

"I'll talk with anybody I want," Stanley bluffed.

"You never learn do you," Starkey pointed derisively at the <u>Post</u> story. "I don't know whatever made me think I could teach you business."

Stanley did not return to his office. He retreated to his Mercedes and defiantly spun his wheels as he raced out of the parking lot.

Two days later a grim faced Stanley dressed in his best blue suit and maroon power tie returned to Astron Computers to chair the extraordinary board meeting. When he entered the room, he found all of the other directors already in their seats. Stuart Miller sat in Stanley's chair at the head of the table. Copies of newspapers from around the country were spread prominently on the long table. Most featured some version of the bozo headline.

"Thank you for joining us, Stanley," Stuart Miller said. "The Board has convened and asked me to chair this meeting as Vice Chairman. Since you are the prime subject for discussion, we unanimously felt that it would be inappropriate for you to preside."

"But I'm the Chairman," Stanley protested weakly.

"We will discuss that," Stuart Miller said ominously. "Please have a seat."

Stanley glanced at the table. Starkey was sitting in Miller's usual position at the chair's right hand. The only empty chair was at the foot of the table. Stanley took it without protest.

"I declare this extraordinary meeting now open," Miller said. He looked at the stenographer sitting against the far wall, indicating she should begin taking notes.

"Do we need to have this on the record?" Stanley asked.

"Yes. Don't worry. This will be a short meeting. Lawrence, do you have a motion?" Miller soberly waited for the company president to begin.

"Yes, Mr. Chairman. I formally request the Board's permission to implement immediately the reorganization plan I have tabled."

"Point of order," Stanley blurted. "What plan?"

"A registered copy was delivered to Board members yesterday," Starkey said coldly.

"I haven't seen it," Stanley protested.

"You were on leave."

"Do I hear a second?" Miller interrupted before Stanley could speak again.

"Second," the sturdy businessman on Miller's left said.

"Motion made and seconded," Miller said.

"We will now vote with a show of hands," Miller said.

"I request that the Board be polled," Stanley said.

"Very well. How do you vote?" Miller turned to the man who had seconded the motion.

"Bozo Number One votes yes," he glared at Stanley.

"Bozo Number Two votes yes," the next Board member said.

"Bozo Number Three votes yes," the next Board member said.

And so it went around the table with only Stanley voting No.

"The motion carries," Miller declared. "I have only one other item and it is for discussion only. Mr. Starkey has requested that I elicit the Board's opinion on an organizational matter. Mr. Starkey please present your issue."

"Thank you Mr. Chairman. The reorganization plan that the Board just authorized includes among other things the amalgamation of the Asteroid and Stella Divisions. I will appoint Mr. Tom Williamson Acting Chief of the

Advanced Products Division. I will not appoint Mr. Stanley Pitts to another operational position. In fact, I plan to issue instructions to security to bar him from access to all Astron Computer facilities. Does any member of the Board object?"

Nobody said a word. A stunned Stanley waited for support that never came.

"Am I still on this Board?" Stanley finally demanded.

"That is a subject we will defer to a subsequent meeting," Miller said.

"How will I get in to attend Board meetings?" Stanley asked.

"We will arrange to have you escorted to the meeting room," Miller said.

"I am still a major stockholder," Stanley blustered.

"Thirty percent. You will be given the courtesy that every stockholder is entitled to," Miller said. "I declare this meeting adjourned."

In the weeks that followed Stanley reassessed his chances of returning and finally decided that they were nil. He then began making plans for starting a new company. He had done it once and was confident he could do it again. He discreetly contacted the Asteroid engineers whose work he respected and probed them for interest in joining him. Several responded positively. Upon learning of Stanley's attempts at suborning key personnel, Starkey erupted. The Board supported Starkey and stripped Stanley of his Chairmanship.

Stanley submitted his resignation and joined Harold Dumbroski as an ex Astron Computers employee.

Chapter 27

By the end of August, David's job as Chief of PC Marketing and Sales had evolved. Despite the lukewarm media reception following the PC's New York introduction, the orders started to trickle in. Word about the new machine, particularly commentary on its reliability, spread. Before long, David's staff was reacting to queries from chains and individual outlets that wanted to feature the PC. Ernie diverted more experienced salesmen to assist David's novices.

As David's anxiety about his professional life ebbed, worry about his private life expanded to fill the emotional gap. Terry assured David during their weekly phone conversations that she had everything in hand, but David was not reassured.

"David," Terry said. "All you have to do is be at the First Methodist Church in Boca Raton at eleven o'clock on the morning of September 4th. It's a Monday and a holiday. You can handle it."

"I'll be there Saturday afternoon for the practice," David said.

"Don't worry. Speedy said he will stand in for you."

"For the wedding?"

"No, for the practice. But Speedy's not a bad idea. He's got a great serve."

"Terry!"

"Just kidding."

On Friday evening, the first of September, at seven o'clock, David signed his last piece of paper for the day, an agreement authorizing a new electronics chain to market the PC. He apprehensively watched the phone as he turned out the light, knowing it would ring to report a disaster that would prevent him from catching his Saturday morning flight to Boca Raton. David planned to rise early, drive to Laguardia, park his car in the long-term lot, and board Southern Airways for Miami at ten.

He was passing through the bullpen when his phone started to ring. He thought about ignoring it but knew that if it were important they would track him to his home in Middleville.

"Howard," David said.

"David, it's Bill Oldham. How you doing?"

David sighed. Since their meeting in New York, Oldham had phoned David once a week trolling for news on the computer industry. The young reporter had protected David as a source, and David liked the idea of having a reliable entre to the media. He had twice put Oldham on the scent of minor Armonk news stories.

"I'm on the way to the altar."

"Someone I know?"

"Me, but that's off the record."

"Not necessary. I'm sure the Post readership doesn't care. Good luck."

"Thanks."

"Shotgun?"

"Of course not."

"You're fortunate."

"What about you?"

"No prospects. Have you heard the news?"

"What news?"

"Astron canned Stanley Pitts."

"That's not news. After your Bozo story what did you expect?"

"Do you have any bozos up there in Armonk?"

"No Comment."

Oldham laughed. "Do you think I would get a reaction if I called your public affairs people?"

"They're not my public affairs people, and no they will not comment."

"I thought all of Stanley's competitors hated him?"

"I don't know about that. We don't think about him much."

"Can I use that?"

"Certainly not."

"What about this headline: 'Armonk says 'Good Riddance' to Stanley."

"Don't you dare. We haven't heard the last of Stanley. He'll bounce back."

"He's a pretty obnoxious character. He almost threw me out of his office, physically."

"Stanley's not all bad."

"I'm afraid his personality got to me. What's your fiancé's name."

"Forget it."

"Don't you think her parents would like to see their daughter's name in the Post Style section?"

"You're kidding me."

"Nope."

"Just a favor for a friend?"

"Yep."

"Terry Mitchel."

"Father's name?"

"Terrence Mitchel."

"Vocation?"

"Golf pro."

"Boca Raton?"

"No, Miami Beach."

"Sounds good. Check the Sunday edition."

"That's it," David ended the conversation hoping he had not been wrong to trust Oldham. The media treatment of Stanley Pitts had offended David.

David's plane landed in Miami on Saturday on time. After he had checked his luggage, he realized he had trusted his rented tuxedo to the vagaries of airline baggage handlers. He waited anxiously at the baggage carousel and grew ever more concerned with each rotation with no sign of his bag. As the flow of luggage from his flight dwindled, David began to anxiously check the growing line at the luggage retrieval counter. A recent article in the Times had noted that fifty bags out of every one thousand checked were lost. Suddenly, the carousel slammed to a stop. A dispirited David turned to join the lost luggage line when someone jerked his sleeve. David turned to complain and found a smiling Terry.

"They've lost my suitcase," David blurted.

Terry kissed him on the lips. "Don't worry."

"It's got my tuxedo in it."

"We can still get married."

"It's a holiday weekend. I'll never find another."

"Lover, in Miami you can find anything anytime. Besides, you could borrow Speedy's in a pinch."

David could see himself wearing Speedy's tux. "I'd look like the scarecrow in the Wizard of OZ."

Suddenly, the carousel started again and David's suitcase popped through the hole.

"There it is," David said.

"See, Moma fixes everything."

David grabbed his suitcase in one hand and took Terry's hand in the other. She led him to David's old Volkswagen.

"I thought you were going to sell this."

"I changed my mind. We'll keep it here so we have transportation when we come down for vacations."

"I didn't know we were."

"There's a lot of things you don't know sweetie."

Terry unlocked the door and climbed behind the wheel. She pulled the lever and popped the trunk. David tossed in his suitcase, closed the lid, then climbed into the passenger seat. "Sure you don't want me to drive?" David hated being a passenger.

"We only have forty-five minutes to get to the church," Terry said.

"We can't keep the preacher waiting," David said sourly.

He rode in silence for several minutes as Terry concentrated on getting out of the parking lot then merging into congested traffic. "Do we really want to do this?" David asked, semiseriously.

"Do what?"

"This whole wedding thing."

Terry laughed. "Cold feet?"

"No, it's just that I'm so young. I'm not sure I'm mature enough for marriage."

"You're not, but don't let it worry you because you'll never be."

"You're talking to a big time executive."

"And I'm a big time fiancée."

"How's the golf pro taking everything?"

"He agrees with you."

"That I'm not mature enough to get married?"

"That you're not good enough for his little girl."

"Then I agree with him."

"Do you have more pictures of the house?"

"Yes."

"Hide them. The picture you sent made it appear like a scrawny little thing."

David smiled. He was keeping the fact that he had succeeded in getting a house in Middleville as a surprise. He had deliberately brought pictures of the most depressing house in could find in Armonk. David privately resented the golf pro's disapproval.

They found Speedy and two others from the Boca Raton office who had agreed to serve as ushers waiting at the church in the company of a flock of chattering bridesmaids. After a flurry of welcomes and introductions, Terry led the group into the church where a thin, balding, pompous clergyman waited. As soon as he entered the church, David's resentment at being forced to go through a ceremony that he considered sheer hypocrisy welled up. Terry, sensing David's discomfort, took his hand and pulled him down the aisle.

"Cheer up. This too will pass," she whispered. Like David, she had never been able to take the required first step of a true believer: suspend critical judgment and accept what you are told on faith.

"This is all to please your mother," David whispered back.

"Yes."

"Then I want a basketfull of credits for tolerating this."

Terry smiled but made no promises. Instead, she whispered: "Mom won't accept our marriage unless it is in the church."

David could think of nothing to say.

The smiling clergyman walked them through their roles. As best as David could determine, his primary function was to stand there, look nervous, say yes, and then kiss the bride. Only the last act interested him. As he fidgeted and impatiently listened to the clergyman indulge his moment of authority, David repeatedly checked his watch. Finally, the clergyman leaned over and asked: "Got something more important to do, son?"

"I'm not your son, father," David cracked back.

The preacher hesitated, obviously debating whether to smile or not, then relented and took the easy path. He smiled. From that point on, he studiously ignored David.

After a thirty-minute experience that David equated with knee surgery, the clergyman dismissed them with a smile and a gratuitous best wish for Terry and a frown and no handshake for David.

"I heard that," Terry punched David in the ribs as they emerged from the church.

"Heard what?"

"That crack about father and son."

"Oh." David regretted his smartass reaction but was determined not to show it.

"You've forfeited some of your credits," Terry smiled.

"Where do you have mom and dad quartered?" David changed the subject.

"At the Hilton."

"And me?"

"I've subleased the apartment and sold the furniture," Terry said. "I'm at home, and you can't stay there, so I put you up at the Hilton too."

"Great," David said and meant it. He intended to minimize the time he spent with Terry's parents. "I'll drop you off and check in with my folks at the hotel."

"OK but remember, we're all supposed to be at the house tonight for a family get together."

David frowned.

"And don't get in a hassle with my father."

"I'll be the best caddy he's ever had." David didn't really mean those words.

361

Saturday night at the Mitchel home proved to be sheer drudgery. David followed Terry about while the older Mitchels and Howards chatted. David could tell that his father had the same negative reaction to Terry's father that David did. David's father had no interest whatsoever in golf—he considered it a silly game—and Terry's father had no interest in anything but golf. Somehow, they got through the evening.

Sunday morning David deflected invitations from Terry's parents. Her mother wanted him to attend church, an effort designed to get her future son-in-law belatedly started on the right path, and her father invited him to a round of golf, planning to whip David's ass thoroughly, establishing beyond doubt who was boss in the family. David earned some credits with Terry's mother when he produced a copy of the <u>Post</u> and casually opened it to the Style page. It had apparently been a slow Labor Day weekend for Washington society; Bill Oldham had succeeded in getting the terse news of David's wedding a prominent position as the leadoff story in the weddings column under a black type subhead: "IBM Executive Marries Daughter of Golf Pro." Terry's father reacted with a grunt. "Big deal."

David took his parents on a tour of southern Florida. Sunday evening David spent with Speedy and his ushers. After a few dispirited drinks, David retreated to the hotel where he watched television and worried about his pending loss of freedom.

Monday morning passed quickly. David prepared his suitcase for flight, donned his tuxedo, then waited uncomfortably for the dragging minutes to pass. Finally, Speedy rescued him and drove him to the church. There, the ceremony proved mercifully brief, and they were soon speeding from the church to Terry's home where the bride and groom changed clothes. After enduring a reception at the golf course, they escaped. Speedy drove them to the Miami Airport.

Fortunately, their plane departed on time. David and Terry each accepted the stewardess' offer of a little bottle of cheap New York champagne, and David toasted his bride: "Thank God that's over," David raised his glass.

"Regrets?" Terry lifted her glass.

"No way. I've got to admit that for the second time in my life, I'm happy to be going to Armonk." David kissed his wife.

"When was the first?"

"My first day."

"So I'm second?"

"No, this is the first day of the rest of my life," David said.

"How long will we be in Armonk?"

"Forever, I'm afraid."

"I wish you had told me that yesterday, lover."

"Too late. I'm all yours, to honor and obey."

"Answer me one question?"

"Yes?"

"Why did you show my father the pictures of that horrid little house when you already have a much nicer one for us in Middleville?"

"Who told you that?" David evaded a direct response.

"I have my sources."

"Damn that Margaret."

Chapter 28

Back at Armonk, the innocent David did not realize at first that he had climbed aboard a corporate roller coaster.

His first three years were comparable to a climb to the top of the first and most formidable peak. On his behalf, David never once succumbed to the ego enhancing temptation to take credit for the phenomenal sales that the PC he had assembled brought to an already dominant IBM. During David's first year as Chief of the PC Sales Group, an eager public gobbled up 200,000 machines. In three years, PC revenues coursed from zero to four billion dollars. This surge lifted IBM's profit line to a staggering seven billion dollars. Inexplicably, Time magazine named PC the Man of the Year. By the time this honor descended, IBM had 10,000 employees working on the little PC.

Quite naturally, David's status and income soared commensurately. After one year, he and Terry moved from Middleville to Seniorville, just one rung down from King's Row. True to her word, Terry found a junior position with the Office of Public Affairs. Here, her tennis prowess stood her in good stead; whenever a visiting VIP or top ranked visitor needed a tennis partner, he found Terry at his or her side. David, whose salary had risen to a level where Terry's modest contribution was not needed, on occasion suggested that his wife might chose a more conventional role, but he learned not to press the issue. Terry obstinately refused to surrender her independence.

From the PC's birth, cautious IBM bureaucrats had worried about two nagging problems that had been inherent since the market transformed the PC from a marginal experiment into a big time product: firstly, impact on the company's primary vehicle, the main frame, and, secondly, cloning. Initially, the impact on main frame sales was marginal; corporate customers preferred to stick with the software and machines they knew and trusted and to rely on good old Big Blue to fix any problems.

Cloning, however, proved to be a huge imponderable. What the Management Group worried about happened. The soaring PC sales and the open architecture made cloning inevitable. Three Compaq executives viewed the PC at a Houston trade show, retired to a nearby restaurant where they worked out the format of a PC clone on a napkin, and shortly thereafter launched their own version of the personal computer. Others followed. IBM at first reasoned that they and their electronic powerhouse would dominate the market with volume sales that let them price basic units lower than competitors. This reasoning failed to compensate for one key factor. PC sales pulled component suppliers into the volume production business, and the cloners purchased their parts at the same price as IBM. This made the cloners competitive. The one exception to this phenomenon was Digital Software. Charles Swift sold aggressively to the cloners, but always at a price higher than that charged IBM. Cloners compensated, however, by operating without the huge overhead that IBM's bureaucracy imposed on the computer Goliath.

IBM and the clones were not the only ones to benefit from the PC's massive success. In Seattle, Digital Software was almost overwhelmed by its quantum leap in business. Digital Software burst through the $100 million dollar sales mark and did not pause. Operating systems, read LDOS, and languages, mainly Charles' favorite, BASIC, were the major money producers. Packaged software for the retail market followed close behind. Not every economy cloner bundled software with its microprocessors, and the fact that millions of machines worldwide could now use IBM compatible software greatly expanded the market.

This phenomenal expansion fueled by IBM's entry into the microprocessor market did not benefit everyone. In Cupertino, Astron Computers, wracked by internal dissension, suffered greatly from market share loss. The Asteroid with its graphical user interface coasted years ahead of the PC, at least technically. A much higher price forced first time purchasers to turn to the PC and its clones, and Stanley Pitt's insistence that the Asteroid architecture be closed and incompatible with any other operating system had devastating effect. The closed system restricted the number of programmers writing software for the Asteroid. PC rocketing sales combined with its open architecture proved too enticing for the software writers to resist. As a consequence, Astron Computers, hamstrung by decisions made years earlier by a now departed Stanley, faced a competitive market against a corporate giant abetted by cloners and eager software programmers.

David's enhanced status gave him access to a better understanding of his company's history and future prospects. Relaxing sessions after golf and tennis and hushed conversation over cocktails and dinners introduced David and Terry to the realties of their corporate family. As a trainee, David had listened to lecturers detail the marvelous story of the 360 Project. Young Tom Watson had courageously bet the company on the massive mainframe and had won. Watson's IBM created a family of computers able to support large numbers of workstations all using a mixture of software programs and peripherals at the same time. The 360 lifted IBM to the corporate stratosphere, an accomplishment whose effect lasted for twenty years. However, technology advances, and IBM had to plan for the future. Adjustments like the 370 and 380 gave the company time to manage the next break through.

The Management Committee in the mid 1970's opted to try another generational leap using a project labeled F/S. Quite ambitiously, F stood for Future and S for Systems. Future Systems. It sounded like Buck Rodgers or Flash Gordon stuff. Years ahead of its time, F/S called for technology able to support multiprocessing and object oriented programming using thin film disk storage, wafer-scale integration and other difficult to comprehend advancements. Multiprocessing managed the performance of many tasks at once. Object oriented programming called for the writing of code in small segments that could be used to perform tasks for a variety of programs freeing programmers from the onerous task of writing the complete code for every task for every program. For a variety of reasons, F/S failed, costing IBM a full five-year stage developmental cycle.

A world economy upturn enabled 370 sales to cover the F/S failure, but in-house the result was felt. Never again would the Management Committee bet the company on a generational breakthrough machine, particularly one that challenged a successful product. F/S had cost billions of dollars. A heavy, cautious bureaucracy appeared to provide top management with safe choices. Administrators took over the company at every level, and strategic thinking and risk taking became the subject matter of past pride.

David finally understood that the IBM he had joined was no longer the company that the Watsons had created. At the time, he had not realized how heavy was the tide that Ernie Hendricks with the support of James Crane had challenged. This new understanding frightened him retroactively. Their selection of a young, technically deficient trainee (David) to establish a new product line that challenged the 360 was a decision that defied comprehension.

Once, David asked Ernie: "Why me?"

"That's easy," Ernie replied. "We would never have gotten the Management Committee to approve the project if we had appointed a serious engineer to head it. They concurred because they did not think it important enough to challenge the company president."

"But still, why me?"

"Because you were eager, intelligent and had already proved your ability able to face challenge?"

"I had?"

"You played running back for Cornell didn't you?"

Following a heavy weekend of competitive mixed double tennis—he and Terry had won the company club tournament—David dropped Terry off at the door to the annex, which housed staff offices including those of the public relations division.

"If I have to stay over, I'll give you a tinkle," Terry smiled. Terry had several appointments later in the day on Madison Avenue.

"How can I reach you?"

"You know where I'll be."

David nodded. IBM maintained a block of transient apartment for executives forced to overnight in New York City. "What'll I do for dinner?"

"That's up to you sweetie. It's your night to cook."

"I get credits if you don't show up."

"No way."

David nodded. This was a familiar refrain that the spouse whose turn it was to cook tried. He knew it was doomed to failure, but he felt obliged to try.

"I'll bet I'm the only group chief who has to cook dinner." This too was a familiar ploy.

"Get promoted to division chief and hire a cook."

"I've been thinking about that. Could I have a personal assistant instead?"

"A personal assistant?"

David smiled. This new idea had just occurred to him.

"Sure. I'm thinking about a newer model, about twenty, blonde...."

"What about a Paul Newman type?"

"No. My tasking requires a girl type."

"Not on your life, buster."

David parked in his reserved space located some fifty slots from the front door and entered.

"Morning David," his mature secretary greeted him. "Ernie wants to see you."

"Why?" The highly successful PC marketing group had no problems pending that required division attention. In fact, things were running so well that David was a little bored with the tedium of group meetings and retail chain negotiations.

"No sir. He just said you were to come up as soon as you arrived."

David shrugged. "Anything I should focus on here first?"

"A few phone messages and the usual memos. Nothing that can't wait."

"Hi Margaret," David greeted Ernie's secretary, his former girl friend. Margaret had finally surmounted her pique over David's marriage. She and

Terry were now good friends, so good in fact that David sometimes felt excluded. He had trouble understanding the difficult female mystique. "Ernie in?"

"He's waiting for you."

"You're looking great," David leaned across Margaret's desk.

"I don't know what he wants," Margaret smiled. "And whatever it is, I'm sure you deserve it."

Her tone did not make it sound like good news.

"Ta Ta," David waved and entered Ernie's office. Without being asked, he closed the door behind him, not to maintain privacy, but to torment Margaret.

Ernie was reading the New York Times.

"Something in there I should know?"

"Yes. The Yankees lost. As a marketing group chief, you should keep up with things."

"Yes sir."

"Oh well, I guess you won't be caring about news in the Times much longer."

"Sir?" David sat up straight in his chair.

"Been a little bored lately, David?"

"No sir," David blurted then remembered who he was talking with. "Well a little."

"How did the tournament go?"

"We won. Well, to be honest, Terry did."

"That's a given. She should really find a more competitive partner."

"Oh, I'm not so sure about that."

"How's the golf game?"

"Whose?" David recognized the signs. Ernie was playing with him. He always adopted this laconic mood when he had something earth shattering to drop on David. He had been lucky so far, in Boca Raton and then his return here, but both times he had to overcome a job he was not equipped to handle.

"Certainly not Terry's. She's got great form."

"So would I if my father was a golf pro." David was not sure if he were defending himself or Terry.

"Golf's not your game anyway."

"I'm not going to play anymore softball," David said, guessing that was the direction Ernie was heading. Ernie still coached the Sales Division team. David had given up the game last summer after breaking his right wrist sliding into third base.

"No, I don't blame you. You're slowing down anyway."

"I am not," David said.

"How do you like your house?" Ernie asked, referring to David's Seniorville home. David and Terry both loved it. After two years, they finally had it furnished and landscaped to their satisfaction. David had just contracted for a pool that would cost more than his parents had paid for their house.

"Oh shit, Ernie, just tell me the bad news."

"The Senior Personnel Assignments Subcommittee of the Senior Management Committee met last night."

"And?" Now, David was really nervous. Ernie was Chairman of the Senior Assignments Committee.

"Somebody brought up your name?"

"Oh shit, Ernie. Who did?" David tried to think of a Management Committee Division Chief he had recently crossed. No matter how hard he tried to humor his seniors, the simple fact of the PC's growing importance to the company dictated that others would involve themselves in his marketing strategy.

"A senior member."

"Who? Was it Parsons?" After all these years, David still worried that Parsons might decide to take his revenge.

"No. Me."

"You?" David was shocked. Ernie was his patron.

"It's time we broadened your perspective."

That could mean anything. "What are you doing to me Ernie."

"You've had your stints in General Products and in Sales. Now it's time you learned what this business is really about."

"I'm begging. Tell me. Is Terry going to kill me?"

"How does she like cold weather?"

"She hates it."

"Good. This will broaden both of you. What do you think about Wisconsin?"

"I try never to think of Wisconsin."

"Congratulations. You have escaped from my command."

"What's in Wisconsin for me?" IBM had a major mainframe plant in Oshkosh. It fell under the Production Division where David had never served, not even on a trainee interim assignment.

"You are now Plant Manager, Oshkosh."

"I don't know shit about Oshkosh. Or plant managing, or production. I'll screw it up."

"That'll be hard, but you'll have to give it a try. You have one week to get there."

"What about Terry?"

"We have no objection to wives living with husbands."

"I mean. Her job?"

"We don't have a public relations office in Oshkosh."

"Oh shit."

"But it snows a lot. You can take up ice fishing."

"I don't fish."

"Oshkosh, named after a Menomini chief, rests on the lovely shores of Lake Winnebago."

"Seriously Ernie, why me?"

"Why not? You might not have noticed, but we are gradually training you, as difficult as it may seem, for a senior position in this humble organization."

"But production?"

"That's what we do, David."

"Terry'll kill me." David was beginning to adjust to the idea. He liked snow.

"You can tell her Oshkosh is a promotion. How does $200,000 sound?"

"It sounds great." David thought about the money for a few seconds. On that salary, they could live really well in Oshkosh. "What does Crane think about this?"

"He likes it."

"He wants to get rid of me."

"No, he thinks you need broadening, too." Ernie smiled. "Maybe the thought of you and a Oshkosh winter also pleases him."

After Ernie dismissed him, David rushed through Margaret's office.

"Congratulations, David," Margaret called after him, then she giggled. "Better Terry than me."

David realized the news would spread through the Armonk bureaucracy quickly. He rushed to his office, knowing he had to be the first to tell Terry. If he didn't prepare her for the blow, David would suffer for a decision that was not his to control.

"Congratulations boss," David's secretary greeted him.

"I've got to tell Terry," David replied.

He dialed Terry's number, and she answered on the first ring.

"Public Relations," Terry said.

"It's me," David said.

"I'll hate Oshkosh," Terry said.

"They have great fishing."

"Ice fishing."

"We'll build one of those little shacks on the ice."

"I'll freeze to death."

"We'll put a heater in it. It'll be real cozy."

"I'm not going to live on the ice."

"This is just for fishing," David said.

"The damned thing will melt the ice and sink."

"Not in Lake Winnebago," David said.

"When do you have to be there?"

David noticed the stress on "you." "Monday."

"Which Monday."

"Seven days from today."

"I close out here and meet you."

"No you won't. Last time you said that it took two months, and I had to go to church too."

Chapter 29

Charles sat at his desk, alone behind a closed door, and stared at the ceiling. His company was booming; Charles was already richer than he had imagined possible, and he was as depressed as he had ever been in his life. James Clapper, "Clap," Charles' boyhood mate, partner, lifelong friend, had Hodgkin's disease. It had appeared out of nowhere. On a business trip, Clap had noticed some lumps in his neck. Now, he was gone. After weeks of radiological therapy, Clap had tried to work half days, but that had not worked out. Clap had lost interest in computers and had resigned, leaving Charles alone. All Charles' life he had felt special, one of the gifted, invulnerable, then this had happened. Charles still had Mike and his parents, but Clap had been special, always there, even during Charles' exile in Cambridge. They had done everything together. Without Clap, Charles felt exposed. Money and power no longer seemed important.

With Clap's departure, Charles suddenly felt the remoteness that growth had brought to Digital Software. In Albuquerque, they had been Robin Hood and his merry gang. Ironically, home in Seattle, the core group was disintegrating. Several of the originals had departed, now Clap, the constant was gone. Charles was so busy dealing with new product development and traveling to meet with major clients, testifying before Congress, making public appearances, that he had no time to write code or share the fun where the carefree band worked. Gone were the days when every employee had access to Charles, when he had the time to share with friends. Now, levels of essential management isolated Charles at the pinnacle. He liked the power, needed the freedom to address the big problems, but he lost much in his splendid isolation.

Right now Charles had to decide what to do about Panes, Charles' counter to the graphical operating system devised by PARC and implemented by Stanley Pitts in the form of Stella and Asteroid. Charles had known it would not be an easy task to develop Panes, but he had not realized it would take this long. Digital Software had the smartest guys Charles could hire, and still they struggled. Charles was prepared to bet the company on Panes, but he prayed that would not be necessary. He had always had Clap to discuss such problems, but now he was gone. Mike had a good business sense, but he was not technical.

Panes posed not only technical challenges; two of Digital Software's major clients, Astron Computers and IBM, resented Charles' insistence on developing his own graphical user interface operating system. The reason was simple. Neither liked the idea that the software company they had helped grow into an industrial force worked on software that would challenge their own products. Nonetheless, Charles, like Stanley, and now the rest of the microprocessor world, recognized that the future depended on those small icons, and he was stonily determined not to be left behind. Customer angst was not new to Charles. Astron Computers, for example, resented the fact that Digital Software provided the operating system for the PC, Astron's ever menacing competitor. Astron had even canceled their original agreement. This did not worry Charles. He kept the $50,000 advance and was free to market his own products. In Charles' opinion, a vital force had gone out of Astron Computers when they dumped Stanley Pitts. IBM posed a bigger problem. IBM was developing its own graphical user interface and deeply opposed Charles' preoccupation with Panes. IBM views on the subject became so strong that Charles had to dissemble, denying that he was still working on Panes.

The intercom buzz interrupted Charles' musings.

"Charles, Jane's on the line," Charles secretary said.

Jane represented Charles' response to his mother's concern that Charles might be gay. Jane, who had become a millionaire in her own right when she sold a Minneapolis accounting software firm she had founded, had recently moved to Seattle. Charles had casually encountered Jane at software expositions around the country and had responded to her scarcely concealed overtures following her appearance in Seattle. Jane, in Charles' opinion, unfortunately, was renown for having popularized the word vaporware to describe software announced but not yet written. All too often Charles' competitors applied the denigratory phrase to Digital Software's aggressive sales practices. Charles had no aversion to advertising products that were still in the planning stages, particularly when it appeared that competitors might be on the threshold of marketing such applications.

Charles had introduced Jane to his parents, and both were delighted. Charles enjoyed Jane's company, when he had time for it. Charles, in private, made no secret of the fact that he had room in his life for only one love, his company, and Jane accepted the situation. Jane, for Charles, presented the perfect solution. She understood he was not yet ready to abandon his single life style, and as long as she accepted the situation, she served as his romantic shield.

"Hi Jane. What's up?"

"I need a change of scene."

Jane in Seattle had established her own consulting firm and had of late engaged in some modest venture capital investing.

"What do you have in mind?'

"Something in the Caribbean."

"No can do. I have to be in Washington Monday."

"OK. What about dinner?"

"I pick the restaurant."

Jane was a confirmed vegetarian. For a while Charles had indulged her, but he had recently reverted to his preferred diet, cheeseburgers and shakes or sausage pizzas with extra cheese.

"No way. We compromise. Let's go Indian," Jane said. "I'll pick you up when?"

"Eightish," Charles said. "Here."

When Charles hung up, he directed his secretary to get Bill Oldham on the phone. Whenever Charles visited Washington, which was more often than he liked, he touched base with his old college roommate. Charles was still comfortable with Bill despite their diverging paths. They could sincerely discuss Clap's problems, and the huge hole his departure left in Charles' life. Charles decided now would be a good time to make another try at enticing Bill to leave the newspaper world and join Charles' growing empire. Bill always resisted. Charles assumed it had to do with ego, not wanting to end up dependent on his old roommate, but now Charles could honestly say he needed another person he trusted.

Charles still had Mike, but their relationship had become more business than personal with most discussions devolving into heated arguments. Charles knew the responsibility for the situation was his. He encouraged subordinates to disagree with him, and Mike did, to the fullest. The demands of business left little room for personal relationships. Somehow, Charles and Clap had always been able to remain friends first and partners second; the security of his major stock holding had undoubtedly sustained Clap's confidence. Charles understood why Bill felt as he did, wanting to avoid a subordinate position under Charles, but he wondered if possibly Bill might consider some other employment in Seattle, close enough to rekindle their past intimacy. Jane helped, but Charles remained more comfortable with masculine relationships.

"He's on the line," Charles' secretary announced.

"Bill, Charles," Charles greeted him.

"Hi Charles. How's it feel to be one of the richest men in America?"

"We haven't reached that plateau yet," Charles said. Talking about money embarrassed him. Bill knew that, so Charles assumed Bill was merely kidding him. "Keeping busy?" Charles asked.

"Same old, Same old."

"Got some time for an old friend?"

"When do you get here?" Bill asked.

"Sunday afternoon."

"Which airport?"

"I'll take a taxi."

"Which airport?"

"Dulles."

"Good. Construction has National in a mess."

"Why don't they close the damned thing down. It's dangerous."

"Too convenient for the politicians. What flight?"

"Northwestern. We get in at four."

"I'll meet you. Can you stay with me?"

"I've got reservations at the Willard."

"I'll come and stay with you."

The Willard still reeked of its past, and its bar was a favorite hang out for Washington politicians and visiting tycoons.

"I've got a suite. Plenty of room."

"Just kidding. Are you free Sunday night?"

"Yeah. I've got a ten o'clock appearance on the Hill scheduled for Monday. Then, I'll catch a red eye flight back here."

"I'm having a small cocktail party Sunday night. You can stay over, then I'll drop you back at the Willard."

"I'm not sure I want to do that, Bill."

"It's a work session for me, but you might find it interesting."

"I don't want to spend the evening making small talk with a bunch of strangers."

"You might find these interesting, a few politicians and lobbyists."

"I don't need either of those two breeds."

"Indulge me. I guarantee you will enjoy yourself."

"If anyone mentions computers or software to me I'm out of there."

"You're on. I'll even make sure they don't mention the over-the-hill gang."

"What's that?"

"George Allen's Redskins."

"I'll hold you to that."

"See you Sunday."

Charles did not like the idea of wasting a rare free night on a Washington cocktail party, but he wanted to talk with Bill, alone. He decided he would have to pay the price.

Charles, toting a carry-on case and a suit bag, emerged first from the debarkation exit.

"Traveling first class has its advantages," Bill said. The Post makes us go cabin class."

"So does Digital Software," Charles said. "I only indulge myself when I have to make turnaround flights. Let's dig out of here."

Bill grabbed Charles' travel bag.

"I can handle that," Charles protested.

"It won't do for the richest man in America to be seen carrying his own bags."

Charles laughed. "Don't believe everything you read in the newspapers."

"You're telling me," Bill laughed.

Bill expected his old roommate to have changed, but he looked and acted like the old Charles, even down to the unbuttoned white shirt, khakis and Nikes. Bill led the way to the parking lot.

"This place is a mess," Charles said as he surveyed the havoc created by construction. Dulles like National was undergoing a facelift. "Air travel today is worse than taking the bus."

"When was the last time you took the bus, Charles?"

Charles hesitated. "I'm not sure I ever took the bus."

After loading Charles' case and suitbag in the trunk, they climbed into Bill's Mustang.

"This a Mustang II?" Charles asked.

"I've had it a month," Bill said.

"Nice wheels. Reminds me of my red Mustang."

"What're you driving?"

"A Porsche."

"Thought so."

Bill forced his way onto the Dulles Access Road. "It's good to see you again, Charles," Bill said. "You still look like a kid."

"I' don't feel like one. It's been too long. Are you sure you won't come and join Mike and me? It would be like old times."

"Thanks for asking, Charles," Bill said. "But I'm not sure I could stand the strain of working for you." Bill's attempt at humor had a hard shell of truth around it.

"Bill, can I ask one thing?" Charles changed the uncomfortable subject.

"Shoot."

"Everything we discuss is off the record, right?"

"Sure Charles." Bill wondered if he should be offended by the question, but dismissed the thought. If he were in Charles' position, he would ask the same question. "I appreciate any leads you give me though. I am a reporter."

"Working for a newspaper that thrives on muckraking."

"We call it investigative reporting."

"Just so you don't pretend you publish all the news that's fit to print. That's hypocrisy."

"You're thinking of the Times."

"Whatever."

"How's your love life?"

"Christ. You're just like my mother," Charles said. "I don't have time for that stuff."

"Don't want a heir to inherit all that money?"

"Sure I do," Charles hesitated. "Another request, please."

"Shoot."

"Don't keep referring to money. It's not important."

"It is if you don't have it."

"Well I have it. If you join me, you would too. Then, you'd see what I mean. How many pairs of Nikes do I need?"

"You didn't answer my question," Bill laughed.

"I've got a friend. We go to the movies, take vacations together."

"And you take her to your mother's for dinner every Sunday," Bill said.

"Maybe once a month."

"And that keeps your mother quiet?"

"You know better than that. She wants grandchildren. At least she's stopped asking if I'm gay."

Bill had speculated when they roomed together about Charles' passive interest in girls. "Still have your Playboy collection?"

"What about you?" Charles, who was uncomfortable with the subject, changed directions.

"What about me?"

"Are you happy at the Post?"

"It's getting a little boring. Business news is not really a grabber every day. The market goes up, the market goes down, guys like you announce a new program."

"Thinking of a change?"

"Not really a change. I've been thinking about taking a year's sabbatical and writing a book."

"Do it. What about?"

"I can't make up my mind. I always wanted to write fiction."

"I know."

"But I've also been thinking about writing a book about the computer revolution."

"Do both."

"I have to eat."

"I mean write a novel about the microprocessor revolution. It's dramatic."

"A real best seller."

"Why not? But make sure you leave me out of it."

"Of course," Bill said. "Leave out the man who now personifies the computer."

"Hardly. What about your love life?"

"I've been dating around. Washington is a single man's dream. All these girls working for the government."

"Will I meet any at your party tonight?"

"No. It's mainly a working deal for me. Payback of sorts. For all the parties I freeload at." Bill reviewed his guest list. "There is one guest you might find interesting."

"Who is that?"

"Ann Page."

"Sounds like a grocery store."

"No relation. Don't try that line. Everyone she meets uses it."

"Why do you think I 'd find her interesting?"

"She a damned good looker. And she's in your business."

"Computers. I don't recognize her name, and I know a lot of people in my business."

"She worked public relations for Intel and a couple of other Silicon Valley companies."

"Maybe I'll recognize her then."

"If you've seen her before, you'll definitely remember her."

"What's she doing in Washington?"

"She's a lobbyist for Intel now."

"Great, a lobbyist. That's just what I need." Charles immediately lost interest.

That evening Charles, not wanting to appear like a guest of honor, took refuge in Bill's second bedroom, waiting for the other invitees to arrive. Charles hated cocktail parties, particularly those with a lot of strangers who would immediately start fawning over him. He had had strangers ask him for autographs at social occasions. Charles immersed himself in a heavy book that Jane had given him, Molecular Biology of the Gene by James D. Watson.

An hour passed, then, Bill tapped at Charles' door and entered. "It's safe to come out now. Everybody has had two drinks and is talking about himself."

"A typical Washington cocktail party. Any spouses?"

"Several."

"Christ. Don't let me get trapped by them. I've heard all the talk about kids, pets, schools and washing dishes that I can stand."

"We've got a couple of older ones."

"Great. Then I'll get to hear about their last visit to the doctor's office."

"Charles. Why do I get the impression you don't like women."

"Don't start that again. Is this Ann here?"

"You can't miss her. She is the one with all the men around her."

"What's she wearing."

"Something silver and shiny."

"Watch my smoke," Charles said. "You go back, and I'll slip in after you. You haven't mentioned my name have you?"

"Nope. These are Washingtonians. They're only interested who's in, who's out, who's screwed up lately."

"Who's screwing whom, you mean."

"You got it."

Bill shut the door behind him. Charles marked his place in the heavy book by bending down a page corner, a habit that really irritated Jane who tended to treat her books like precious treasures while Charles used them and discarded them like newspapers. Charles slipped on a wrinkled blue sport coat, thought about a tie and decided against it. Bill's guests, none of whom he would ever meet again, would have to take him as is.

Charles was surprised to find Bill's large living room and dining room filled with noise, smoke and loud conversation. He paused in the doorway and thought about retreating.

"Don't run," a voice to his right ordered.

Charles turned to find himself face to face with a pretty brunette wearing a shiny silver dress. Bill had not said that it was also very tight and very, very short. Charles immediately thought of the old cliche: "Thank God he had lived long enough to witness the advent of and then return of the mini."

"Charles, Bill assigned me to serve as your palace guard."

"To prevent my escape," Charles said.

"You don't like cocktails?"

"Not in large doses, like this."

"You don't remember me, do you?"

"Certainly I do." Charles did not, not particularly, but she did appear familiar somehow.

"No you don't."

"No I don't. Where did I meet you?"

"You didn't. But I've seen you countless times."

"I'm sure I would have remembered if I had seen you?"

"Like my dress?"

"How could I not? How do you get in such a thing?"

"With difficulty. I wore it for you?"

"You didn't even know I was going to be here."

"Yes I did, Bill called to warn me."

"I asked him not to."

"Not to warn me? How droll?"

"No, not to tell anybody I would be here."

"The price of fame."

"Notoriety. You didn't answer my question."

"How do I get out of this dress?"

Charles blushed. "I didn't ask that."

"But it was what you were thinking?"

Charles' face turned red. "No I wasn't," he stammered.

"Then I know what you were thinking."

"What?"

"How do I get her out of that dress."

Charles immediately began to search the room for Bill.

"He won't save you," Ann laughed. "He assigned me to you, remember."

Charles finally spotted Bill across the room, watching him. Charles waved for Bill to join him. Bill shook his head negatively.

"Want a drink?"

"Are you Bill's hostess?"

"Sort of. Well?"

"Could I have a screwdriver?"

"A real Harvard man's drink."

"I prefer to think of myself as coming from Seattle."

"Well Seattle man, stand here, and I'll get your screwthing for you. Then, when I get back we can talk about your money. Bill tells me that you love to do that."

Ann turned toward the kitchen where a bar had been set up before Charles could think of a glib answer. He was standing silently apart from the party, being ignored, when Ann returned with his drink.

"I hope it's mostly orange juice," Charles said. "I've got business tomorrow."

"You have to testify."

"Bill again?"

"Keeping track of Congress is my business, particularly when the subject of computers is being discussed."

"You're a lobbyist."

"Yes. Does that offend you?"

"A defender of special interests."

"Yes, the computer industry."

"You're not doing a very good job. Those jerks on the hill don't know the difference between a microprocessor and a minicomputer."

"Who does? Is it important?"

"No."

"That's what I thought."

"What are you going to testify about? Software?"

"The industry and its future."

"I'll be in the audience."

"May I take you to lunch afterward?"

"No. But I'll take you. It'll look good on my expense account. Lunch with the high and mighty."

At that point, two party guests joined them, both women in their fifties. Charles assumed they were spouses.

"How can I reach you?" Charles asked quickly.

"Ask your roommate," Ann said and slipped away.

"Tell us young man, who is your father?" One of the matrons asked.

At nine fifty-five AM Charles, unaccompanied, entered a Capitol Hill hearing room to address a subcommittee of the Senate Commerce Committee. No one met him at the door, and no one directed him to his place. The guard at the door passed him through, assuming he was another teenage nerd present to hear their idol, Charles Swift III, speak. Charles glanced about. There was only a handful of spectators in the audience. A bright-eyed Ann Page stood out. She waved. She wore a beige, knit dress, and Charles wished she would stand up so he could better admire it. Charles waved back, and proceeded to the front of the room where he took a desk facing a raised platform where the Senators would sit behind polished tables that flanked a podium.

"May I help you," a page, who was clearly younger than Charles looked, asked.

"My name is Swift. Is this where I sit?" Charles asked.

"Yes sir," the page, obviously impressed, answered.

Charles impatiently waited. After a few minutes, staffers began to drift in, chattering with each other, ignoring Charles and the negligible audience. Then, the Senators arrived, alone or in pairs. Finally, a white haired red faced man with a big nose took the seat to the right of the podium. He glanced at the podium and then turned and whispered something to the staffer sitting immediately behind him. Immediately, two young men jumped up and moved the podium to the rear of the room. The Chairman then declared the subcommittee to be in session and nodded at Charles.

"Welcome Mr. Swift. We thank you for taking the time from your busy scheduled to come to Washington to share your views on the current state of our technical industry with us."

Charles nodded. He sat erect with his hands folded on the empty table in front of him. He resisted the urge to rock from side to side, something he liked to do when talking.

"Are you waiting for any colleagues?" The Chairman asked.

"No sir. You have only me."

The Chairman smiled. "This indeed is a precedent setting occasion. We are accustomed to having our visiting dignitaries flanked by an assortment of potted plants."

Charles smiled at the weak attempt at humor. The phrase "I am not a potted plant" had been used by Colonel Olly North's lawyer when he had been admonished by the IranGate subcommittee to keep quiet.

"We normally proceed with an opening statement by our witnesses," the Chairman explained. "Do you wish to follow this precedence?"

"I'm here at your behest," Charles said.

"Our interest today is in your views on this technical phenomenon which we understand has been labeled the information highway. Could you please share any insights you may have acquired during your awesome career in the computer industry. I am sure that everyone in this room is aware of your impressive contributions."

Charles waited for the Chairman to continue. When he did not, Charles assumed it was his turn.

"I find this opportunity daunting," Charles said. He did not really, but he felt obliged to pretend that he did. In reality, he considered most of the Members of Congress he had met as second rate. He had yet to meet a single one he would consider hiring for Digital Software. In any case, writing software was a young man's business.

The Chairman smiled indulgently, indicating he understood that Charles felt the opposite of what he was saying.

"Many describe the change that has been wrought by our nation's nascent technical industry as a revolution. I do not. We are only at the threshold of a technical revolution that will change how we work, play, live. I don't have to describe for this committee the remarkable advances that our engineers and scientists and businessmen have given us. We are not yet at the stage where we have a microprocessor on every desk and in every home. The day is not far away when we will. The personal computer one day will control our homes, setting our individual climates, managing our appliances, storing our information, paying our bills, ordering our groceries, providing our entertainment on demand. These computers will all be interconnected within the home and without.

"Today we have what we call the Internet. The Internet was first conceived in the early 1960's. Then, the Department of Defense's Advanced Research Project Agency, ARPA, developed a small network known as ARPANET, that is Advanced Research Project Agency Network, to connect researchers across the United States. In 1969, ARPANET was established on four university campuses, enabling researchers at those four institutions to use computers to communicate. These hubs were sited in Stanford, UCLA, UC Santa Barbara, and the University of Utah. Thus was born electronic mail, e-mail. By 1971, ARPANET enabled twenty-three hosts in government research centers and universities to share their projects. By 1981, ARPANET had 213 hubs. In 1982, the term Internet was used for the first time. Today, the Internet has more than a million hosts located throughout the world. This may sound like a lot, but the day will come within the decade when over two hundred million computers will be talking to each other on this Internet. What began as a system for a handful of elite government and academic researchers to communicate, has become an everyday tool for the common man. Just imagine, today's desktop computer is more powerful than a ten million dollar IBM mainframe of just twenty years ago. If the computing power of today's desktop doubles every eighteen months while cost drops, you can just imagine what this will mean to our generation, not just future generations.

"This Internet, or information highway, will place the libraries of the world at the fingertips of tomorrow's schoolchild. Tomorrow's worker will be freed from the confines of today's office. They will telecommute from homes in the same city or across the country. They will read electronic books that they download from the Internet. Libraries as we now know them will become obsolete. We will shop in the electronic marketplace for everything we need, including food supplies and clothing and transportation. If you want to take a trip, you will consult your electronic travel agent. Pay your bills, your electronic bank with take care of that. And for those of you who invest in the stock market, the Internet will give the individual investor access to the information you need, and electronic brokers will buy and sell your stocks and bonds. Our homes will become self-contained entertainment centers. Like movies? We will be able to order movies on demand, watching what we prefer when we want. Concerts? Share favorite tapes, records, and CDs? Use the Internet.

Change will be traumatic. Entire industries will disappear, and new ones will take their place. People will worry about the proliferation of personal information that this easy access provides, and troublemakers will create security challenges. Our present day telephone wires will have to be replaced with fiber optic cables. An immense infrastructure filled with servers will have to be built. There will be social repercussions. Many fear that tomorrow's netizen will become isolated, lose his or her interpersonal skills. Yes, that is a risk, but imagine the opportunities that e-mail gives us to communicate with others. Our elder citizens with declining physical abilities will be able to use the computer and the Internet to talk, share, and play with friends and strangers.

When Charles had finished, the Chairman opened the floor to questions, and the first was asked by a female senator from Utah with a known bias against the changes being wrought by the technical revolution.

"Mr. Swift. The change that you allude to frightens me for one. Obviously, the social repercussions will be immense, and I am not sure how all of our citizens will share in this marvelous New World. Many will not be able to afford to buy the expensive equipment that will make it possible. Also, there are those of us in my generation, myself including, who are not computer literate. What will happen to us? Are we to be simply cast aside."

"The social..." Charles began.

"No please, I was speaking rhetorically. I do not expect a big businessman like yourself to address these social issues," the Utah Senator interrupted. "I'm more interested in your response to another related issue. You are reputed to be the richest man in the United States. You derived your good fortune from the so-called marvelous technical advances that you have described. Aren't you really telling us you want a good thing to continue for your own selfish gain?"

Charles frowned. He was tempted to tell the rude, headline seeking Senator to piss off, but he decided to answer with forbearance.

"Madame Senator," Charles said. "Or should I call you Ms. Senator."

The small audience laughed.

"Either will do."

"It's true Digital Software stands to gain from this technical revolution. We work hard to earn our profit. That what America is all about."

"You don't have to tell me about capitalism," the Utah Senator interrupted.

"I wasn't intending to," Charles snapped. "We write software. I am proud to say that our program language is used on virtually every computer manufactured in the United States."

"Some describe you as commercial pirates," the Utah Senator interrupted.

"I think another computer company, or more precisely another division of another company, has flown the skull and crossbones." Charles tried to distract his adversary by referring to Stanley's flag over the Asteroid Division.

"Some claim that you have built a virtual software monopoly by forcing IBM and others to buy your operating systems," The Utah Senator snapped.

"We gave IBM a very good deal. For a low one-time fee, IBM is free to use our operating system on as many computers as it can sell. I assure you our profits from our IBM contract are small. There was no force involved. They saw a good deal and jumped on it."

"If you are so altruistic, how do you make your spectacular profits?"

"It's no secret. We do not make money from IBM, but we do on sales to the companies that manufacture the PC clones. They have duplicated the IBM PC and therefore need PC Compatible software which Digital Software gladly provides."

"I bet you do."

"Is there something illegal about that?"

When the Senator did not respond, Charles continued.

"You raised a very important issue when you referred to the social consequences of this revolution. There will be social consequences. Some of our citizens will not be able to afford computers. This is a problem we must address. Henry Ford and his peers revolutionized the transportation industry.

Every one of our citizens was not immediately able to put his horses out to pasture and share in this marvelous new invention, the automobile. The same was true with the invention of the telephone or the electric light bulb. Time was required for society to adapt. Our businessmen worked hard to make their products affordable. Our government funded programs that spread the wires across the land even to our most inaccessible rural areas. We in my industry will do everything we can to reduce the price of our products. Others, like yourselves here in this room, must join with us in guiding this technical revolution into channels that make its fruits available to the largest possible numbers of our citizens. I assure you, the revolution is underway, and neither you nor I can stop it. We must work together to ensure it develops in a manner hospitable to our nation's citizens, all of them."

The Senators succeeded in dragging Charles' appearance out for an additional hour with questions, most of which Charles considered downright silly. After the Chairman finally thanked Charles for his presentation, Charles turned and found Ann Page standing in the aisle near his chair.

"Nice job," she said.

"What was that all about?" Charles asked.

"Congress in action."

"That was action?"

"You ought to see them at play. I could introduce you to the Senator from Utah."

"I'm not sure I like your friends," Charles said.

"Who said she was a friend. She hits me up for a free plane ride to California every once in a while."

"You are also in the travel business?"

"If you can buy someone for a plane ride, then I'm in the travel business."

"You're going to have to tell me how this lobbying business works," Charles said.

"Do you have a free three minutes in your world class schedule?"

"Three minutes?"

"Yep. I'll give you the long version."

"We'll do it over lunch."

"OK, but I'll pay."

Charles gave Ann a quizzical look.

"Business expense," Ann smiled.

As they walked up the aisle, Charles asked: "What is the short version?"

"Money, money, money."

"I thought it was illegal to bribe legislators."

"Who said bribes? Campaign contributions. Gifts from friends repeat friends."

"Where do we find a taxi?" Charles said as Ann guided him toward the rear exit of the Capitol.

"No need. I've got my car."

Ann led the way into the parking annex. About midway down the lane, Ann stopped behind a Mercedes 220. "This is it."

"Nice wheels," Charles said.

"We lobbyists have to keep up appearances," Ann said. "We're just like realtors."

"How do you rate a parking spot?"

"Friends in high places."

"One hand washes the other right?"

"Right."

After they had gotten into the car, and Ann had maneuvered into the street, Charles said: "Digital Software doesn't have a full time lobbyist."

"I could take on another client," Ann said.

"I said full time," Charles said.

Ann maneuvered around a bus before replying. "You don't need a full time lobbyist, at least not yet."

"Meaning you think I will?"

"Then you didn't understand what was going on in there?"

"No."

"The Senator from Utah just fired the first shot across your bow."

"Please explain that."

"Digital Software is getting too big. When you belong to a political party that doesn't collect much money from big corporations, there is a lot of mileage to be gained by taking on a big boy. Remember Ma Bell."

"No, no, no," Charles said. "There's no comparison."

"Yes, yes, yes. You had better start preparing to defend yourself."

"And how do I do that?"

"By lining up your friends on Capitol Hill. It's all politics."

"And how do I do that."

"You don't. You hire someone to do it for you."

"A lobbyist? Or CIA?"

"Whatever works for you."

"Just kidding."

"I'm not," Ann said. "Charles when a politician smells big money, there is no middle ground. You are either their friend or their enemy."

Charles, despite himself, was impressed with Ann Page. He had been jousting with her as a form of flirting, but her warning about the Senator from Utah had a ring of truth to it. He too had been wondering how long it would be before the politicians took an interest in the computer industry. Charles did not follow politics, but he knew human nature well enough to understand that phenomenal success, earned or not, attracted unwanted companions.

"What do you think of Bill's plan to take a year's sabbatical?" Charles changed the subject.

"You've known Bill a lot longer than I have. We help each other out now and then. If I pick up an interesting lead for a story, I pass it along."

"And what does he do in return?"

"Ask him that."

Charles wondered if he might be poaching on Bill's turf. Bill had encouraged him to pursue Ann.

"We're not lovers, if that is what you want to know," Ann said.

"Yes, that's what I wanted to know," Charles said.

"Do you ever get out to Seattle?"

"I visit there now and then."

"Do you think you might find the time to write up a political game plan for me?" Charles asked.

"I don't come cheap."

"I didn't expect that you would."

Ann pulled up in front of a Georgetown restaurant with the odd name "1812." She handed her keys to a parking valet. "Thank you Ms. Page," the college boy said.

"This one of the in places?" Charles asked.

"It's a power eatery. All the right people will see who my lunch partner is and fall out of their chairs trying to figure what's going on."

"What is?"

"You tell me," Ann smiled.

After they were seated, Ann ordered for both of them including the wine. After the waiter had departed, Ann spoke: "You know I once had lunch with Stanley Pitts."

"One of my clients," Charles said, signaling he was not interested in gossiping about Stanley. The Astron Computers contract was too important to Digital Software to risk for the sake of a little gossip.

"I met Stanley when he was stilling working out of his father's garage."

"Really?"

"Yes. I was checking him out. He took me to a hole in the wall fast vegetarian food place. You know, he was a Hare Krishna."

"Stanley is frequently underestimated," Charles said.

"Not by me," Ann agreed.

"A lot of people are making a mistake in thinking Stanley's out of the business for good. Astron Computers took a big misstep doing what they did. Stanley doesn't have an academic degree in engineering..." Charles paused and smiled. "Who does? Stanley is very intuitive about computers and has a great eye for design."

"Cosmetic or technical?" Ann expected Charles to reply design.

"Both" Charles said.

The waiter served the wine, filling Ann's glass first. She offered it to Charles: "Prefer to pass judgment?"

"I'm not big into the wine stuff. Diet coke suits my taste."

Ann tasted the wine, smiled, and nodded to the waiter. "Very good."

The waiter frowned at Charles when he poured.

"I guess he doesn't like diet coke," Charles said when the wine waiter and withdrawn.

"You're threatening his livelihood."

"I'll have to apologize," Charles said.

"He wouldn't accept it. He's already classified you as a philistine."

"The world wastes too much time talking about wine," Charles persisted.

Ann sipped her wine and smiled.

"How was Stanley's vegetarian fast food?" Charles asked.

"Terrible. And he made me walk for miles to get there. I don't think he owned a car."

"Did he give you a tour of his garage?"

"No way. I dumped him fast."

"You made a mistake."

"I know. Do you really think he'll make a comeback?"

"Astron Computers needs him."

"Were you serious about hiring me full time?" Ann asked.

"Write up a game plan for dealing with Congress and bring it out to Seattle and we'll discuss it."

"You better worry about more than the politicians. They're ambidextrous. They swing either way. If the Democrats get in office, you better worry about the Department of Justice."

"Really?"

"Really. They'll bring in a herd of young lawyers, all eager to make a name for themselves as prosecutors. Guess who they'll target?"

"Work on that game plan, and I'll give you a first class tour of Seattle."

"I've been to Seattle."

"What about the Seahawks?"

"What about them. I'm a Redskins fan."

"Then we'll have to figure out when the Redskins play the Seahawks."

"A deal."

"You never answered my question about Bill."

"What was it?"

"Can he be enticed away from the <u>Post</u>?"

"To write a book?"

"Sure."

"Only to write a book and then he'll be back at the <u>Post</u>."

"What's so great about the <u>Post</u>?"

"Nothing. But a lot of uninformed people think there is."

Chapter 30

Stanley's departure liberated Lawrence Starkey, and few employees mourned Stanley. Many, particularly in the Astron 2 Division, cheered; most, however, worried about the pending changes. Lawrence was ready. Within the month, he fired 1,200 employees, mainly in marketing; he announced the closing of three of six factories by the end of the year, and he reorganized the shaky management structure. Lawrence gave the departing one month's severance pay, and to those who remained he made it clear that Astron Computers was one company, not a conglomerate of competing, backstabbing baronies. Stockholders cheered, and on Wall Street Astron stock moved upward.

Lawrence created two key groups: a division responsible for engineering, manufacturing and distribution, and a division overseeing all marketing and sales. The two new vice presidents were specifically charged with bringing the company together.

"Let no one misunderstand. I want a team effort," Starkey declared.

On Starkey's recommendation, the Board authorized the filing of a law suit against Stanley alleging that he had pirated company trade secrets, suborned key employees, and was entering into direct competition using Astron developed technology. Stanley retaliated by selling twenty-one million dollars worth of Astron stock. The media speculated that a brain drain of sizeable portions would ensue, but this did not occur. Harold Dumbroski returned to Astron. Harold cryptically observed: "It was time." He refused to link his return to Stanley's departure. Harold purchased 100,000 shares of stock, presumably picking up many of the shares his former partner had sold.

The lawsuit was eventually settled out of court. Stanley promised not to hire any more Astron employees for three years and to eschew developing products that would compete with his former company. Stanley defiantly sold his remaining shares in Astron, keeping but one.

Lawrence faced a considerable task. Stanley had bet the company on one product, the Asteroid, and as remarkable as its technology was, the high priced, underpowered machine with no hard drive—Stanley had not liked the noise caused by the requisite fan—met with a cool reception. Stanley had predicted a sale of two million machines during the first year; disappointed consumers bought only 500,000. Astron's market share plummeted to eleven percent, down from twenty. Before the slight surge precipitated by Starkey's cost cutting actions, Astron Shares fell to a meager fourteen dollars, down from its peak of sixty-three.

Stanley had created the computer for the future, but its present proved dismal. Lawrence named the Frenchman, Pierre Dupris, his Director of Product Development. Dupris reported directly to the Vice President for Engineering, Manufacturing and Distribution. For his other appointments, Starkey turned to experienced managers and bureaucrats. He needed creativity, however, not business school managers in Product Development. Starkey intended that Dupris be his Stanley, with emphasis on "his." Starkey assigned Dupris the task of correcting the Asteroid's shortcomings. Asteroid 1 operated on 128,000 bytes of memory, not enough to power their graphics programs. Stanley knew this, but he had convinced himself that he had to reduce cost in order to make Asteroid competitive. He argued that if users wanted more memory, they could buy it. Dupris committed himself to giving Asteroid 2 a full megabyte of memory.

But, Dupris could not correct all the shortcomings himself. The Asteroid had launched with a limited software arsenal. Disappointed Asteroid loyalists had only to window shop at a computer store to compare their meager options with the full shelves of packaged software available for the PC. Stanley had insisted on closed architecture, denying the nation's programmers the opportunity to write code for the computer of the future. Starkey hired a marketing executive named Cordell Haynes to persuade software developers to write for the Asteroid. This was a daunting task because most developers could see for themselves that Astron's market share was falling while the PC share was rising, both dramatically. Their decision was preordained: sell to the rising market.

Hayes styled himself an "evangelist." He and two others armed themselves with Asteroids, which they knew the programmers would covet, and embarked on a nationwide campaign to convince the developers to write for the Asteroid. One of his first successes was the discovery of a group of Seattle engineers who had developed a program that could deploy text and graphics on the same page. The Asteroid was a natural for this program. It was made for businesses, providing them with the ability to prepare their presentations and newsletters without resorting to scissors and paste layouts. The evangelists touted this new technique to corporate customers under a catchy name: "desktop publishing."

Lawrence changed Stanley's marketing strategy. Stanley had focused on schools, home users, and small businesses. By default, this left the most lucrative market of all to IBM and its Personal Computer. Stanley's stress on the little man left Astron Computer sales hostage to the whim of the nation's economy. In bad times, their target market had no money for luxuries like the Asteroid. Lawrence found it ironic that the more expensive microcomputer, the Asteroid, was aimed at the portion of the market less able to bear the higher cost while the cheaper PC sold to the fat corporate customers.

"No wonder we're getting shellacked," Lawrence complained regularly at his management committee meetings.

With the advent of desktop publishing, Sharkey directed his Vice-President for Marketing and Sales to widen his market targeting to include corporate offices. They immediately encountered corporate resistance. The big boys trusted IBM. The quality care lavished by Big Blue on its corporate leasees had carried over to the PC. IBM was a proven quantity; Astron was a Johnny come lately with a cute product with no hard drive and limited memory. No self-respecting business could risk storing its critical information on floppy disks. This was another Stanley directed shortcoming that Dupris was more than happy to have the opportunity to correct.

With his team in place, Starkey took upon himself the burden of repairing his company's severed relations with other components of their industry. In his wake, Stanley had left a littered battlefield. Stanley had insisted that Astron Computers did not need the others; his Astron Computers was a go it alone outfit. The skull and crossbones flag that he flew over his Asteroid Division symbolized his attitudes.

A shy man with no technical background, Starkey undertook the difficult job of erasing Stanley, Mr. Charisma, from Astron's corporate image. Despite Stanley's arrogance and personality shortcomings, people admired him for having lifted Astron Computers from the garage to corporate superstardom. Stanley, like Starkey, had a limited technical background; he talked a good game and made people believe him. Starkey forced himself to learn. Over time, he refined his message: "Astron Computer no longer wants to go it alone."

By the end of Starkey's first year at the helm, he succeeded in stemming the corporate hemorrhaging. His books showed a modest profit, largely the result of his cost cutting, but he had his marketing team in place; Product Development had the Asteroid 2 ready for launch, and Starkey could in all honesty tell his growing audiences that the future for Astron Computers was indeed rosy. They had turned the corner. Starkey did not, however apportion credit and blame. How could he without demeaning Stanley, the man who had created Astron Computers with his vision and sheer will while simultaneously fomenting disharmony and self-defeating restrictions in his wake. Starkey concentrated on the positive. Starkey liked to play the successful manager telling his audiences that Astron Computers now had all the pieces in place: Asteroid 2 with a hard disk and enhanced memory, new software featuring desktop publishing, and a new image.

Charles who was renown for his close monitoring of industry developments—some paranoid souls even claimed he had spies in every company—followed developments at Astron Computers closely. He regretted the fact that the new management had found it necessary to jettison Stanley Pitts. Charles had even called Stanley one week after his firing and patiently listened while Stanley vented his anger. Charles counseled patience and encouraged Stanley to look forward not back. At the same time, Charles meticulously filled every contract he had with the new management. Over one third of Digital Software's programmers were employed on Astron projects. There was no doubt in Charles' mind that graphical interface was the system of the future and that Asteroid represented a quantum leap forward.

After the turmoil surrounding the Astron reorganization subsided into routine, Charles made his move. He phoned Starkey in Cupertino, congratulated him on his downsizing and focus of resources, and requested an appointment. Since promoting the development of software for the Asteroid ranked high on Starkey's agenda, he responded to Charles' request with alacrity.

Three days after his call, Charles flew to San Francisco, hired a rental car, and drove to Cupertino.

"Good morning Mr. Swift," Starkey's executive secretary greeted him. "Please have a seat, and I will inform Mr. Starkey that you are here."

Charles studied the vista from the window. Located on the top floor of a new building, Starkey's headquarters had an impressive panoramic view of the Astron campus. Charles calculated that four buildings had been added during the past year.

"You may go right in Mr. Swift," the secretary said, rising to lead Charles to the President's office.

"Charles, welcome to Cupertino," Lawrence Starkly greeted Charles at the door.

Charles fought a smile. The contrast with previous receptions was amusing. Stanley always played a game of one-upmanship, and Charles appreciated Stanley's devious ploys. On more than one occasion, Stanley had greeted Charles with: "What do you expect me to do with this latest shit you are sending me?" Charles always responded with excessive courtesy that tended to provoke Stanley into more outrageous reactions before settling down grudgingly to seriously discuss business. Dealing with Starkey would make Charles' life a little easier if less entertaining.

After a modicum of small talk and pleasantries, Charles turned to the subject he had interrupted a busy scheduled to broach. Charles had concluded that Digital Software could best profit from an Astron Computers with open architecture. Charles had reaped considerable advantage from his relationship with IBM. If he could persuade Sharkey to license the Asteroid to other companies, Charles would be in a position to dominate software sales in an entire industry.

"I must confess that I have had this conversation with Stanley in the past," Charles admitted, assuming that Lawrence Starkey was aware of those conversations.

"Please, don't let that inhibit you," Starkey said. "I assume that Stanley rejected whatever you had in mind."

Charles smiled.

"Stanley tended to reject any ideas that were not his own."

"Today," Charles said. "Astron Computers, you, have a window of opportunity that will not stay open for long."

Lawrence concealed his impatience. He wished the boy genius would get to the essentials of his pitch. Lawrence assumed Charles had come up with another way for him to corner a market and reap money.

"You have a well deserved reputation for innovative technology." Charles did not say that this was obviously based on Stanley Pitts' vision. "The time is ripe for you to establish Asteroid as a standard for the graphical user interface, GUI." Charles pronounced gui "Gooey." Starkey's lack of a technical background worried Charles. He was not sure how much of the technical jargon Starkey understood. This placed Charles in the difficult position of possibly talking down to a client and insulting him.

"And how do we do that?" Starkey asked.

"A good question," Charles said, knowing the answer was obvious to most people in the industry. "You need to open your architecture."

"Cloning," Starkey said. "That is something that Stanley stubbornly resisted."

"I know. As I noted, I've had this conversation with him," Charles said. "If you license other companies to manufacture the Asteroid, the size of your market will increase geometrically. IBM has proven that."

"By doing so we create our own competition. Look at IBM today. It runs the risk of becoming a marginal player in the marketing of their own product, the PC."

"One of the reasons that the average user selects the PC over the Asteroid is the availability of software. A closed architecture discourages software writers. Naturally, they will write for the machine that has the biggest market. Today that is the PC."

"I understand all too well. We need more software to sell more machines, and we need to sell more machines to get more software written for the Asteroid. A neat dilemma."

"A Gordian knot. The solution is simple. Slash the knot with your sword, open architecture."

"Easily said. Harder to do. Tell me something. Don't you think clones are a real threat to IBM?"

Charles hesitated. He had tried to avoid answering this question. "The solution rests in IBM's hands just as it would with Astron Computers. If you and they remain content to produce the same machine—if you insist on marketing only the Astron 2 for example—the answer is yes. But IBM and Astron Computers are the industry's technical innovators. Technology does not stand still. IBM and Astron computers have the resources to outpace the competition. While the clones are selling the Asteroid 1, Astron Computers must be selling the Asteroid 2. Your superior and more advanced technology will even support your higher price structure."

"We will sell more advanced machines than the clones at a higher price," Lawrence mused.

"Exactly," Charles agreed. He could not tell if Starkey understood or simply was parroting Charles' words. "And the bigger market will attract the software writers."

"It sounds simple."

"There are other advantages. The PC has established itself as the industry standard. This dominance literally invites increased financial investment which produces enhanced software and better peripherals while funding advanced R&D."

"And Astron Computers by itself has no chance of establishing itself as a standard."

"You have already lost that battle."

"Cloning will let us recoup?"

"It gives you a chance."

"I can see what you stand to gain out of this," Starkey smiled. "You write software for the Asteroid now. Cloning means more customers for you."

"Certainly," Charles said. "We're in business to make a profit. One third of my programmers are writing code for Astron Computers. Why shouldn't we want to sell our product to others? Unfortunately, right now there are no others. Open your architecture, and we both will benefit. That's good business."

"You also write software for the PC, IBM and its clones."

"That's right."

"So you want a monopoly of the software business?"

"Not a monopoly. That's bad business. We all need competition to do better. There will always be competition in the software industry. That's the nature of the game."

"Some kid will always be coming up with the next monster application."

"Exactly."

"Tell me why you think Asteroid has not become a gui standard. The PC is five years behind, technologically speaking."

"That's why I am here. Now is your time of opportunity. You must act now."

"I understand that," Starkey said, "but please Charles, humor me. This is all new to me. Why has Asteroid not become a standard?"

"Because you have failed to attain the critical mass necessary for your gui to become a long term contender. Since you have no competition from Asteroid compatible manufacturers, others are afraid to lock in to the Asteroid. Your company's internal turmoil frightens others away from making the necessary investment. Nobody is sure how long Astron Computers will retain the internal will to remain the dominant technological leader."

"And if we solve our internal problems," Starkey without stating it referred to Stanley Pitts and his free swinging operating style.

"You still have been slow in producing hardware and software improvements."

"Such as?"

"A hard disk, more memory, bigger monitor, better keyboard. Need I go on."

"We're working on those things," Starkey went on the defensive.

"But not fast enough. You also need a bigger corporate sales force."

"Anything else?" Starkey was really peeved at this kid coming in and telling him how to run his company.

"All the publicity coming out of Astron Computers for the past six months has been negative. Buyers read too and become uncertain about Astron's future. For them, paying the bigger price for an Asteroid becomes a risk. They worry that you might suddenly fold and leave them with an expensive gadget without support."

"This company will not fold. Not as long as I sit at this desk," Starkey said.

"I know that. I'm just telling you what the average buyer worries about."

"And how did you become an expert?"

"By studying the market. We retail software too." Charles recognized he had hit Starkey too hard, but, having come this far, was not about to back down.

"If we license Asteroid, who should we consider as cloners?"

"Go with the established companies. I could name a dozen off the top of my head."

"What should we expect from them?"

"Established companies would add credibility to the Asteroid's future. Select licensees who will broaden your base by offering additional Asteroid compatible lines of computers, peripherals, software."

"I understand that Digital Software is working on its own gui operating system." Starkey decided to shift the focus of their conversation.

"We are. We call it Panes. We have a long way to go. We're no threat to the Asteroid."

"Really. You're not re-engineering the Asteroid?"

"No sir. That's insulting. We're writing our own system."

"An icon based system?" Starkey did not venture into more details because he did not understand them.

"Yes."

"Then you are duplicating the Asteroid."

"No sir. We're both taking our inspiration from PARC's Alto and Star. The entire industry is indebted to Xerox."

"What does IBM think about your Panes research?"

Charles smiled. Starkey was shifting to familiar territory. "IBM is one of our major clients. We are working closely with them in developing what they call OS2, Operating System 2."

"And they don't resent that you are working on your own graphical user interface operating system at the same time?"

"That's our prerogative."

"Mr. Swift, I thank you for taking the time to have this discussion with me. I will certainly take your views under advisement. As a valued business collaborator, would you be so kind as to send me a memo on the subject of what cloning would mean to Astron Computers?"

"Certainly," Charles said, not positive that he had made any headway. The request for a memo seemed positive, but it might mean only that Starkey wanted an argument to use with his engineers that would not embarrass him.

Charles stood up, prepared to leave, then he had one further thought. "Lawrence, I feel obliged to tell you that I intend to have lunch today with Stanley Pitts. Common courtesy requires me to tell you this."

Starkey hesitated before answering. "Certainly. This company owes its past to Stanley, and he remains one of our stockholders."

Charles smiled. He had read the story that Stanley had sold all of his stock but one share. "Stanley's an old friend and former business collaborator," Charles said.

"Give Stanley my best," Starkey said.

"I will," Charles said, knowing that was the last thing he would do. "I'll send you a brief summary of my thoughts on cloning," Charles promised, then departed.

Charles, driving his compact rental car, found his own way to Stanley Pitts' new mansion located a short distance from Astron Computers. Charles followed a long paved driveway abutted by freshly mowed lawn. Obviously, Stanley was spending some of his money. Charles parked behind a shiny black Mercedes bearing the license plate "Again." Charles smiled. Stanley might live in a two million dollar mansion that dwarfed Charles' much more modest two hundred thousand dollar home on the lake, but he was still the same arrogantly defiant Stanley.

Charles rang the doorbell and listened as chimes responded inside. He waited. He pushed the button a second time, and again the chimes pealed. Charles waited. Finally, the door opened.

"Yes sir," an attractive brunette in a thigh high black maid's uniform smiled.

"My name is Charles Swift. I have an appointment with Mr. Pitts," Charles said.

"Please come in, sir," the brunette's bright blue eyes sparkled.

Charles decided that Stanley selected his own staff. He studied the perfectly formed legs as he followed them down a long hallway to a room on the right. Charles wondered how Stanley got any work done with that kind of distraction always on the premises.

"Please have a seat, and Mr. Pitts will join you shortly," the brunette said. Her amused eyes seemed to ask if Charles had enjoyed the view.

"Thank you," Charles said. "May I ask your name?"

"I'm Lola."

"Glad to meet you Lola," Charles said, offering his hand.

Lola returned Charles' firm grip, letting her hand linger a brief bit longer than normal, just enough to signal interest.

Charles watched as Lola spun about and walked from the room with exaggerated hip movement. Charles immediately decided that Stanley might have been fired, but he had not lost his sense of style.

Charles appraised the room. It was filled with overstuffed leather furniture that lined the walls much like the waiting room of a prosperous society doctor. On the coffee table, Charles found only a two-year-old copy of Time magazine with a smiling Stanley on the cover. Charles ignored the magazine and went to the window that faced to the rear of the mansion. To Charles' surprise, he saw Lola talking with a solitary Stanley who sat reading a newspaper at a table near an Olympic sized pool. Lola apparently announced Charles' arrival. Stanley nodded, dismissing Lola, picked up a cup that set on an umbrella covered table at his side, and sipped. He then turned a page of the newspaper and sat with his back to the window where Charles waited.

"Bullshit," Charles declared. There was no way he was going to let Stanley get away with his constant games of oneupsmanship. Charles leaned over and picked up the magazine and slipped it under the cushions of the couch. He then left in search of a door that led to the pool. In the hallway he met the smiling Lola.

"This way to the backdoor?" Charles asked.

"Yes sir. End of the hall and turn right."

Lola did not seem the least bit discomfited by the fact that Charles was disobeying her command to be seated and wait.

"Would it possible to get a cup of tea?" Charles asked.

"Yes sir. English breakfast or lemon?"

"Lemon," Charles smiled.

This time Lola watched as Charles made his way to the pool. He followed the flower-lined walk and approached Stanley from behind.

"Some people know how to live," Charles announced his presence.

"Hi Charles, have a seat," Stanley said, waving to a chair on the other side of the table. He did not rise or offer his hand. He still acted as if he were the chief executive of one of Charles' major customers.

Stanley pretended to continue to read the newspaper.

"Cut the shit, Stanley," Charles said, reaching across the table and pulling the newspaper from Stanley's grasp. Charles noted that Stanley had been reading the comics and chuckled as he folded the paper before dropping it to the table.

"Sorry about that," Stanley smiled. "We unemployed have to keep up our pretenses.

"Unemployment insurance must pay good these days," Charles said, waving his hand to include the pool and surrounding garden and poolhouse.

"One makes do. Can I offer you some coffee, since I must be hospitable."

"That's all right," Charles said. "Lola is bringing me some lemon tea." Charles nodded in the direction of the approaching Lola who carried a silver tray that held a shining teapot, cup and saucer, and silver sugar bowl.

"Damn it, Lola, wait until I call for you," Stanley complained.

"Yes, your highness," Lola smiled.

"Good help is hard to get," Stanley complained.

"Help? Is that what you call it?" Lola countered.

Stanley grimaced and the ever-smiling Lola departed.

"She's still learning first floor duties," Stanley said.

"Normally her post is the second floor?"

Stanley smiled. "How's the software business?"

"Tolerable. Just visited your old office," Charles said.

"Is it still there? I heard those shits are going out of business."

"Things seem to be going along just fine," Charles needled Stanley. "They had a lot of problems to solve, but the new management seems to be getting on top of things."

"Damned flavored water salesman," Stanley complained. "I taught him everything he knows about computers which isn't much."

"A hard disk and a lot more memory, and the Asteroid is a formidable machine," Charles said. "You gave them a product that is years ahead of its time and will save the company."

"How's your graphical interface coming?" Stanley asked. "What do you call it, broken windows or something?"

"Panes."

"I knew I should not have given you those four machines," Stanley referred to the Asteroids he had loaned to Digital Software so that Charles' programmers could write software for them. You pirated my operating system."

"No way, Stanley. We both stole it from PARC."

"I stole only the idea. I created the Asteroid."

"Then, I too stole only the idea from PARC. Don't throw stones," Charles' tone indicated that he was serious.

"What do you want here?" Stanley asked ungraciously.

"Just thought I would drop in on an old friend and see how he was taking early retirement." Charles could not resist putting the barb in a true statement.

Stanley twisted sidewise in his chair, turning his back to Charles. "Can you see it?" He asked. "Is the knife still there?"

"Doesn't show," Charles said.

Stanley laughed and turned back to face Charles. "I'm starting a new company. I'll put those bastards out of business."

"Are you referring to Astron Computers? That will be a big job."

"I started from nothing before. I can do it again."

"This doesn't look like nothing. What did you pay for the house? Two million I hear."

"Chump change," Stanley laughed. "Seriously, I'm going to build a new computer."

"What's the name of your company?"

"Again."

"Thought so. Already advertising?"

"I thought the license plate was a nice gesture," Stanley said.

"Will I be able to sell you software? Need a good graphical operating system?"

"No way. We'll write our own software. Think I'll drive you and Astron Computers out of business."

"Good luck.

"Whose money are you going to use Stanley? Your own?"

"Oh I'll let a few small time investors in on the action."

"Such as?"

"A little guy from Texas named Perot, know him?"

"I may have heard the name."

"You interested?"

"In what?"

"Investing. I hear you just made <u>Fortune's</u> five hundred list."

"Must have been some other fellow."

"I thought so. You're too young to be on the list."

"Unlike you."

Stanley smiled. "Are you interested in investing? Is that why you're here? Want to get on the bandwagon with old Stanley?"

"No thank you Stanley. I'll sell you an operating system and software when you need it."

"No thank you," Stanley sneered. "We'll write our own."

They sat in silence for a few seconds while Charles debated whether to commiserate with Stanley for his recent loss of employment or not. He had come to cheer Stanley up, but he decided that was unnecessary. Stanley was the same old arrogant Stanley ready to take on the world.

"I met an old friend of yours recently," Charles changed the subject.

"Who? I have so many."

"Ann Page."

"Nice looking tootsie. You banging her?"

Charles thought about turning the question back to Stanley but decided against it. "She said you took her to lunch once. A vegetarian takeout."

"What's this takeout shit. I taught her what a good Nicoise salad should be. Then, she dumped me."

"Why did she do that?"

"Guess I wasn't big enough for her. You banging her?"

"I met her in Washington."

"What's she doing there?"

"Lobbyist. Think I should hire her?"

"Get a good man. You're going to need one."

"Why do you say that?"

"You're getting too big. The politicians smell money."

Three weeks after his visit to Cupertino, Charles faxed a terse three-page memo to Lawrence Starkey as promised. After scanning it, Lawrence decided to present it to his management committee. Rather than submit it as Charles' recommendation, Lawrence summoned a young man who headed Astron's Investor Relations Office. The fact that Starkey favored the young Economics major from Stanford and that he frequently used him as a catspaw to test ideas that Starkey did not want to advocate himself was well known. This transparent ploy did little to enhance the young man's status with his peers and superiors, but it served Starkey well.

"Chris, I have a little project I want you to undertake for me," Starkey smiled at his willing pawn.

"Yes sir." Privately, Chris Stuart worried. Shit, not again, he said to himself.

"Take this memo. Read it carefully. I want you to present it to the management committee as your own proposal."

"Yes sir." Christ took the memo, glanced at the title. "This is from Charles Swift?"

"Obviously, but I don't want anyone else to know that."

"If I present it, they'll assume the proposal comes from you." Chris delicately tried to evade the task.

"That's all right. You just act as if the idea is yours."

Christ scanned the first paragraph. "It advocates cloning."

"I read it, Chris."

"They'll eat me alive," Chris said.

"I'll be there to defend you."

"Yes sir." Chris knew how that would work. Starkey would silently listen to his deputies as they orally pummeled Chris for coming up with such a stupid idea. The technical people automatically rejected any proposal coming from the administrative staff, particularly one on such an emotional issue as opening the architecture of the company's primary product. Chris could already anticipate Pierre Dupris' emotional response.

"Make this your priority task. I scheduled an ad hoc management committee meeting for tomorrow morning."

"I'll need more time, sir."

"Do the best you can," Starkey ordered.

At exactly seven thirty the next morning, Lawrence Starkey entered the company conference room to find his senior managers deployed around the conference table and a fidgeting Chris Stuart waiting at the podium.

"Good morning, gentlemen," Starkey said, taking his place at the head of the table where he faced the podium.

A chorus of weak good mornings greeted Starkey. Starkey smiled. He knew the others resented the early timing of the meeting, but Starkey proceeded on the assumption that the pre-work hour sessions conveyed a message to the company work force.

"Mr. Stuart has an interesting concept that I think the management committee should consider. To a man, the senior division chiefs deployed around the table frowned. They did not like the idea of listening to the president's toady front another impractical proposal.

"Please go ahead, Chris," Starkey said.

"Gentlemen," Chris said. "Our subject today is licensing."

Before Chris could continue, Dupris groaned loudly.

The sound discomfited Chris, and he stood silently wondering whether he should continue or say something about the obviously negative reaction.

"Gentlemen," Starkey said. "Common courtesy requires us to give Mr. Stuart a chance. Please save the theatrics and your comments until he is finished. Chris," Starkey nodded to Chris, indicating he should continue.

"I need not tell you that Astron Computers is respected throughout the industry as an innovator. Our Asteroid user interface is indeed revolutionary. We have yet to achieve our goal of establishing the Asteroid operating system as an industry standard. I don't have to tell you gentlemen what this would mean to our profit line. However, no matter how advanced our technology may be, we cannot achieve this lofty goal on our own. No computer company, not even IBM, can create a standard without independent support."

With this introduction, Chris presented Charles Swift's argument as his own. He concluded: "We must license Asteroid technology to five significant manufacturers. And the best way to establish our standard would be to put it on an Intel platform."

When Chris finished, Starkey spoke: Thank you Chris. You have given us much to think about. You may leave now."

Chris gathered up his papers and departed.

After the door closed softly behind him, Starkey said: "As unpalatable as this proposal may be to most of you, it is something we must evaluate. IBM and its licensees are isolating us in the marketplace. Despite the fact that we have the superior technology, software programmers are not writing applications for us. If we do not do something immediately, we face a downward slide to bankruptcy."

"May I say something, please," Dupris interrupted.

Starkey smiled and nodded affirmatively. Dupris had barely been able to contain himself during Chris Stuart's presentation. Dupris had turned red in the face and had pounded the table in frustration. Starkey assumed it was Dupris' Gaelic blood showing.

"This cannot be done," Dupris shouted. "Establishing a standard may sound like salvation to non technical ears, but as Director of New Product Development I must point out that we are discussing nonsense."

"Could you be more specific and less emotional," Starkey tried to calm Dupris down.

"No Way. It is not possible. Our computers use a Motorola chip. The PC uses an Intel chip. Our Asteroid operating system is linked to the Motorola chip. It is not possible to run on an Intel based machine."

"Pierre," Starkey said. "Are you saying it is impossible to adapt our system to run on an Intel machine?"

"It might be possible," Dupris spoke more calmly, "but it is not practical. Our system is totally integrated into the Motorola chip. Our software and hardware are inseparable. We could write a software program that would let a PC run our applications on a PC, but why should they want to do that? They have more and better software than we have. Besides, our software and hardware

403

are so closely integrated, our software will not run smoothly. Technically, it is ridiculous. And to talk about licensing, that is nonsense. We license our beautiful machine, and we simply invite others to steal our sales. Of what good is this so called standard then? It's all bullshit."

"Could we have a little less emotion and more objective discussion," Starkey said, waiting for someone to debate the other side.

Nobody spoke. Starkey waited. Finally, Dupris started up again:

"You want objective discussion. I will give you objective discussion Lawrence. We sell 700,000 computers a year at three thousand each. Our revenues are approximately two billion dollars. If we were able to write the software and sell it to every single one of the four million customers who buy Intel computers every year, we could earn about one hundred dollars each. To me that ideal figure would total four million in sales. Our sales would fall, and we would find ourselves a one billion-dollar company. Is that what we want? Is the board prepared to cut our revenues in half with this licensing?"

"I am not sure that your figures represent the final outcome," Starkey equivocated. "Does any one else have any views?"

No one responded.

"Very well, I take it that the management committee rejects the concept of licensing."

"That's right," Dupris said.

"Very well," Starkey said. "I will take your views under advisement."

Starkey returned to his office convinced that licensing was a dead issue. Dupris felt so strongly that he was sure to take the matter to the Board if Starkey tried to implement it without unanimous Management Committee support, and that he did not have.

While he pondered the next step, the market intervened. The changes they had made in the Asteroid began to have their impact. With a hard disk, increased memory, and a cleaner bug free system, sales began to rise dramatically. Stanley's baby, the Asteroid, replaced the Astron 2 as the company's primary profit producer. This phenomenon led Starkey to commence worrying along parallel lines. Rather than thinking about licensing, he began to worry about protecting their graphical user interface and his first point of concern became Digital Software and its "Panes" project. Starkey consulted his Chief of Product Development, the outspoken Dupris, who reinforced Starkey's insecurity.

"They are stealing from us," Dupris shouted. "Stanley should never have given them our lovely Asteroid."

"They needed the machines to write software for us," Starkey said.

"No matter, they are stealing from us. Sue them!"

Starkey consulted his lawyers who of course salivated at the thought of a multimillion-dollar copyright infringement suit. It could drag on for years, a virtual cornucopia of billing hours.

Without advance warning, Starkey dispatched a lawyer to Seattle to fire a shot across Digital Software's bow. His instructions were clear: accuse DS of corporate espionage and copyright infringement; warn Charles Swift to cease and desist or face a lawsuit. The visit accomplished little other than to anger Charles. Charles countered by noting that the father of the Asteroid was Xerox not Astron Computers. DS and Astron Computers had only one thing in common; they had both learned from Xerox. Charles rejected every Astron Computers allegation.

Charles, after the lawyers departed, phoned Lawrence Starkey directly.

"Lawrence, I'm very disappointed in you," Charles began the conversation without polite preliminaries. "Your lawyer was just here, and I'm very angry."

"Why is that?" Starkey tried to dissemble.

"He accused us of crimes that we did not commit."

"You are infringing on our copyrights. You took advantage of our good will."

"Good will my ass," Charles said. "If anyone stole anything, it was both of us. We both learned from PARC."

"Nothing will be gained from such talk."

"We're on a collision course," Charles said. "Back off."

"Don't threaten us."

"Continue with the law suit, and I won't threaten you. I will stop all development of Asteroid products and that includes our new spreadsheet program."

At the time, Astron was responsible for over one third of Digital System's business. A termination of the business relationship would hurt Digital Systems, but not fatally. Burgeoning sales of the PC by cloners would eventually pick up the slack. A cancellation of the spreadsheet program would impact dramatically upon Astron Computers effort to access the corporate market.

"Let's not be hasty," Starkey cautioned.

"Let's meet and settle this once and for all," Charles said.

"Agreed."

Charles flew to Cupertino the following week.

Before meeting with Charles, Starkey consulted his management committee. The group's advice was unanimous: don't back off. Nonetheless, Starkey worried about Charles' threat. He did not think Charles was bluffing. Losing critical software at this point in Astron's resurgence on the back of the sturdy Asteroid could cripple the company.

Starkey, flanked by his lawyers and key executives, including Dupris, met with Charles and his single lawyer in the Astron conference room. Starkey expected the lawyers to dominate the meeting with their legalistic haggling, but Charles took command early.

"We don't need your permission to develop a graphical operating system," Charles spoke firmly but without heat. "I told Stanley in clear and uncertain terms that we would continue to work on Panes. You," Charles addressed Starkey directly," have known about our plans from the first day. We have not hidden anything from you. We've done everything we can to make your Asteroid project a success. Now, you threaten us with this bullshit lawsuit. How do you explain that?"

Before Starkey could respond, his lead lawyer intervened. "Let's put this in a legal context. Digital Software has infringed on Astron Computers copyrights."

"Bull shit," Charles erupted. "We haven't infringed on anything. If you persist on that line, this meeting is over." Charles stared at Starkey.

"We're all friends here," Starkey retreated. "There must be a gross misunderstanding."

"There is no misunderstanding if you threaten us with copyright violation. Tell the truth. Stanley was as impressed by PARC's technology demonstration as much as we were, and we both set out to duplicate it. Your Asteroid is as much a copyright infringement as Panes. As a matter of fact, investigation may prove it to be more so." Charles paused and let the threat of a possible countersuit sink in.

"Can we reach a middle position?" Starkey avoided looking at Dupris who he was sure was about to erupt.

"Sure. I'll give you a middle position," Charles said. "We'll continue to work on Astron software including the new spreadsheet which is as good as anything written for the PC, but only under a clear contract."

"Which should include?"

"The usual legal bullshit plus a clear license so this nonsense dispute does not come up again."

"Now wait a minute," Dupris interrupted.

"Not now, Pierre," Starkey shut Dupris down.

"What do you mean by license?"

"To employ the look and feel of the Asteroid operating system in any software we write for you."

"I see no problem with that. What about Panes?"

"Panes are not your concern. We have our own license from PARC."

Starkey and his lawyers immediately saw the viability of a lawsuit slipping away from them. Astron Computers would be vulnerable if Charles had involved PARC.

"Let us draft an agreement."

"That's what you should have done in the first place before sending this lawyer," Charles glared at the culprit, "down to threaten us."

A draft agreement was written, submitted to Digital Software, rejected, re-written then signed by both parties. The contract acknowledged that DS was free to develop its own digital technology as long as its conception originated with Xerox not Astron Computers. In the rewrite Charles had personally drafted a key paragraph: it granted Digital Software a "non-exclusive, worldwide, royalty free, perpetual, nontransferable license to use derivative works in present and future software programs and to license them to and through third parties for use in their software programs." These few words authorized DS to use virtually any features borrowed from Asteroid in Pane 1.0 and all future versions. In his rush to dampen the dispute he had created, Starkey had surrendered the future.

Chapter 31

The "Fasten Your Seatbelts" and "No Smoking" signs flashed on, and Ann Page heaved a deep sigh of relief. She turned, raised the blind, looked out, and suddenly wished she had not. She had expected a panoramic view of Seattle and the Puget Sound. Instead, all she could see was a thick blanket of dark angry cloud, streams of water flowing across the wings, and raised flaps fighting the weather as the United 747 forced its way down. Ann turned and pushed against the seat back, tightly grasping the arms of her seat. She felt the huge plane shudder as the wheels locked into place. She looked out of the window again, hoping to see an approaching runway. In the distance thunder rumbled.

"Why do I do this," Ann mumbled. This was a question she asked herself each time she crossed the breadth of the United States. She enjoyed her work in Washington and believed she was good at it. Certainly, the salary and generous expense account exceeded that she could have earned elsewhere, but the frequent cross-country flights were a heavy price to pay, and this one was of her own doing. She had contrived a reason to come back and confer with her employer, Intel, mainly because she wanted to make a side trip to Seattle to find out if Charles had been serious when he proposed that she undertake to represent Digital Software in Washington full time. If Intel learned of her unannounced side trip to Seattle, she would be in serious trouble.

Suddenly, the plane lurched to the right and all of the passengers took a deep breath as one. The pilot jerked the plane back in line and suddenly they broke through the overcast. Ann exhaled. Now that she could see the ground, she relaxed. The plane slowly drifted downward. They cleared the boundary fence and the wheels reached for the runway. A sharp shriek announced touchdown, and then they were racing along on the ground. The pilot reversed the engines and plane fought inertia. The man in the seat in front of Ann leaped up and jerked his bag from the overhead compartment.

"Please remain seated until the plane comes to a complete stop at the terminal," the stewardess announced on the loudspeaker.

The man maneuvered back into his seat, but others ignored the stewardess' warning and were soon crowding the aisles in a desperate attempt to retrieve their belongings. Ann watched and waited. She was not in that great of a rush. She checked her watch. She had three hours before her scheduled appointment with Charles at his Digital Software office. Ann was familiar enough with the Seattle airport to know she had ample time to check into her hotel and take a quick shower before presenting her memo to Charles and his staff. Ann opened her purse, took out her compact and ran her comb through her hair. She added some lipstick and decided that would do until she reached the hotel.

Finally, the huge airplane snuggled up to its arrival gate. The aisle filled with anxious passengers, waiting for the stewardesses to open the doors and free them. Ann shared their desire to escape. She felt like she had been locked up in a cage. She hated airplane travel for that single reason, the loss of freedom.

Ann waited patiently. Finally, the aisle cleared, and she was able to retrieve her carry-on bag. She checked her watch. She still had plenty of time. She was among the last of the passengers making their way up the covered ramp. An experienced traveler, Ann knew that it would do no good to hurry. They would all wait at the luggage carousel where democracy and chance reigned. The first eager arrival could be the last to claim luggage.

"Ann," someone called her name. This surprised her because she had not expected anyone to meet her. She had only talked with Charles. Ann turned and found a beaming Charles jumping up and down waving his hand.

"Charles, what are you doing here?" Ann asked. She knew that the chief executive of a Fortune Five Hundred company did not have the time to spare for meeting lowly lobbyists.

"Let me carry that," Charles grabbed her carry-on bag. "Welcome to Seattle."

Ann did not know how to respond, so she said nothing.

"Follow me," Charles said and began pushing through the crowd.

"I have a suitcase," Ann said.

"We're on our way to Carousel 3," Charles said.

After a long walk, they reached their destination. The carousel was already turning. Miraculously, Ann's bag appeared. She reached past Charles and retrieved it. Charles grabbed it from her grasp. "Follow me," he said.

"Charles," Ann stopped. "This is not necessary. You have more important things to do than carry my luggage around."

"No I don't," Charles smiled. "Follow me."

Again, Charles took off. Ann followed. Charles led the way to the arriving passenger exit and turned left. He ignored the waiting taxis and stopped at a green Porsche parked just beyond the line of taxis. A security guard stood to the rear of the car and waved the taxis around it.

"That's us," Charles said.

He threw Ann's bag into the trunk, discreetly slipped something into the security guard's open palm, then rushed to unlock the door for Ann.

"I'm not used to such service, Charles," Ann said. She wished she had taken more pains with her appearance. "I'm a mess," she said. "I planned on a quick stop at the hotel before our meeting."

"No problem," Charles said. "We're not due at my office for forty-five minutes.

"Charles!" Ann protested. "I just got off a six hour flight and need some time to regroup."

"I had to move our meeting up a little. We have an engagement tonight."

"What kind of an engagement?"

"It's sort of a surprise."

"Please, Charles. I need some time."

"You've got it. Take a half an hour."

Charles gunned the engine, darted between two passing taxis, then passed another on the right before they reached the main highway. Ann fastened her seat belt and half wished she were back in the dark angry clouds. "Is it always so overcast?" She asked.

"Overcast? This is a clear day for Seattle. You ought to see it when it rains."

"Isn't this rain?" Ann referred to the light drizzle that had begun as soon as they turned on to the main highway.

"Nah. Just a little morning mist."

"It's three in the afternoon."

"We have long mornings."

Ann decided to keep quiet and let Charles concentrate on his driving. She had heard stories about Charles and cars but had discounted them as media hype. Now, she was not so sure. She watched with the concern as Charles darted around fast moving traffic. The needle on the speedometer reached eighty.

"Charles, slow down, please," Ann said.

Before Charles could respond, a siren sounded behind them. Ann leaned forward, peered in the window mirror, and was relieved to see a helmeted cycle cop motioning with his thumb for Charles to pull over.

"Ahh shit," Charles said as he slowed and pulled to the side of the highway. He rolled down his window and waited with his left hand raised.

"Good afternoon Mr. Swift," the cop said as he approached the opened window. "Out for a weekend race?"

"Just picking up a friend at the airport," Charles smiled.

"Lady, let me give you some advice. Next time take a taxi."

Without asking to see Charles' license or registration, the cop scribbled on his pad then handed Charles his citation. "You're lucky," the cop said. "I caught you this time before you broke eighty."

"Thank you officer," Charles said.

"Hold it down, Mr. Swift. If I have to stop you again today, I'll take you in and have your nice Porsche towed."

The cop returned to his cycle.

"This happen often?" Ann asked as Charles reached across her, opened the glove compartment door, and tossed the citation inside.

"Are all those tickets?" Ann asked as she peered at the jammed compartment.

Charles flipped on his left turn signal and jumped the Porsche back into traffic. "You said you were in a hurry."

"Yes, but I want to live long enough to make my meeting."

At the Four Seasons Charles drove up to the front entrance and stopped directly in front of the door. He hopped out and retrieved Ann's luggage from the trunk. Two bellhops appeared. One grabbed the suitcase and the other the carry-on bag. "I can take that," Ann retrieved the carry-on bag. Charles flipped the car keys to the disappointed bellhop.

"Good afternoon, Mr. Swift," the bellhop snatched the keys out of the air.

"We'll only be a few minutes," Charles said.

"Take your time Mr. Swift."

"Why do I get the feeling you've done this before?" Ann asked.

Charles smiled. "This is my home town."

After Ann signed in at the receptionist's desk, she turned to Charles. "Wait here and try not to get arrested. I be down in ten minutes."

Thirty minutes later Ann reappeared. She had showered, changed into a clean suit, and applied fresh makeup. She found Charles slumped in a chair in the lobby nervously tapping his fingers on the arm of his chair.

"Now that wasn't bad, was it?" Ann surprised herself with her casual attitude toward one of the richest men in the country who just might become her next employer. She assumed it was his boyish appearance and cavalier manner that invited familiarity. She warned herself to behave more formally in the presence of others.

The Porsche waited exactly where Charles had parked it. Charles slipped the waiting doorman some folded green and retrieved his keys.

"Come back soon, Mr. Swift," the doorman smiled.

A fast ride and ten minutes later they reached the Digital Software campus.

"I'm impressed," Ann said as she surveyed the expanse of green grass. Even on a Saturday afternoon yard crews were busy.

"Not as big as Astron Computers, but we're growing."

Charles ignored the parking lot and stopped in a clearly marked no parking area in front of the first building. Ann said nothing. Charles could clearly park wherever he chose here. She got out before Charles could get around the car and open the door for her.

"Lead the way, sir," Ann smiled.

Charles guided her through a busy lobby to the elevator. Young men and a few young women, all dressed informally, were standing around in groups chatting.

"Get to work," Charles shouted good-naturedly.

Nobody paid much attention. "Personal time Charles," a longhaired youth in dirty T-shirt and jeans answered.

Ann noticed that several of the young female employees were carefully assessing her. Ann was glad that she had chosen her lawyer suit and regretted she had not brought a briefcase with her. She did not want to create any misimpressions about her relationship with Charles.

They took the elevator to the top floor, and Charles led the way to a corner office.

"Everybody's waiting in the conference room," the mature secretary greeted them.

"I hope they haven't been waiting long," Ann said, offering her hand to the gray haired secretary. "Hi, I'm Ann Page."

"June," the secretary smiled back. "Don't worry. Somebody saw Charles pull up and alerted me. I summoned the others."

"Does everybody in Seattle work on Saturday afternoon?" Ann asked.

"Only here," June smiled. "And Saturday night and all day Sunday."

"It's the high pay," Charles laughed.

"Tell me about it," June said.

Ann was impressed. Everyone seemed to treat Charles informally.

"Would you like to chat a bit in my office or should we go directly to the conference room?" Charles asked.

"I didn't prepare a presentation," Ann said. "I wrote my memo and came to talk about it."

"No problem," Charles said. "If I could have your memo, I am sure June can find a way to make some copies for the others."

Ann retrieved her three-page memo from her purse, opened it, and handed the paper to June.

"Make enough copies for the others, then bring that to me," Charles ordered.

"Yes, master," June smiled.

Charles entered a corner office to June's left. It was nicely furnished, but small. Ann had expected something far grander. Charles waited for Ann to take a seat on the two-seater couch and then dropped into a chair on her left. The furniture was Danish Modern, nothing fancy.

"Do you want me to make my pitch now?" Ann asked.

"I want to hear it, but why don't we wait until we join the others," Charles smiled. "No sense repeating the same thing twice."

"You might not want to waste the others' time."

"I doubt that we will."

"You're the boss." Ann could see that Charles was accustomed to having his way. She had heard, however, that he expected his employees to speak up when they disagreed with him. Stories of Charles' shouting matches with Mike Hamilton were the stuff of legends.

They sat in silence. Ann was not quite sure how to respond. She had come to Seattle to make a formal pitch that Digital Software badly needed representation in Washington, but something about Charles' casual style made her feel like they were on a date.

"You mentioned an engagement tonight," Ann decided to break the silence.

"If you're free."

"Certainly, I'm free. I hope it's nothing formal. I didn't bring anything appropriate with me."

"No problem," Charles smiled. "We're all informal here."

Ann noticed that he did not elaborate on their "engagement." She wondered if it would be prudent to press but decided against it. She calculated she might pick up a hint from one of the others.

"Here is your memo back," June said as she entered the office. She gave June the memo and a copy to Charles. "The others already have theirs."

"Thanks June," Charles and Ann said in unison. Ann chuckled and June winked.

"Let me scan this, then we'll join the others."

Ann waited while Charles read quickly.

"Great, let's go," Charles said.

Ann could not believe he had read what she had written. Charles had barely taken the time to scan her sub headings.

"If it's not what you wanted..." Ann began.

"No. It's great. Let's join the others."

Ann could not tell whether he liked the proposal or not. If not, she had taken a long flight and wasted a whole weekend to boot.

Charles jumped up and led the way back to the hallway and then into a small conference room. Ann was surprised to see that "the others" consisted of two men. She recognized Mike Hamilton but not the other.

"Mike," Charles said. "I think you know Ann."

"Hi Ann," Mike said. "Good to see you again."

"Hi Mike." Ann did not recall having previously met Mike Hamilton, Charles' Number two.

"And this is John Henry Perkins," Charles introduced the second man. "John is our staff counsel."

"Hi Ann," Perkins, a slender man with black hair and a prominent nose, offered his hand.

"Pleased to meet you Mr. Perkins."

"Call me John."

Charles dropped into a chair next to Mike and waved his hand to indicate that Ann should take the seat at the head of the table.

"The meeting's yours," Charles said.

Ann sensed she was being tested, so she dropped the informality.

"Did you have a chance to go over my memo?" Ann asked.

"Yes," Mike and the lawyer responded. Charles did not.

"As I make clear, I believe that Digital Software is going to badly need Washington representation in the very near future."

"Why?" Charles asked bluntly.

"The Democrats do not like big business, and Digital Software is on the threshold of becoming the biggest."

"We donate to both parties," Charles said.

"Doesn't matter. The voters expect the Democrats to hate you so they have to appear to do so whether you donate money to them or not."

"Then we're wasting our money."

"I didn't say that." Ann felt Charles' attitude forcing her into a confrontational style. She smiled to soften her words. "Donations to individual campaigns will buy you votes but don't expect the Democratic Party as a whole to endorse you."

"Why not?" Charles demanded. "The computer industry is now driving our economy, and a booming economy is good for everybody."

"But no one likes a monopoly."

"Digital Software is not a monopoly."

"Charles I didn't fly across the United States to debate the question whether you are a monopoly or not. You're big enough that people will believe you are if the politicians claim you are."

"I can't help that."

"And you have a reputation for unfair business practices."

"That's bull shit," Charles blurted.

Ann looked at Mike for assistance, but he merely smiled.

"No it isn't Charles. If I were a staffer on the Hill I would have no difficulty finding witnesses willing to testify that Digital Software forced them to bundle software in return for licenses to sell machines using your operating system."

"That's because the PC is the industry standard."

"Again I'm not here to debate the point. I'm talking what people will believe, not reality."

"What do they hope to gain by attacking Digital Software?"

"Votes. Politicians need issues. Who defends big businesses like yours? Your competitors? How is your relationship with Astron Computers or Motorola?"

"We can defend ourselves."

"Exactly my point. You need representation in Washington to keep pressure on the Congress."

"How do you do that?"

"Threaten and cajole your opposition and treat your friends nicely."

"Buy votes?"

"Campaign contributions. You're playing in the big time now."

"Any questions?" Charles turned to Mike and John.

"I agree with Ann," Mike said. "I don't see imminent peril, but who knows what waits down the road."

"I take that as a yes," Charles said. "John?"

"This may sound like heresy, coming from a lawyer, but I hate politics," John smiled at Ann. "I don't like the ring of this anti-monopoly talk coming out of Washington. Someone is clearly preparing a game plan."

"Why us?" Charles asked.

"Because, as Ann so aptly pointed out, nobody likes the big guys," John said.

"Then the Republicans are our only hope?" Charles said.

"They're the party of big business," Ann said, "but you can't concentrate on them exclusively. You have to work the Democrats and the media."

"Fuck the media," Charles said.

"Fuck who you want," Ann replied, not intimidated by Charles' language. "But you need friends in the media too."

"How do we start?"

"Hire yourself a good lobbyist," Ann smiled. "Know any."

"We're going to need more than one," Mike said. "A whole office full."

"We'll have to give this some more thought," Charles said. "Do you think you could work out a budget for us?"

"For the office? No problem. But I am not sure how much of this you want to put on paper."

"Everything has to be above board," John cautioned. "Nothing illegal."

"Precisely," Ann said. "You are going to have the Justice Department all over you and your files. Everything must be within the letter of the law."

"We're not a bunch of thieves," Charles blurted.

"But you do pressure clients," Ann said.

"That's just business," Charles said. "If we were a bunch of patsies we wouldn't survive."

"Why don't you give us a ball park figure, Ann," Mike suggested.

"If I were setting up an office for you, for Digital Software exclusive, I would start with a million for overhead, salaries, representation. Campaign donations, media costs, that would have to be figured out later."

"What do you mean by media costs?" Charles asked.

"You need an outlet that constantly sings your praises."

"I've been telling him that for years," Mike said. "We need a magazine that focuses on the microcomputer industry."

"The PC," Charles said. "Astron Computers has their own outlet, AstroWorld."

"You should have a quality magazine that will promote your image while it sells your products," Ann said. "You need a professional for that."

"I might have a candidate for that job," Charles said.

"I bet I could name him," Mike said.

"I think I might also," Ann said.

Charles checked his watch. "OK Ann, you've given us something to think about. We'll get back to you."

"I'm ready when you are," Ann said. She hid her disappointment that Charles had not pressed her to take on the job of representing Digital Software in Washington. She was not sure she wanted the job because it would require her to terminate her relationship with Intel which had always treated her well. It would have been nice, however, to have Charles as a suitor.

"I think I'll call it a day," Charles said.

"It's only six o'clock," Mike said. "Are you feeling all right?"

"I don't believe what I'm hearing," Ann laughed. "It's six o'clock on a Saturday afternoon and you guys are acting like you're quitting early."

Charles paused at June's desk while Ann shook hands with Mike and John.

"You'll be hearing from us," Mike confided. "I've been pressing him on a Washington office for some time now. Charles hates to spend money."

Ann laughed. "You guys are printing it on your own presses. Anyway, a little investment now will prove invaluable in the long run. I'm serious when I say the Democrats have you guys in their sights."

"I know," Mike agreed.

"Anything that can't wait?" Charles asked June.

She held up a stack of messages. "Nothing that can't wait until tomorrow. Have a good evening, boss," June nodded meaningfully at Ann.

Charles blushed.

As they made their way back to Charles' Porsche, Ann was amazed at the number of people still busy at their desks.

"Is it always like this?" Ann asked.

"Work is what we do," Charles said.

416

"You mentioned you had an engagement, Charles," Ann said. "I can take a taxi back to my hotel."

"No you can't," Charles said. "I said we have an engagement."

"But I don't have any formal clothes with me."

"You won't need any."

"I can't go anywhere in this outfit. I look like a butch lawyer."

"I hadn't noticed."

"Do we have time to go back to the hotel?"

"Again? Don't women do anything but change clothes?"

"Will a blouse and skirt do?"

"Perfectly."

"Mind telling me where we're going?"

Charles hesitated. "I guess you'll learn soon enough."

"Like when we get there."

"My parent's house. It's my dad's sixtieth birthday, and I promised I would be there."

"Oh Charles. I don't want to intrude. You go without me. This should be for family."

"My sister can't make it. We need a fourth."

"A fourth?"

"You play bridge don't you?"

"I haven't since college."

"Good. You'll do."

Charles waited in the lobby of the Four Seasons while Ann changed. He gave her twenty minutes and she took forty-five. She showered, took pains with her makeup, and gave thanks that she had brought a new blouse and skirt. She particularly liked the silk scarlet blouse with a tailored collar. The white linen skirt was a little tight, but the two went together well. She debated heels and flats but settled on the white pumps.

"I hope we're not late," Ann said when she rejoined Charles.

"No problem," Charles said.

"It is if they're holding dinner."

"It'll just be some steaks on the grill," Charles said.

"Outside?" Ann asked. "Should I change these shoes?"

"They look great," Charles said, taking Ann by the arm and guiding her towards the lobby exit.

"This is just like a date, Charles," Ann said.

"Don't worry," Charles said. "You won't have to write it off your expense account."

Ann let the dig pass. Charles again tipped the smiling doorman who opened the Porsche door for Ann.

"My parents live on Lake Washington," Charles said.

"Do you still live at home?" Ann asked.

"No, I have a house across the lake."

"That's unusual."

"What is?"

"Living so close to your parents."

"It's convenient," Charles said. "I spend most of my time at the office."

"No girl friends?"

"Don't have time for them," Charles said.

"What do your parents think about your bringing a stranger to their party."

"It's not really a party. I couldn't get out of it. They'll be glad to see I have a girl friend. You don't mind pretending do you?"

"Why Charles?"

"My mother thinks I'm gay."

"Are you?"

"Of course not." Charles frowned.

"I didn't mean to insult you."

"You didn't. I just don't have time right now for a social life."

Ann smiled. Charles whipped the car into the opposite side of the road and passed two cars before he jumped back narrowly missing an oncoming trunk.

"Charles?"

"Yes."

"I wouldn't mind if you slowed down a little."

Charles lifted his foot from the accelerator and allowed the Porsche to slow down to a peaceful sixty miles an hour.

Charles parked in the driveway behind a shiny green Cadillac DeVille.

"There's mom," Charles said.

Mrs. Swift, an attractive mature woman with stylish gray hair and wearing an obviously expensive housedress, waited in the doorway.

"Charles, you're late," she admonished, tapping her wrist.

"Sorry mom," Charles said. "We ran over a bit at the office."

"And who is this?" Mrs. Swift greeted Ann with a warm smile.

"Mom this is Ann. Ann, Mom," Charles said.

"Welcome Ann," Mrs. Swift said. "This is indeed a pleasure. Charles seldom brings his friends to visit."

"I apologize if I'm intruding," Ann said.

"Not at all. The pleasure is mine," Mrs. Swift beamed. "Charles, where have you been hiding this lovely person?"

Charles shrugged.

"Mr. Swift is waiting for us on the patio," Mrs. Swift said as she held the door for Ann to enter.

As Ann passed, Mrs. Swift gave her son a sharp inquiring glance that he ignored.

"Straight done the hall and out the back door," Mrs. Swift said.

Ann glanced at the living room on the right and then a study/library. "What a lovely house," she said, sincerely impressed.

"Thank you," Mrs. Swift said before stepping past Ann and opening a door that led to a covered patio.

"Charles," Mrs. Swift said. "Your son has brought a guest."

The handsome man with gray hair, rimless glasses, and slightly flushed face pushed himself up from his chair. Mrs. Swift made the formal introductions.

"I didn't realize you were a junior," Ann smiled at Charles.

"He's not," Mr. Swift laughed. "I'm junior."

"I'm trey," Charles said.

"Charles Carleton Swift, III," Mrs. Swift explained.

"We didn't really want to do that to him but his grandmother, my mother, insisted," Mr. Swift said.

After they were seated, Mrs. Swift took charge. She delegated Charles to serve them each a glass of white wine and dispatched Mr. Swift to light the charcoal and alert "Marsha" that they would be dining in forty-five minutes. Then, she turned to Anne. "Are you from Seattle?" She asked sweetly.

"No, Mrs. Swift."

"Marjorie," Mrs. Swift said quickly. "I'm only Mrs. Swift at my Board meetings. I'm Mom or mother to Charles, depending on my son's mood, and otherwise I'm plain old Marjorie."

"Yes, Marjorie," Ann said, thinking that Mrs. Swift was anything but plain old Marjorie. She was a formidable woman accustomed to having her own way. "Right now I'm from Washington, D.C., not the state, but I grew up in Palo Alto."

"You went to Stanford then?"

"Yes."

"Major in education?"

"No, computer science."

"That explains it. Where you met my son, then."

"Actually, I met him a month ago at a cocktail party given by a mutual friend."

"William Oldham."

"Yes. How did you know?"

"William was Charles' roommate at Harvard. We talked frequently on the phone. Charles is not an easy person to keep track of."

"I'm beginning to learn that."

"What do you do in Washington?"

Ann was tempted to resist. She did not like being interrogated, and Charles' mother came on with a heavy hand. "I'm a lobbyist for Intel."

"Really!" Mrs. Swift responded quickly. "That sounds interesting. Do you find it difficult working in a man's world?"

"Times are changing."

"Too slowly. I'm a token female on several Board of Directors," Mrs. Swift said.

Charles returned and served the wine, but his mother ignored him.

"And what are you doing in Seattle," Mrs. Swift continued.

"Business, mother," Charles tried to interrupt the interrogation.

"What business, Charles?"

"Digital Software is considering opening an office in Washington D.C. and Ann is advising us."

"Really? Why are you opening an office there? Are you in trouble."

Charles' father rejoined them. "The coals will be ready in about twenty minutes. How do you like your steak Ann?"

"Rare, if that's not a problem."

"I'm well done," Mrs. Swift said. "These two are rare, so you'll fit right in."

"Charles," Mrs. Swift addressed her husband. "Ann is a Washington lobbyist and Charles is considering hiring her to represent him."

"Oh," Charles Jr. studied Ann with intensified interest. "I'm a lawyer. What's a nice girl like you doing in that shady business?"

"Mom. Who told you I was going to hire Ann?" Charles frowned. He kept his mother as far from his business affairs as possible.

"Not me," Ann smiled.

"I'm your mother. I usually can figure out what you're up to." She turned towards Ann. "I've had years of practice keeping track of trey."

"Let's not get into that, please," Charles said.

"I'll only tell one story," Mrs. Swift said.

"Please do," Ann encouraged her.

"When Trey was in the sixth grade he was so difficult that I took him to a psychologist to find out why he was so stubborn. He wouldn't do anything I wanted."

"Mom, please not that old chestnut," Charles frowned.

Charles Jr. leaned back and smiled indulgently. He had long ago given up trying to control either his son or his wife.

"After a full year, the psychologist told me that my son was an independent character who would do exactly as he pleased and that I should just give in and enjoy it. Well, we're proud of him, but I haven't enjoyed everything that he has done."

"You should be proud of him," Ann said. "He's created a wonderful company."

"But he did not graduate from Harvard," Mrs. Swift persisted.

"I didn't have time, Mom. You know that. I could have, but I could not let the opportunity pass me by."

Mrs. Swift did not respond. With Charles Jr. present and skillfully guiding the conversation, they moved on to more neutral ground. They discussed the gloomy Seattle weather, the lovely green countryside, and the political situation in Washington. The latter subject allowed Ann to share the latest political gossip. They switched to red wine when Charles Jr., assisted by Charles, served the steaks, and Marsha, the maid, added the French fried potatoes, stiff green beans, and hot rolls.

"Charles, I thank you for your present," Charles Sr. shook his head when he addressed his son. "I think you got carried away."

"You deserve it. Anyway, it was Mom's idea."

"It was not," Mrs. Swift said quickly. "I told him not to be so extravagant." Charles shrugged.

"What do you think, Ann?" Charles Sr. asked.

"What was the present?"

"You may have seen it on the way in," Charles smiled.

"The green Cadillac," Ann said. "It's beautiful. That was very kind of you Charles.

"I'll enjoy it," Charles Jr. said.

"It was too extravagant," Mrs. Swift said.

"It makes up for that red Mustang you gave me," Charles said. "I loved that car."

"And I'll love this one," Charles Sr. said.

After dinner, they retreated to the family room where they took seats at a green beige covered table. Mrs. Swift directed Ann to the chair across from hers.

"I'm so glad that Charles found us someone who plays bridge," Mrs. Swift declared as she shuffled the cards.

"Would you prefer poker?" Charles asked Ann.

Mrs. Swift ignored him and dealt the cards. Ann realized she was in trouble when Charles looked at his father and said: "All right dad, we'll whip their asses."

"Charles, watch your language," Mrs. Swift remonstrated.

Charles Jr. smiled at his son and nodded.

They played serious bridge. Ann did not really care who won, but it was clear that the three Swifts did. Ann did her best and made only two serious errors. Both were enough to cost Mrs. Swift and Ann two hands. Mrs. Swift frowned, making her disapproval clear, but she said nothing. Finally, Charles and his father won. They both smiled, and Mrs. Swift glowered. "Don't you dare say it, Charles," Mrs. Swift stared at her son. She threw the cards down angrily.

"We whipped their asses," Charles said and jumped up and touched the ceiling.

Charles Sr. and Ann tried not to smile.

Mrs. Swift glared at Charles. "You must excuse my son's manners," Mrs. Swift spoke to Ann. "I obviously failed in his upbringing."

"Ahh mom. You taught me to win," Charles laughed.

"It has been a lovely evening," Ann said. "But I should be getting back to my hotel. I have to catch an early flight."

"Oh, you're staying at a hotel," Mrs. Swift picked up quickly. "Which one?"

"The Four Seasons."

Mrs. Swift nodded approval.

Ann turned to Charles as they sped down the deserted road: "Your mother thought I was staying with you."

"She was disappointed you aren't."

"I doubt that."

"She liked you."

"I'm not sure of that."

"Take my word for it. She did."

"She didn't seem too happy when we left."

"That's because she lost. Want to see my house?"

Ann considered her response before answering. The temptation was great, but she decided no, knowing where it would lead. She was determined not to mix business with pleasure. "I would like to sometime, Charles, but not tonight. I'm exhausted from the flight and have to check in with Intel then leave from San Francisco tomorrow night."

"OK. You've got a raincheck."

"I'll take it." Ann had trouble reading Charles' reaction.

Chapter 32

Despite the fact that he knew better, Charles was excited. He and Ann, Bill Oldham and Mike Hamilton, even David Howard and his wife Terry were all in San Francisco along with 4,500 of their friends, celebrities, and media stars to attend the dramatic unveiling of Stanley Pitt's new computer, Starbright. After three years and the expenditure of a two hundred million dollars, Stanley planned to stage a sensational show. He hired Davies Symphony Hall and orchestrated the presentation himself. Charles, whose on again off again relationship with Stanley was in the off position, wanted to see for himself what Stanley had wrought this time. As Charles told Ann, the show itself would compensate adequately for the arduous crosscountry flight.

Charles and Ann occupied a penthouse suite in the Mark Hopkins Inter-Continental, Charles' choice because he liked the famous lounge, Top of the Mark. Charles and Ann made no secret of their relationship. Whenever they were in the same city, they shared everything. With Charles in Seattle and Ann heading up Digital Software's lobby operation in Washington, they did not see as much of each other as both wished. Ann refused to surrender her career for a housewife slot, and Charles did not press, primarily because he was not ready for marriage. His empire required his full attention, and Charles was smart enough to recognize he did not have enough available time to share with a family.

Friends were another matter.

Bill Oldham had finally succumbed to Charles' entreaties. He had taken a sabbatical from the Post and written his book about the evolution of the microcomputer. The book sold well, a modest success, but the royalties were not enough to sustain Bill while he wrote. Bill with Charles' unlimited support had started Charles' magazine, "Computer News." Bill served as Digital Software's beard. He had full editorial control; Charles discreetly subsidized the startup costs, and CN, as "Computer News" was known in the trade, had overcome a slow start and developed into a commercial success. CN won praise for its terse objective reporting and glossy illustrations. Like most magazines, CN served its advertisers, foremost among whom stood Digital Software. Based in Los Altos, from which he covered the Silicon Valley, Bill had a short distance to travel to attend Stanley's gala.

David Howard and Terry flew in from Armonk. David, who now headed IBM's Personal Systems Division, had important business to discuss with Charles and had been delighted to accept Charles' suggestion that they meet in San Francisco. Charles knew why David was coming, and he was not looking forward to their formal meeting. Obviously, David's superiors were trying to take advantage of their old friendship. David was not responsible for the development of IBM's new operating system, OS-2, Operating System 2, but he would be charged to insist that Charles desist in his development of Panes. This Charles could not do. He had already released Panes 1 and Panes 2 to a lukewarm market, and Charles was determined that eventually Panes would surpass the Asteroid and become the standard for graphical operating systems. The meetings would come after today's celebration and socializing.

"Are you ready Ann?" Charles called. "We should get there a little early. It will be a mob scene and I don't want to risk missing Stanley's show."

"Ready," Ann popped out of the bathroom holding her hands wide. "How do I look?"

"Jesus Ann," Charles said. Ann looked ravenous in her tight blue silk dress with a high hemline.

"That bad, huh? Should I change?" Ann twirled about.

"They'll think I'm a cheap gigolo with a movie star."

"Which one?" Ann smiled.

"How about Loretta Young?"

Ann retreated to the bathroom then reemerged mimicking Loretta Young's patented entrance routine. "How's that."

"Great."

"I wish you had said Marilyn Monroe," Ann said.

"Don't set your sights too high," Charles laughed.

"Look at you," Ann said.

"What's wrong? This suit cost me five hundred bucks."

"You're the cheapest billionaire I know. Check your tie. Didn't your mother teach you how to tie one properly?" Ann adjusted Charles' knot.

The phone rang. Ann answered: "Hello."

She listened. "We're ready. Meet you in the lobby."

"Bill?" Charles asked.

"Yes. He, Terry, David and Mike are waiting in the lobby. The limo's here."

When they reached the lobby, they found Bill Oldham pacing.

"Christ Charles, what have you been doing?"

"Not what you think," Ann laughed.

"We don't have to worry," Charles said. "Stanley will make the audience wait."

"For dramatic effect," Bill agreed.

Four blocks from the Davies Symphony Hall, they reached a traffic backup.

Bill squirmed anxiously on the jump seat for a few minutes, then reached for the door handle.

"You guys will have to excuse me. I'm working today. I'll see you inside."

Bill jumped out and jogged down the street.

"It's good seeing the media take their jobs seriously," Mike laughed.

"You guys aren't fair," Ann said.

The traffic gradually inched forward and finally, twenty minutes after the presentation was to start, they reached the front of the Hall. They climbed out and started up the stairs. To their surprise, Bill met them at the top.

"The crowd's so big they're only letting a media pool inside," Bill complained.

"Don't worry," Charles said. "Come with us."

Bill slipped between Charles and Ann. At the door, a security guard stopped them and held out his hand for their formal invitation that served as their entrance ticket.

"That's all right," Some one behind the guard called. "That's Mr. Swift and his party."

"Sometimes it's useful to be famous," Mike whispered to Bill.

The guard stepped back, and they entered the crowded hall.

"Let me escort you to your seats," the man who had recognized Charles said. He said he represented Again Corporation's Public Relations Office.

"See you guys later," Bill stayed behind.

The Public Relations officer led them down the aisle to the front of the crowded hall. He unhooked a velvet rope that protected the first four rows. "Stanley wants Mr. Swift in the front row center where he can see him," he instructed the usher.

Charles assumed Stanley would somehow try to exploit his presence but did not protest.

They were barely seated when a blast of rock music filled the hall then subsided into silence. The lights darkened, and the audience waited expectantly.

"The master showman is at work," Charles whispered to Ann.

The huge curtains slid back on the darkened stage. Suddenly, the music resumed, and the audience stared at the inky black tableaux. They waited, at first amused. The silence lengthened, and many shifted in their seats in nervous anticipation. Two minutes passed.

"What is happening?" A woman seated behind Charles asked.

Suddenly, the music erupted in full volume. The audience waited, then the music stopped abruptly. A single spotlight pierced the darkness and focused on a table holding a single vase of black flowers and a shrouded object. The audience murmured. The spotlight went off, and again the audience stared into the blackness. The rock music blared, and a second spotlight highlighted Stanley Pitts standing before a microphone on the right. Stanley wore black: a shiny black suit, black shoes and a black turtle neck shirt. He stood motionless, smiling silently at the crowd. Someone in the back began to clap and within seconds the entire audience was applauding and cheering. Stanley waited silently, clearly aware that he had the audience under his complete control.

Stanley suddenly stopped smiling. He touched his palms together in front of him, an act symbolizing prayer, and bowed his head until his chin touched his fingers. The audience gasped. Stanley waited a few seconds then raised his chin high and began to talk softly:

"Speaking for everyone at Again" Stanley paused for effect, then he shouted loudly, "I say its great to be back!"

Stanley hesitated, and the audience clapped enthusiastically. He waited for the applause to subside then began again:

"We are about to share with you one of those occasions that occurs only rarely, a time when a new product appears that changes the future. Thomas Edison gave us the electric light. Alexander Graham Bell gave us the telephone. Again Corporation gives you Starbright."

The second spotlight pierced the dark stage and highlighted the table holding the vase of flowers and the black shroud.

The audience cheered, and a smiling Stanley waited.

"We have worked on Starbright for three years," Stanley spoke softly. "And it has turned out incredibly great," Stanley shouted. He pronounced "in-cre-di-bly" in syllables, and the audience loved it.

Then, with the audience comfortably hanging on his every word, Stanley gave his spiel, speaking with the subject of his paean still shrouded. Stanley informed his audience that computer architectures had ten-year cycles. He cited the Astron 2 that reached its technical peak in 1982, then he referred to the IBM Personal Computer and finally the Asteroid which currently, until this moment, had the momentum.

"And here," Stanley pointed at the shroud, stands the architecture of the nineties waiting to carry us into the next millennium."

Everyone expected the shroud to rise dramatically to disclose the Starbright, but it did not.

"We consulted the leaders in educational computing from research universities such as Stanford and Carnegie-Mellon, liberal arts schools like Vassar and Reed, as well as state universities from California, to Michigan, to Maryland. We asked them not for their list of specifications, but for a list of their dreams; not to extend what computers have been, but to imagine what they could be. Only then did we begin to work on our partner in thought."

"They told us they wanted a microcomputer that thinks like:" Stanley turned and raised one hand toward the dark screen behind him. A picture of a huge mainframe computer appeared, and the audience cheered.

"We quickly realized that higher education wants a personal mainframe."

After describing the characteristics that such an odd device would need, Stanley announced: "And thus we give you a one foot cube containing the entire system."

The audience waited breathlessly for the shroud to lift. It did not move.

Charles leaned across Ann and whispered to Mike: "This is all bullshit."

"But its great bullshit," Mike whispered back.

"Give Stanley credit," Ann whispered. "The audience loves it."

Stanley continued his tease by describing the factory that produced the Starbright "completely untouched by human hands."

He turned again and watched while a clip showing robots at work appeared on the screen.

When the film ended, the lights dimmed, and Stanley shouted: "I give you the computer that will carry us into the millennium. The spotlight focused on the rising shroud that lifted to disclose a jet-black cube.

"I will now let Starbright take us on a tour."

Stanley walked to the table and pressed a button on the front of the black box. A picture of Starbright flashed on the screen. Stanley stepped back. "Entertain us," Stanley commanded, and Starbright filled the auditorium with the heavy beat of rock music. The picture of a shooting star exploded on the screen.

The audience cheered again.

"Now, Starbright take us back in time," Stanley ordered.

A picture of Martin Luther King appeared on the screen while King's "I have a Dream Speech" boomed from the auditorium speakers. This was followed by John Kennedy declaring "Ask Not What Your Country Can Do for You."

Stanley and his machine performed for another dramatic half-hour and ended with a simple statement from Stanley:

"We are going to charge higher education a single price of $6,500 for this marvel." The audience applauded. "And our printer that everyone will want in their office will sell for the great price of $2,000."

"He is out of his damned mind," Charles spoke loud enough to be heard by those in the row behind him.

"And those who wish to supplement the capabilities of the optical disk drive with conventional hard drives will have to pay a mere two thousand more." More applause followed.

"In other words," a shocked Charles said. "This marvel will cost the average college student over $10,000."

That evening before Charles and Ann were to join David and Terry for dinner, Charles and David met in the Top of the Mark lounge.

"What did you think of Stanley's presentation?" David asked.

"Pure Stanley," Charles replied, "from top to bottom."

"Stanley knows how to put on a good show," David said.

"What about Starlight?"

"It's a beautiful machine," Charles said honestly. "I even like the dramatic black box."

"But?"

"It's a machine I would like to own," Charles sidestepped the question.

"And I. But?" David smiled.

"We both know the answer to that," Charles said.

"Stanley's built the machine that he wants to build ignoring his target market."

"If he consulted with higher education as he claimed, Stanley chose to ignore their advice. He doesn't stand a chance at $10,000 a machine," Charles shook his head and sipped his Budweiser.

"Why does he do it?" David asked.

"For the same reason he does everything," Charles said. "Stanley lives for Stanley."

"Well, we all live for ourselves, to a degree," David said.

"Not when we're building a product to sell," Charles said. Charles thought of his struggle to underprice the competition. Charles' dogma was: dominate the market at whatever cost, then reap the profit when you own it. "They love him, though. Did you hear the applause when he announced those ridiculous prices?"

"Even the media at the press conferences served up softballs," David said. "Are you going to write software for Stanley?"

"Are you kidding? Stanley insists Again will write its own software."

"He's repeating history," David said, referring to Stanley's restrictions on the Asteroid.

"We have our own problems," Charles changed the subject.

"That's why I'm here," David said.

"Shoot."

"You've got our people worried," David began.

"They've got no cause to be."

"The bureaucrats are convinced that you're not working on OS-2 like you should."

"That's bull shit, David. We have a full team working on OS-2."

"Then you're not working on Panes?"

"Certainly we are. I've made no secret of the fact that we're going to produce our own graphical operating system. You want to replace LDOS and so do we. We think you're making a mistake with OS-2, but we will meet the terms of our agreement. We're working together with you to build it if we can." This was an issue that Charles was determined not to compromise. "We've committed a lot of resources to OS-2. Frankly, it pisses me off when I hear this crap."

"Calm down, Charles," David said. "Astron Computers with Asteroid has finally gotten it right, and they're winning back market share. Not only that, the clones are starting to take our sales."

"And the microcomputer is eating into your mainframe market," Charles said.

"So you can see why we're worried."

"You had a good ride with the PC," Charles said. "We both did well by it."

"Now we need a new graphical operating system. Without one, we're hurting big time."

"That's the nature of the game," Charles said.

"You seem to be doing all right," David said. "I've got to speak frankly. There are a lot of people in Armonk, senior people, who resent the fact that Digital Software plays all sides. You license software and operating systems to us. You do the same for the clones, and you write software for Astron Computers. They really do believe you are stringing us along with OS-2 while you are working full time on Panes."

"Do you blame me for selling to everyone I can?"

"I don't. I envy you."

"Then, why complain."

"They say they can't trust you."

"Bull shit. When I say I'm working on OS-2, I am. If they think they can do better without me, fine." Charles glared at his friend.

David sipped his beer and tried to think of a middle ground.

"Besides, I'm hearing things I don't like," Charles counterattacked.

"Such as?"

"Let's get everything out on the table, David. I hear that IBM is talking with Unix about standardizing their operating system. And what about the rumor you're thinking about licensing Stanley's Starbright operating system. Where does that leave me?"

David did not know how to respond. It was true that IBM's Corporate Technology Team was exploring these other options. In typical IBM fashion, different divisions competed until a final decision was reached. The difficulties that the OS-2 team were encountering only served to fuel the competition.

"If I worked exclusively on OS-2 as you seem to want, what happens to Digital Software if IBM suddenly drops OS-2 in favor of another system?"

"I know, Charles," David said. "We're all beginning to understand what a miracle Stanley wrought when he developed Asteroid. It's not a simple matter of copying PARC's technology."

"Tell me about it," Charles laughed wryly. Panes was proving to be a much greater challenged than he had envisioned. He had already devoted six years to developing Panes and still was not close to matching Asteroid, which Stanley had engineered in three years.

"Our engineers tell me that your OS-2 people are not even talking with ours," David said.

"Mine tell me the same thing, only in reverse order," Charles said. "What's this I hear about IBM considering two versions of OS-2?"

"You're better informed than the CIA," David replied.

"And one of those versions would be sold exclusively on IBM machines."

"We have to protect ourselves from our clones somehow. You can't blame us for that."

"And you can't criticize me for working on Panes to protect myself. We are in the operating system business. We will develop new products, and the graphical interface operating system is the system of the future."

"Can you assure me that you will continue to work with us to produce OS-2?"

"Yes. I have never stopped."

"Then let's get our two teams back on track."

"Agreed."

"I guess that is the best we are going to achieve here today."

"Just cut out this complaining about Panes."

"Agreed."

"We've got other problems, you know," Charles said.

"Such as?"

"My Washington people tell me that staffers at the Federal Trade Commission and lawyers at the Justice Department are talking about an investigation into our collaboration on OS-2."

"What in the hell for?" Charles' words surprised David. IBM's legal eagles had not mentioned this subject in the Management Committee.

"Monopolistic restraint of trade. Price fixing. You name it. Sherman Antitrust Act, Clayton Act and all that good stuff."

"Can you be more specific?"

"Just between us friends, right?"

"Right."

"They think we have a secret agreement to divide up the market. You will use OS-2 on IBM machines, and Digital Software sells Panes to the clones and everyone else. Between us, we will have a monopoly."

"And they think we're scheming to fix prices?"

"You got it."

"What ever happened to Bush and Reagan?" David asked. Both Republican administrations had reined in the trustbusters.

"These are civil service bureaucrats making plans for the next Democratic administration."

"I know two special interest groups that had better get busy and make sure we don't have a Democratic administration elected next time," David said.

"So do I, but let's not discuss it," Charles said.

"Is there anything we can do publicly to highlight the nature of our agreement on OS-2 to negate any talk of monopolistic conspiracy?"

"Let's make the whole thing public," Charles said.

"Agreed," David said. "I'll have our Public Relations people get in touch with yours.

"Tell them to talk with Ann."

Chapter 33

Laurence Starkey's Astron Computers rose with the tide of new Asteroid purchasers. With a graphical operating system five years ahead of its peers, thanks to Stanley, Astron Computers had the time to reconsolidate after the disastrous years when they lost market percentage to the PC. Having lost their creative drive with Stanley's forced departure, Starkey's team concentrated on internal organization and maneuvering for position while Starkey turned outward where he thrived in the role of industry spokesman. He spent his time on the road giving speeches, making public appearances, and hobnobbing with politicians like the Clintons.

Starkey took time from his public appearances to give Charles a severe case of heartburn. Some two months after the appearance of Panes 2, Starkey directed his lawyers to file suit against Digital Software alleging violations of copyrights on both Stella and Asteroid.

Charles was seated at his desk when Bill Oldham called: "Charles, have you heard the news?"

"What news? You know I don't have time for that nonsense." Charles made no secret of the fact that he avoided the ranting and raving of the media.

"You better find the time. I predict that your phone will soon be ringing."

"Who will be calling?"

"The media."

"Tell me the bad news," Charles sighed, leaning back in his chair and putting his feet on the desk. Charles was in a good mood. Panes 2 was selling better than expected. "Don't tell me Panes 2 has developed an unknown bug." Charles expected this. Every new application had bugs, and the same applied to operating systems that were far more complicated. A few fixes and everything would be fine.

"Astron has filed suit," Bill reported.

"Against whom?" Charles dropped his feet to the floor and sat up straight. Lawsuits were not an unknown phenomenon in the industry, but Charles hated them. Although they were usually settled out of court, the first wave of publicity was bad for the markets, stock and consumer.

"Digital Software."

"For what? I talked with Starkey on Monday, and he said nothing about a lawsuit," Charles complained.

"Astron claims that the visual displays and images in Panes 2 constitute an 'illegal copying of the Asteroid audiovisual works.' Astron asks for damages of $50,000 per infringement and the destruction of all Panes 2 systems."

"Those shits. What audiovisual works?"

"Lawyerspeak."

"Starkey's desperate. He knows we have a license. Just a minute, I've got it here." Charles called to his secretary and told her to bring him the file with the Astron Computers license. She did with the file opened to the exact page. "Listen to this," Charles told Bill:

"They granted Digital Software a 'non-exclusive, worldwide, royalty free, perpetual, nontransferable license to use these derivative works in present and future software programs and to license them to and through third parties for use in their software programs.'"

"That sounds pretty clear. A lawsuit doesn't make any sense."

"I'll fax you a copy of the signed agreement," Charles said. "Use it as you wish."

Bill understood that Charles was ordering a lead article on Astron Computers' bad faith for the next issue of "PC News."

"Do you have anything more?" Charles asked.

"No, but I'll check things out in Cupertino and keep you informed."

As soon as Bill hung up, Charles dialed Starkey's private number.

"Hello," Starkey answered.

"Lawrence what in the hell is going on?"

"I don't know, the lawyers...I don't know much about it," Starkey stammered.

"I thought we had an agreement, that we put all this nonsense behind us," Charles challenged.

"Oh don't worry this is lawyer stuff," Starkey regained his composure.

"The hell it is. You knew we were working on Panes and didn't object."

"Your license was only for Panes 1. Our lawyers tell me that subsequent versions are not covered," Starkey fought back.

"The hell there're not," Charles raised his voice. "I just talked with you two days ago, and you said nothing."

Starkey did not reply.

"I take this personal," Charles said. "You stabbed me in the back." Charles bit his lip to keep from adding: "Just like you did Stanley."

Robert L. Skidmore

"Don't take that attitude, Charles," Starkey said. "Let's leave everything in the hands of the lawyers and see what the courts say."

Charles slammed the phone down.

Immediately, the intercom buzzed.

"Yes," Charles said gruffly.

"The Los Angeles Times is on the phone," Charles' secretary said.

"Tell him..." Charles started. "I'll take the call and then no more media. Tell them I am in conference."

"Yes," Charles said after punching the flashing button.

The reporter identified himself.

"What can I do for you?" Charles asked.

"Have you a reaction to the Astron lawsuit."

"Yes, it's garbage. We have an agreement that Astron signed two years ago. Want me to read it to you?"

"Please."

Charles complied.

"Astron claims that applies only to Panes 1."

"We'll see them in court," Charles said.

"They allege you are not licensed to reproduce the 'look and feel' of the Asteroid."

"What the hell does 'look and feel mean'?" Charles shouted.

"Don't shoot the messenger," the Times correspondent laughed.

"Astron doesn't license 'look and feel' to developers. Nobody has to have a license to do Asteroid software."

"They claim that Panes 2 infringes on their copyright because it uses overlapping windows, moveable icons and other exclusive interfaces."

"That's baloney," Charles said. "And you can quote me. We got our inspiration the same place they did, from PARC. Call PARC and find out if they are going to sue Astron Computers for this touchy feely nonsense." Charles hung up.

Charles then called his staff lawyer, ordered him to get on top of the lawsuit and handle all media queries, and to file a countersuit charging Astron with slander and the intent of wrongfully inhibiting Panes development.

Charles, aroused, called David Howard at IBM.

"David, Charles. Have you heard about this shit that Astron is pulling?"

"I've got our lawyers in my office now. What does this all mean to Panes?"

Charles hesitated. He assumed that IBM was worried about a similar lawsuit being filed against them. If Panes violated the look and feel of the Asteroid, then certainly IBM's OS-2 would also. This placed IBM in a delicate position. They would not be unhappy if Panes went down the tube, but they needed OS-2.

"It doesn't mean shit," Charles said firmly. It's just legal nonsense. I don't know what's gotten into Starkey. What do you hear?"

"We've had no indication that they are filing against us," David said. "Have you heard anything?"

"No. If I do, I'll call you. I hope you'll do the same."

"Certainly. Our lawyers don't know what look and feel means. They're researching to see if there are any precedents."

"Mine tell me there is nothing in computer law."

"Astron must be desperate," David said.

"Starkey is grasping at straws. They've lost market share and can't decide which way to jump on licensing and price."

"They've already missed the boat on licensing. The PC and its clones have set the standard."

"We'll let the courts decide. This thing is passing news. We'll tie them up on the case for years," Charles said.

"You're pushing ahead on Panes?" David asked.

"Damned right. We're already working on Panes 3."

"You know what we think about that."

"You'll change your minds. How are OS-2 sales going?" Charles needled. He knew the answer to his question.

"Not too well," David said honestly. "We might well need Panes 3."

"The clones seem to like Panes."

"I know."

"We still have a long way to go before I'll be satisfied," Charles said.

"You'll never be satisfied, Charles," David laughed.

"Keep in touch," Charles said and hung up.

Subsequently, Digital Software's legal team moved to have the case dismissed. When the presiding judged demurred, they moved to bifurcate the case with one suit focusing on whether DS had been licensed to use Asteroid features in Panes 2 under the original agreement, and secondly to determine whether DS had violated Astron's copyrights. The judge subsequently granted the bifurcation.

The lawsuit created impediments to Lawrence Starkey's public relations campaign, disturbing in particular his association with industry rivals. At the same time, however, internal Astron problems caused considerable heartburn for the earnest president. During Starkey's long absences, two rivals positioned themselves as Starkey's successor. One was Pierre Dupris, who had become a major baron with product development, marketing and manufacturing under his aegis: the other was Han Wifler, the arrogant chief of Astron Europe. To the outside world, Starkey described his imposition of a bureaucracy on Astron

Computers as a maturation process, the addition of standard business structure on what had been a playground for young adults. The young mavericks who had created the Asteroid did not like what was happening to their company. The managers, the suits, had taken control. Placid forms and memos imposed control, dampening the excitement that the clash of ideas had created. They had a saying: "If you don't like this reorganization, don't worry. There will be another in ninety days."

Starkey liked to say that his goal was a flatter organization that would allow an expanding Astron Computers to become more flexible and innovative. It sounded good but did not mesh with reality. Lacking a captain at the helm with vision, Astron Computers drifted during the good times in search of a strategy. Starkey frequently observed that he sought increased market share with a reasonably priced superior product. Pierre Dupris, for one, vigorously resisted. Dupris argued that premium machines commanded premium prices, and the Asteroid deserved this respect. While Starkey tried to lead from the outside with his grand speeches and public appearances, the company drifted. The Board of Directors concerned itself exclusively with the bottom line.

An internal debate developed, led by the Sales Department. They argued that IBM's operating system problems provided an opportunity. By dropping prices, Astron Computers could recoup market share in greater quantities. Starkey favored the idea, but Dupris bitterly opposed. Dupris went so far as to claim that their superior machine did not need to have market share. Premium prices would guarantee fair profit. Dupris even once suggested—some thought in jest—that Astron raise prices.

Trouble awaited. At the end of the fiscal year, the accountants reported that the unfocused R &D budget had risen a staggering forty- percent. This coupled with a twenty-five percent rise in administrative expenses, largely the result of the expanding bureaucracy which Starkey denied even existed, required counteraction. Unable to overcome Dupris' resistance to price cuts and inhibited by a Board majority unable to understand how cutting prices would increase profits, Starkey acceded to his management team's acceptance of the recommendation that they raise Asteroid prices by thirty percent.

The Sales Division howled in anger, and Dupris watched the furore with a satisfied smile. Astron users, normally a loyal band of forgiving believers, protested with the one effective tool at their disposal. They boycotted the Asteroid over the critical Christmas sales season. Sales and profit lines plummeted. By the end of January rumors of another Astron reshuffle began to circulate in the Valley. Bill Oldham with a weekly deadline approaching phoned Charles in Seattle.

"Charles, there's a covey of rumors about Astron circulating down here. Have you heard anything?" Bill asked.

"I haven't heard anything. What rumors?"

"There's talk about a major reorganization. Some big heads may roll."

"Sounds like the same old stuff to me," Charles said. "They won't make the hard decisions. They won't license, and they refuse to cut prices. They've got their heads in the sand down there."

"I think I'll check them out," Bill said.

"Keep me informed," Charles said.

Bill immediately dialed Lawrence Starkey's number and was surprised when Starkey's secretary put him through.

"Hi Bill, what can I do for you?" Starkey greeted him.

"I'm going to be in Cupertino tomorrow and wondered if you might fit me in for a few minutes," Bill said.

Starkey laughed. "The sharks are circling. You've been listening to the rumor mill."

"I've heard you're about to announce a major reorganization," Bill said.

"Why don't you drop by tomorrow when you get here," Sharkey countered.

"Care to set a time?"

"Can you be here by ten?"

"It's a date," Bill said. He was about to hang up when Sharkey apparently had an afterthought.

"You can tell Charles he's got nothing to worry about."

"I suggest you tell him yourself," Bill said. He hated it when others virtually accused him of being Charles' eyes and ears in the Valley.

Bill arrived at Starkey's office at exactly three minutes to ten.

"Please have a seat. Mr. Starkey will be right with you," the friendly secretary greeted him.

Bill sat down in a chair facing Starkey's office. To his surprise, he could hear raised voices emanating from behind the closed door. Starkey was having an angry argument with someone. Bill gave the secretary a quizzical look. She shrugged and glanced at the door with a worried expression on her face. She obviously did not want Bill to hear what was being said but was unsure how to handle the delicate situation. Finally, she picked up the intercom and pressed a button.

"I'm sorry to interrupt, Mr. Starkey but Mr. Oldham has arrived."

Bill could not hear the response.

"Mr. Starkey apologizes but he is running late. He will be with you shortly," the secretary said.

"No problem," Bill said, wishing he could make out what was being said inside.

The argument continued for another five minutes, then the door burst open and a red faced Pierre Dupris charged past Bill without speaking. Bill watched as Dupris hurried through the outer door and down the hall.

"Hi Bill. Sorry for the delay," a smiling Starkey appeared in the doorway to his office. "Please come in."

Once they were settled inside, Starkey leaned back in his chair. "I suppose you wonder what that was all about."

"I'm sure it's none of my business," Bill said.

"Well, it'll be common knowledge pretty soon. Can we speak off the record?"

"Certainly."

"I suppose you saw our fourth quarter results. We released them last week."

"I did." Bill refrained from observing that they were terrible. Sales were up only a sickly six- percent for December and consequently profits for Astron Computers were down eleven percent. Astron's stock had fallen twenty percent.

"It wasn't very good news for the company and worse for me. I've been here almost five years and was looking forward to keep my promise to my wife that we would move back east after five years. But I can't leave with Astron in this condition."

Bill did not know what to say, so he said nothing.

"We've got to make some changes. Internally, the company is not running the way I want it to. While I concentrated on representing Astron to the world, I expected my management team to pull together and continue our astonishing success."

"There've been rumors about continuing internal dissension," Bill said.

"They're true, but I want to thank you for not printing them."

Bill wished he had.

"I need a strong inside manager. The team approach is not working. This afternoon I will announce that I am appointing Hans Wifler as our chief operating officer. Hans has done an outstanding job managing Astron Europe. In addition to his operating duties Hans will also assume responsibility for manufacturing and marketing."

Bill understood immediately what the shouting match had been about. Pierre Dupris had not been happy to learn that two thirds of his empire had been taken from him.

"I take it that Pierre was not too pleased with the changes," Bill fished for a response.

Starkey smiled. "Pierre will continue to direct our R&D effort. I don't have to tell you how important that is to the company. We're going to have to get leaner and meaner here."

"Will these changes impact on your marketing strategy?" Bill asked. Bill assumed that Dupris' apparent demotion would mean that that the fiery Frenchman's opposition to licensing and lowering prices would be devalued.

"That's a good question. Now that Hans is assuming a bigger share of the burden, I'll be free to readdress all strategic issues."

"Have there been any developments in your law suit against DS?" Bill asked.

Starkey smiled. "I don't think that's a subject we should discuss." Starkey was fully aware of Bill's relationship with Charles.

Bill then asked a few token questions, and Starkey answered in generalities. Having learned what he had come for, Bill decided to cut the interview short.

"I know you're very busy," Bill said, "and I appreciate your taking the time to chat with me."

"Any time Bill."

That evening Bill Oldham retreated to the bar of the Quality Inn for a beer and a few minutes respite from the tribulations of Astron Computers. He had taken his first sip of his first Budweiser when Jake Owens, the LA _Times_ columnist who specialized in the Silicon Valley, appeared at his side. Bill knew him from his days in a similar position at the _Post_.

"Hi Bill," Owens said, nudging Bill with a friendly elbow. "You buying?"

"A beer for the working journalist," Bill called to the bartender.

"What's a PC guy like you doing in Cupertino?" Owens asked.

"Same thing you are."

"Want to compare notes?" Owens asked.

"Why not?" Bill said, fully aware that anything he said would reach print before the next issue of Computer News.

"You first," Owens said as he raised his glass.

Bill recited the gist of his session with Starkey.

"Not much there," Owens said.

"I couldn't get a thing out of him on the lawsuit," Bill said.

"Best as I can tell, the lawsuit's all Starkey's idea. His lawyers tried to tell him it was a loser, but Starkey insists he has to slow Charles down while he revs up his own technical shop."

"The PC and the clones have him in a difficult position," Bill opined. "If Panes ever catches up with Asteroid's graphic operating system, Starkey's lost his only advantage. He and his high prices will go to computer heaven."

"Any idea what he plans? Dupris thinks the world belongs to Asteroid and won't budge on price or licensing."

'I wonder how long Dupris will be around?"

"Or Starkey," Owens said.

439

"Another?"

"Why not? I've filed for tomorrow's edition?"

Bill ordered two more Buds.

"What can you tell me?" Bill asked.

"I can pass along a little gossip I picked up," Owens replied after a pause.

Bill smiled. Charles liked gossip, especially when it came from someone with reliable sources like Owens.

"I heard there was a confrontation last week between Starkey and Dupris," Owens paused. "This is off the record, right?"

"Yeah," Bill smiled. "Who reports gossip?"

"I'm told the computer magazines do," Owens smiled. "I wouldn't know myself. I never read them."

"I only read one of them," Bill smiled back. "Give."

"Well, Starkey came back from one of his trips—that guy spends more time out of Cupertino than he does here—and learned that Dupris had been badmouthing Starkey behind his back."

"So what else is new?"

"This time Dupris went further. He blamed Starkey for the poor last quarter and suggested that all the division chiefs should as a group go to the board and demand that Starkey be removed."

"That could get on your nerves."

"Starkey invited Dupris to a face to face at a local restaurant."

"How do you know what happened?"

"Starkey took another guy along."

Bill did not ask who. He wanted to hear the story.

"Starkey started out by asking Dupris what he thought of him."

"That sounds plausible. Starkey is the kind of guy to confront problems at their source."

"That's where Dupris made his mistake. He told him straight out. For the better part of an hour."

"What did he say?"

"The gist of it is that Dupris told Starkey he was the wrong person to run Astron. He said Astron is Asteroid and Starkey didn't know enough about Asteroid and its technology. He said you are a great outside man, but the engineers don't respect you."

"How did Starkey react?"

"He asked what Dupris thought of him and the jerk answered him honestly. He said he didn't respect Starkey either."

"That's a great line to use with your boss. I'll see if I can remember it," Bill laughed.

"There's more. Dupris told Starkey that feeling was running so strong against him that a lot of people were plotting to dethrone him."

"Good move."

"Yeah. When Dupris finished telling Starkey how bad he was for Astron, Starkey spoke frankly too. He said Engineering was taking too long to bring new products to the market. They could not live forever on the Asteroid just as they couldn't with the Astron 2. Starkey said they had to go for market share."

"How did Dupris take that?"

Owens laughed. "You want a lot for two lousy beers."

Bill ordered a third Bud for Owens but passed himself.

"This is the good part," Owens said. "Dupris said to hell with the market. That's not what this is about. You Lawrence Starkey, don't know enough about the products for me to respect working for you."

"Dupris is finished," Bill said, amazed at the Frenchman's lack of discretion and good sense.

"Dupris knows that. Afterward he told a couple of engineers that it was his Last Supper."

One week after Starkey' reorganization, Pierre Dupris submitted his resignation. To the surprise of all and the dismay of Dupris' engineers, Starkey appointed himself Astron's Chief Technical officer. Starkey, a marketing man, frankly admitted his lack of technical expertise, but he believed vision was more important in a leader than a knowledge of logarithms. The engineers protested but they buckled under. Starkey instituted a daily meeting every morning at seven thirty. For years, the Astron engineers had been a privileged lot free to pursue whatever projects they chose. Despite his hostility to everyone outside the engineering department, Dupris had given the engineers and researchers their heads. Starkey instilled discipline. While Starkey ran engineering and provided the public front, Hans Wifler with Germanic efficiency ran everything else.

The following year Starkey introduced the Asteroid Classic. With Dupris, the main opponent to lowering price in a bid for market share, gone, Starkey took a chance and brought the Classic out at the unbelievable, for an Astron computer, price of $999. The market responded. Demand was so high that Astron struggled to meet orders. Quarterly reports showed double digit growth, and Astron stock rose to $73 a share. Unfortunately for Starkey, the profit margin fell despite the increasing sales volume. Lower priced machines simply returned less profit to the company. To compensate, Starkey sacked ten percent of his employees, cut salaries by ten percent, and eliminated many of the perks that had made Astron a unique place to work.

Chapter 34

David sat alone in his office behind a closed door and solemnly listed in his mind the various factors that were making his life as Chief of the Personal Systems Division so miserable. It was more than the fact that jealous peers considered him a Lone Duck who refused to fly with the rest of the formation. Naturally, they considered him the recipient of outright favoritism. David still did not understand what had motivated Ernie Hendricks to take David under his wing, and he acknowledged that Ernie had given him the opportunity to fly on his own in Boca Raton. David liked to believe, however, that he had responded to the pressures and had played a key role in launching the Personal Computer. Then, Ernie had manipulated David's transfer to Oshkosh where again David had distinguished himself. Now, David had his own division and honestly believed he had earned it. At least, he, Terry, and possibly Ernie thought that, but none of David's peers did.

The others resented David's rapid advancement in a non-traditional career track. He was a non-engineer in a technical company. Most considered David too independent. David enjoyed the luxury of being his own boss in Boca Raton and Oshkosh, and he despised the management by committee style of the IBM bureaucracy, but David tried to fit in and not antagonize others. This assignment to work with Nick DiGenova was threatening to undermine everything that David was trying hard to accomplish.

Ernie's suggestion that David be used to mediate in the dispute between DiGenova and Charles Swift placed David in an untenable position, exactly in the middle between Charles and DiGenova. David could not see any way he could win. He disliked DiGenova and admired Charles. This placed him on the wrong side. In David's opinion, IBM had mishandled the OS-2 development program and was now trying to put the blame on Digital Software. Charles had recognized an opportunity and seized it. If the decision were David's, he would cancel the OS-2, which cost IBM one hundred and fifty million dollars a year, and license Charles' Panes. David had more confidence in Digital Software's ability to develop a viable graphical interface operating system than he did in IBM's massive effort under Nick.

Even the personalities involved troubled David. DiGenova was a caustic egotist determined to be the chief operating officer of IBM before his time. Gracious to his superiors, disdainful of peers and subordinates, DiGenova was admired only by his spouse. A graduate of the mainframe division, DiGenova had acquired all the biases and attitudes of that traditional hierarchy. No one could write code, build machines, manage a customer base better than a mainframe man. David considered it preordained that DiGenova and Charles would clash.

The two disparate personalities crossed swords at their first meeting. DiGenova considered Charles a smart aleck and his company little better than a college fraternity. They wrote hasty code, kept irregular hours, dressed indifferently and acted like children at recess. Charles rated DiGenova as a pompous snob, a Colonel Blimp with an outdated knowledge of technology, and a disastrously limited technical background rooted in the mainframe base. DiGenova did not understand the personal computer revolution and the changes it had wrought. In short, Charles considered DiGenova a dinosaur living in the shadow of the twenty-first century.

David recognized that he had few options. Caught between these two antipathetic personalities, he had to find a mutually acceptable middle ground. Charles, through skillful maneuvering, had IBM by the short hairs. After Charles perfected Panes, and David had no doubt that he would, IBM would have no option but to fold OS-2's tent. The clones and software companies would jettison the clunky OS-2 and climb on the Panes bandwagon. Already, the software companies were complaining. They could not afford to write software for both OS-2 and Panes. One or the other would have to be selected as the software standard for the PC. For now, they scrutinized Digital Software and IBM with nervous eyes.

The clones had already dangerously eroded IBM's domination of the Personal Computer market. Companies like Dell and Compaq threatened IBM's once dominant market share. For now, the clones continued to use LDOS, the operating system standard imposed by IBM. David considered it ironic that LDOS was owned by Digital Software who reaped a license reward every time a clone sold a PC. David was responsible for that situation. David and the lawyers, in their haste to assemble IBM's PC, had allowed Charles to retain the rights to sell LDOS to others. DiGenova blamed David for this, and David suspected that Ernie had suggested David for a mediator's role in order to give him a chance to recoup.

Fat chance, David thought. DiGenova's engineers were faltering and attempted to mask their problems by attacking Digital Software. Charles had Panes under intensive development; he also held a contract to assist IBM in developing software for OS-2. IBM had given their rival the keys to their own product. Charles insisted that Digital Software had its Panes team and its OS-2 team compartmented. As best as David could determine, the Digital Software OS-2 team was seriously pursuing their assignments. It came as no surprise, however, that the IBM engineers and DiGenova used the apparent conflict of interest inherent in Digital Software's situation as an excuse. The different corporate cultures did not help, and to make matters worse, the IBM contingent to a man was jealous. When Digital Software went public, Charles became a billionaire over night. The lords of IBM resented this, but the fact that many of Digital Software's employees became millionaires at the same time was too much to tolerate. The IBMers could not understand how this bunch of unruly kids working for a two-bit company could deserve such wealth.

Depressed by his assessment of the situation, David turned to the one person who he knew had the power to cheer him. He phoned Terry at home.

"Hi Terry, what's up?"

"Nothing changed here in the past two hours. What's up with you?"

"This place gets on my nerves."

"Take the rest of the day off, and we'll play a round of golf."

David seriously considered the invitation. In Oshkosh, both he and Terry had seriously taken up golf. David with the skill of a natural athlete had gotten his handicap down to eighty-five after a year. To his disappointment, eighty-five is where it hung. David knew he could do better, but his job kept him from putting in the time he needed on the course. Terry, also a natural athlete, had the time David did not and when she played regularly she was a scratch golfer. David hated to lose when they played head to head, but he was also proud of his wife's skills. Their last year in Oshkosh Terry had won the club championships in both golf and tennis.

"I wish I could," David said. "I've got a meeting with DiGenova at ten."

"That's why you called."

"I guess."

"Don't take any shit from that Mafioso," Terry said.

"Don't talk like that. You don't know who might be listening."

"You think he might bring an EEO suit against me for defaming a minority?" Terry laughed.

"He wouldn't dare," David laughed. "You would whip his ass."

"You betcha, lover."

"How would you like to go to the meeting with me?" David joked.

"I can be there in ten minutes. I would have to wear my tennis outfit."

David could picture his wife in her short skirt breaking up DeGenova's meeting.

David's intercom buzzed.

"Just a minute, Terry," David said. "Yes?" David answered the intercom.

"I'm sorry to interrupt," David's secretary said. "DiGenova's assistant called. They've moved the meeting up by half an hour, and they want you there forthwith."

"I'll bet that's what DiGenova said."

"I don't doubt it," David's secretary said.

"Thanks," David clicked the intercom and returned to Terry. "Sorry, DiGenova wants me."

"Then let him come to your office. He doesn't outrank you."

"I know. But he thinks he will."

"No way lover. You're always tops."

"With you anyway."

"Wait and see. He's afraid of you. You're his main competition."

"Thanks but I doubt it. I'm not in any race."

"You are whether you know it or not. Besides, I would like to be the CEO's wife. All the girls at the club would let me win all the time."

"You do anyway."

"Who cares," Terry said with the inconsistency that David loved.

"I've got to go," David said. "See you later."

"Whip his ass," Terry said before hanging up.

"They're all waiting for you in Mr. DiGenova's office," DiGenova's secretary said when David appeared.

"I'll bet they are," David said irreverently.

The secretary frowned. David smiled. The woman's lack of humor amused him. The fact that secretaries tended to adopt the personalities of their bosses always baffled David.

"It's about time," DiGenova growled when David entered without knocking.

"Good morning, Nicholas," David said. He nodded without speaking to the other four members of DiGenova's staff who were present. Given the overwhelming numbers aligned against him, David assumed that DiGenova intended to propose another unworkable stratagem designed to bring Charles and Digital Software to heel.

"Our situation is untenable," DiGenova immediately launched his tirade. "We are in our seventh year of the program cycle with OS-2. It costs us one hundred and fifty million dollars a year, and we still have problems."

"Why don't you kill it?" David played the devil's advocate. He knew that DiGenova could not do that. Thanks to IBM accounting procedures, they had not written off all the costs over those seven years. The bookkeepers liked to defer the costs and then spread the write-off over the years when the product is sold. This made good sense to the beancounters and the bottom line, but it put DiGenova in a difficult position. If he canceled the project, his division would have to take the hit for the accumulated costs all at once. Undoubtedly, he feared this would destroy his career because it would have a bad effect on IBM's bottom line and impact on the stock price arousing stockholders.

"Don't be funny," DiGenova snapped. "We inherited this problem, and we're going to solve it. Charles Swift and his friends have taken this company for a jolly ride," DiGenova glared at David as if he were solely responsible for their current situation. All present knew that David had had nothing to do with the slow development of OS-2, but they were aware that David had negotiated the original PC agreements with Digital Software, thus bore some responsibility for the relationship between the two companies.

David smiled, determined that he would not be provoked by the rude DiGenova.

"I've decided to terminate the contract with Digital Software," DiGenova announced.

"That might present certain difficulties," DiGenova's software team leader spoke softly.

"We can do anything they can do, better," DiGenova smiled at his feeble attempt at humor.

His subordinates laughed weakly, David not at all.

"I can't speak for the others," the software chief said. "But termination of the contract would present insurmountable problems for software."

"And why is that?" DiGenova demanded. "We write our own software for the mainframes don't we?" A mainframe man, DiGenova knew this to be irrefutable.

"Yes sir."

"Are you telling me that we can write mainframe software but cannot handle a comparatively uncomplicated task like software for a little computer?"

"Are you sure you need me for this discussion?" David interrupted, giving the harassed software man time to formulate an answer.

"Are we boring you David?" DiGenova asked.

"No, but it seems to me that you are discussing internal OS-2 team matters of no relevance to me," David replied calmly.

"We'll soon be discussing how to handle your friends at Digital Software," DiGenova said. "As the Management Committees mediator, you will carry the message."

David let this pass even though he did not like DiGenova's tone.

"We can write the software, but it will set us back two years at least," the software man said.

"And why is that? Are Digital Software's programmers that much better than ours?" DiGenova snapped.

"No sir, but the way the tasks were allocated when the project was devised, key elements of the code were deferred to Digital Software. If we break with

DS, we will have to rewrite the code from scratch. It might take longer than two years."

DiGenova's face flushed a bright red. He pivoted in his chair and stared out the window, ignoring David and his staff. After several uneasy minutes, he turned back. He slapped the desk with his palm. "All right. I can't fire these guys because the bean counters have locked me in with their accounting games, and Swift outsmarted us when we set up this project." DiGenova glared at David as if he were accusing him of being responsible for their position.

"Are you trying to blame me?" David challenged.

DiGenova ignored the question. "These guys are taking us for a ride. We can at least make them pay their fair share." DiGenova grabbed a file from a stack on his desk and consulted it. "We have more than 1,000 people working on OS-2 and Digital Software has less than two hundred. Is that equable?"

No one responded.

"I assume that with eight hundred more programmers working we write more than our fair share of the code," DiGenova continued.

The software leader squirmed in his seat.

"Well?" DiGenova demanded.

"We discussed that with Digital Software last year. We even had a committee study the problem. There are mitigating circumstances."

"Inform me."

"The DS programmers break all the programming rules and write much tighter code than we do."

"So the kids are careless?"

"Our committee concluded that IBM procedures require seven programmers to equal what one DS programmer produces."

"Bull shit," DiGenova shouted. "I know programmers working on mainframes that make DS programmers look like amateurs."

"The committee even cited an example."

"Share it."

"A DS programmer reduced 30,000 lines of IBM code to 200 lines."

"I don't believe that. Those kids?"

Despite himself, David let a slight smile creep across his face.

"Do you have something to say about that Howard?" DiGenova called David.

"I suggest you concentrate on fixing OS-2 before Charles gets his upgrades to market," David refused to be intimidated.

"You've been talking with the software companies," DiGenova accused.

"That's part of my job."

"Well I do too. Make sure you tell them that there is no way Digital Software is going to beat IBM."

"If you will excuse me," David rose to his feet ignoring DiGenova's comment. "I have work to do. I've got PC's to sell."

In Seattle one of Charles' smart guys solved a problem that bedeviled the Panes Workgroup. At a cocktail party, a Panes team member casually discussed his project with a physics professor from Arizona who was spending his summer working at DS on a debugger he had designed. The professor, who had recently adapted his debugger to run on LDOS, opined they could use the debugger to make Panes run in a protected mode. Without permission from management, the engineer and the professor experimented until they had Panes running in the protected mode. This meant that with adjustments Panes could run on existing machines and DS would not have to discard all of its old applications. After three weeks of work, they demonstrated their discovery to Mike Hamilton and Charles who immediately recognized the import of the breakthrough. They now had the memory, speed and graphics that users wanted. Charles immediately established Panes 3 as DS' number one priority.

The news eventually leaked to IBM where DiGenova erupted. He stormed into David's office.

"Did you hear what those bastards have done now?" DiGenova demanded.

David, who had been concentrating on his own problems, had not been in contact with Charles for several weeks, not since Charles had agreed to give OS-2 priority. He raised his palms to indicate he did not know what DiGenova was talking about.

"They've been lying to us all along, your friends. They never stopped work on Panes and now they're working on it full blast."

"Why are they doing that?" An astounded David asked.

"Because they never intended to meet their commitments on the OS-2. They took us for a ride, just like they did on the original licensing deal."

"I can't believe it," David said.

"Believe it. It's true.

"Charles assured me that Panes was only a temporary product designed to hold the market until we have OS-2 ready to go."

"He lied to you."

"How do you know?" David asked.

"Because the word is out. All the little software companies are lining up to write applications for Panes. They think Panes 3.0 will come out soon and will sell for half of what we charge for OS-2. That bastard has torpedoed us."

"I'll call him right now and ask what he's up to," David offered.

"Do it," DiGenova ordered.

David was tempted to tell DiGenova to take his orders and stuff them but decided against it. IBM would be best served if he called Charles.

David and DiGenova sat in glowering silence while David's secretary called Charles on his private line. Fortunately, maybe unfortunately in David's opinion, Charles came right on line.

"Hi David," Charles greeted him.

"Charles, I'm putting you on the speaker phone," David said. "Nicholas is here too."

David could visualize Charles' negative reaction, but Charles concealed it. "Fine," Charles said. "What's up?"

"That's what we want to know?" DiGenova shouted.

"David?" Charles said.

"We understand that you have prioritized Panes 3 over OS-2," David said.

"Oh, I understand," Charles said. "There's nothing for you guys to get upset about. We had a lucky break here. One of our engineers was demonstrating Panes 3 and our problems with its memory to a temporary hire and he suggested a way to get around our problems."

"You're deliberately undercutting OS-2. You set us up," DiGenova hollered.

"That's not true," Charles replied firmly. "I resent your implications."

"Resent them all you want, but you lied to us," DiGenova shouted.

"David, talk reason with Nicholas," Charles said.

"Does your breakthrough change your commitment to us?" David asked calmly.

"No," Charles said. "Panes 3 remains an interim solution until OS-2 comes on line then it will fade away."

"I don't believe him," DiGenova said to David.

"Will you make a public statement to that effect?" David asked.

"Sure. Why not?" Charles said. "If DiGenova apologizes right now for his intemperate remarks."

David looked at his irate colleague. "It's up to you."

DiGenova started pacing. "All right. I apologize. I got excited. But I want a public statement from DS immediately."

"You've got it," Charles said reasonably and hung up.

DiGenova left David's office without another word.

Two days later in Seattle Charles issued a press release:

"DS will cease development of Panes software after the release of Panes 3.0. Panes then will be left for the low end of the market while IBM's OS-2 will become the main PC operating system for the 1990's."

Charles had worries other than IBM, OS-2 and Panes to occupy his spare time. The Astron Computer lawsuit came to the fore with a decision by the San

Francisco District Court. The Judge ruled that the 1985 agreement did not give Digital Software the right to develop future versions of the graphical user interface as it pleases. He ruled that Digital Software received a license to use the visual displays named "as they appeared to the user in November 1985." Consequently, faith in Digital Software stock wavered. In a demonstration of resolution, Mike Hamilton spent forty-six million dollars to purchase 945,000 shares of DS stock.

The legal battle turned to Pane 1 and Pane 2, both of which the Judge had ruled were "fundamentally different." At Charles' direction, the DS lawyers commenced a legal wrangle to define the differences between what Astron had authorized and what they had not. The hairsplitting began. Under pressure from DS lawyers, the Judge ruled that 179 out of 189 cited infractions were authorized.

David scheduled a meeting between DiGenova and Charles at a trade show in Las Vegas. During a series of indirect negotiations, conducted with all the sensitivity of a heads of state summit meeting, issues were negotiated. Finally, they reached an agreement. DiGenova would endorse Panes; Charles would reiterate that Panes was an interim system designed for low end personal computers while OS-2 targeted high end users with more powerful PCs. In a direct meeting to discuss their scripts, both DiGenova and Charles hedged their positions. DiGenova toned down his endorsement of Panes, and Charles removed his commitment to cease Panes development after Panes 3.0. At the dinner in the Hilton, both men presented equivocal statements.

One developer stood up and asked Charles a direct question: "Charles, a while back you told me I should be developing applications for Panes. Are you now telling me I should develop applications for OS-2 instead?"

Charles in a longwinded answer waffled, finally concluding that developers should do applications for both Panes and OS-2. Asked which he recommended they should work on first, Charles replied that was a decision for each company to make for itself.

Dissatisfied with the results of their first efforts, Charles and DiGenova after a full day of subsequent negotiations held a final press conference. DiGenova offered a lukewarm endorsement of Panes as a temporary system, and Charles opined that OS-2 was a system of the future.

Both men departed Las Vegas angry. Subsequent media coverage stuck with the equivocations. The software developers departed with a sense that IBM had capitulated to Charles Swift. DiGenova had embraced Panes, signaling that they should abandon work on applications for OS-2.

David gradually withdrew from the negotiations and fortunately for him his position as mediator had been dropped when Charles announced the release of Panes 3.

Emulating the showman Stanley, Charles earmarked a million-dollar budget for the Panes 3 extravaganza. Over a hundred software vendors announced Panes 3 applications, and DS hosted presentations in eight major cities. Users and media responded, and Panes 3 to everyone's surprise emerged an immediate winner. Capitalizing on the excitement, Charles authorized a ten million-dollar follow up campaign. In five weeks, Panes 3 sold close to 400,000 copies. Charles and his DS colleagues were ecstatic, but fateful reactions occurred beyond their ken.

In Armonk, DiGenova was furious, so much so that he refused to travel forty miles from Armonk to New York City to attend Charles' gala. DiGenova phoned David:

"Now tell me about your friend," DiGenova shouted. "This is what he calls an interim system."

DiGenova hung up before an embarrassed David could respond.

In Washington D.C. a routine meeting at the Federal Trade Commission began modestly. Commissioner Chairperson Irene Boggs presided.

"Can anyone tell me what's behind yesterday's Post story?" Bogg's indifferent manner concealed the Chairperson's white-hot anger. The Post, apparently short of real news on a slow day, had used innuendo and conjecture to manufacture a hostile attack on the FTC and Chairperson Irene Boggs personally. The opinion piece, masquerading as a news report, had savaged the FTC. No one knew better than Irene Boggs that the FTC slumbered as a deliberate result of the Reagan/Bush hands off policy. The Post story was grossly unfair. Its allegation that Chairperson Boggs was in the pockets of big business skirted libel.

"Typical liberal bullshit," one of Irene's fellow commissioners noted.

Irene was tempted to reply that the speaker would not be so dismissive if he sat in Irene's chair. There was no way she was going to sit still and allow the charge that the FTC under her stewardship had lost its sense of purpose.

"There's no doubt that we are vulnerable to this kind of attack," one of Irene's closest supporters on the commission said.

"What do we do about it?" Irene demanded.

"We need a showcase investigation." Irene's friend said. "Nothing serious. Just something to make a few headlines, enough to distract our enemies."

"Anyone have any suggestions?" Irene asked.

An embarrassing silence descended. Finally, Irene tabled her own solution. "What about this IBM/Digital Software thing. I see Digital Software has released a new operating system. Didn't we once investigate the possibility of antitrust violations?"

"If I recall correctly," Ernestine Bailey, a FTC staff investigator who Irene had elevated to serve as her special assistant, volunteered. "A couple years back we investigated the possibility of collusion between IBM and Digital Software. Something to do with the two dividing up the PC market. IBM reserved the high end users for its OS-2 and gave Digital Software the dregs for its Panes operating system."

"I don't understand all that computer stuff," Irene admitted.

"Neither do I," Irene's supportive colleague said. "I don't think we want to get too deeply involved in an investigation of IBM. They pull a lot of water at the White House."

"We need something to divert the damned <u>Post</u>," Irene said. "Take a look at it Ernestine, and see if there is something we can use."

After the meeting, Ernestine diligently checked with the enforcement office to get the details of the previous investigation. After searching their files, a staff investigator produced a clipping that reported on Charles and DiGenova's public statements regarding their uneasy truce.

"I really think there is something there," the investigator opined.

"Why don't you take a look at it. Irene wants a headline grabber to show that we still exist."

"An investigation of IBM and Digital Software would do that," the investigator said.

"Then do it," Ernestine said, exceeding her authority.

"We would need approval from the Department of Justice to do something this big. And the White House might get involved," the investigator, a hesitant bureaucrat, cautioned.

"You prepare the paperwork and let Commissioner Boggs worry about policy," Ernestine said haughtily.

Ernestine then briefed Commissioner Boggs on the action she had initiated on her own authority.

"Good," Irene approved, satisfied that a simple feint in the direction of an investigation would serve her public relations purposes. She knew nothing really serious would be mounted against IBM or Digital Software under a Republican Administration.

Using clippings from FTC files as his sources, the investigator prepared a low-key memorandum for the Department of Justice requesting authority to re-open the old investigation. After checking with Ernestine, the investigator's branch chief forwarded the routine request over his own signature. One month later, the memo came back to the FTC with an approval line signed by an Assistant Deputy Assistant Secretary of Justice, a political appointee, who had initialed the request which had been buried in a stack of other papers submitted routinely for his signature minutes before he was due to tee off at Congressional.

Armed with this authority, the FTC investigator directed his staff to begin. They started by screening the nation's most influential newspapers. Although not one of the assigned investigators had a technical background, they easily accumulated a mass of information in stories inspired by the phenomenal success of Panes 3. Digital Software sold over two million copies of Pane 3 in six months. This success frightened competing companies like Lotus and WordPerfect who leaked stories alleging collusion between IBM and Digital Software. Most alleged that Digital Software had only pretended to support OS-2 as a part of a nefarious scheme by Digital Software and IBM to abandon OS-2 and work together to establish Panes as an industry standard. The subsequent press play fed the growing files of the FTC investigators.

In Seattle, Charles ignored the complaints as sour grapes. The unanticipated success of Panes 3 had disrupted Charles' long range strategy. He ordered his programmers to begin work on a successor to Panes 3, which had proved buggy despite its phenomenal sales success, and to de-emphasize the development of OS-2 despite his previous assurances to David and DiGenova. The market place had spoken, and Digital Software strategy obediently followed its command.

In Armonk a furious DiGenova, ignoring the rumors of a pending FTC conspiracy investigation that he knew was groundless, decided he had been right all along. Charles Swift had lied to him and had manipulated IBM. Despite his assurances to the contrary, Charles had had no intention of seriously promoting OS-2. The assertions that Panes 3 was only an interim system until OS-2 came on line were part of a devious strategy.

DiGenova informed DS that IBM had decided to go it alone on OS-2. Charles greeted the information with a smile and assigned the remaining personnel from the joint OS-2 team to the Panes group. Charles briefed Tom Oldham, and Computer News broke the story. The breakup of the IBM/Digital Software joint effort gave the FTC investigators heartburn; it undercut their basic contention of conspiracy. With relief, for to a man they feared taking on IBM, the investigators refocused their investigation exclusively on Digital Software. By then, Commissioner Irene Boggs had almost forgotten why she had reopened the investigation. Nonetheless, almost a full year later, details of the investigation leaked, and the probe became public. This stimulated Digital Software's competitors to promote follow-up stories detailing Digital Software's "illegal" business practices. Nothing, however, impaired Panes 3's success. By year's end, five million users had purchased Panes 3. Charles had for years recognized the appeal of a graphical interface system; he had not realized, however, how many PC users felt as he did. Panes 3 let the genie out of the bottle.

At the FTC, investigators finally decided on why they were investigating Digital Software. Charles' and James Clapper's little company had grown so big that it had become a monopoly. Allegations now surfaced charging that Digital Software had built secret interfaces in Panes that gave an unfair advantage to Digital's own applications. For these duplicitous reasons, they argued, Digital's operations ran better on Panes than those of competitors, giving birth to the charge of unfair competition. FTC investigators armed with warrants descended on Seattle where they demanded copies of every Digital document, electronic and paper.

For three years, the investigation dragged on. Administrations in Washington changed. Irene Boggs left office, and finally, a new set of FTC Commissioners considered the Digital Software case and the simplified question: did Digital Software use its monopoly position to exclude others from the market?

One commissioner recused himself on grounds of conflict of interest, and the subsequent voting resulted in a 2-2 tie. The FTC thus ended its investigation, but it did not free Charles from the government's bureaucratic grasp. A Democratic administration was now in power, and the liberals held sway. Rather than surrender, the Department of Justice took over the investigation.

Chapter 35

The new Democratic Administration swarmed into office with a cast of characters that the President vowed would look like the American people. After several false starts, President Clinton selected an intimidating female prosecutor from Florida, Janet Reno, as his Attorney General. Ms. Reno in turn selected another female lawyer, Julie Winthrop, to lead the near dormant Antitrust Division. Julie, an experienced Washington lawyer, breezed through her Congressional hearings and reported to her dingy third floor office to learn that the first case she faced required her to decide what to do with the Digital Software investigation.

Julie did not hesitate. She decided to prosecute. Julie was determined to show her political sponsor, First Lady Hillary Clinton, that she, like the Attorney General, could be twice as tough as any man. Julie had cemented her relationship with Hillary while raising money for the Children's Defense Fund, one of the current First Lady's pet projects. Julie was not daunted by the fact that Digital Software's yearly revenues of close to five billion dollars had made Charles Swift the richest man in America.

Julie's staff patiently explained that it was perfectly legal for a company to monopolize a market if that position was acquired by legitimate means.

"Bullshit," Julie exploded. "No one can tell me that Charles Swift got all that money legally."

"If they acquired their monopoly position through legitimate competition, if they maintain that status through the use of superior skill and reasonable industrial practices, they are not violating the law," the ranking civil servant lawyer in the division insisted.

"When is monopolization illegal?" Julie asked.

"When it can be proved that the company intended to establish a monopoly and used anti-competitive and predatory conduct to achieve same."

Julie studied her staff and realized they were intimidated by the formidable task of taking on Digital Software. "Re-open the investigation," she ordered.

Although she did not state it openly, Julie frankly admitted to herself that her decision was a political one. This did not trouble her. Julie was a political appointee and was confident that the activists who now occupied the White House would back her completely. They would love the case.

Department of Justice investigators descended on Seattle armed with subpoenas that required Digital Software to surrender every document. Upon orders from Charles, the company responded enthusiastically. Soon the halls of the Antitrust Division were filled with paper. The fact that the suit dragged on did not trouble Julie; she and her sponsors were more interested in the media coverage than the outcome. For two years, the investigators researched their case, and finally the best they could come up with was to file suit charging Digital Software with antitrust actions and then to offer a consent decree which if accepted by Digital Software's combative army of lawyers would result in an agreement to desist from certain specified actions. Since this was a political case, Julie believed she could obtain her objective if she could force Digital Software to tacitly admit wrongdoing by submitting to a consent decree.

Julie's legal staff decided that they should try for a comprehensive decree. They opted to force Digital Software to refrain from using per processor licenses which required contracting companies to pay a royalty for each computer shipped with Digital Software operating systems and to substitute instead a system where customers purchased specific numbers of licenses. This would prevent Digital Software from collecting royalties for computers sold with operating systems other than DS systems. The lawyers also intended to force Digital Software to impose a time limit on their licenses rather than leaving them open ended. The proposed decree also prohibited volume discounts to companies who purchased licenses in great numbers, and, most importantly, the decree prohibited DS from linking sales of products. They could no longer require companies that purchased DS operating systems such as Panes to also license DS's word processors and other applications.

After grabbing headlines by deposing Charles Swift, Julie submitted a draft of a consent decree to Charles' lead lawyer, Jeffrey A. Potter, a former member of Charles' father's law firm. Then, the legal wrangling began. After two weeks, agreement was reached, and Jeffrey Potter signed on behalf of Digital Software.

After all the talk, charges, counter-charges, and tons of paper, the results seemed miniscule:

Digital Software agreed to stop licensing their operating systems by the processor and to substitute instead licensing by the number of operating systems sold. Also, Digital Software agreed not to link the sale of one product to another.

The White House declared victory, and the Department of Justice withdrew from the field of battle. The public and the media were confused. The government's victory appeared very minor indeed. No penalties were imposed; Digital Software admitted no wrongdoing, and the word monopoly was used not at all.

The night of the signing Charles, Ann and Bill met in Bill's Crystal City apartment for a modest victory celebration. While Bill fetched the drinks, white wine for Ann, a screwdriver for Charles, and a Budweiser for Bill, Ann sat on the couch and Charles selected the nearby-overstuffed chair.

The couch and the chair faced the full floor to ceiling windows that looked out on the Potomac and the presidential memorials in the distance.

"I love this view," Charles said, meaning it. "Washington from afar is a beautiful city."

Bill served the drinks, then joined Ann on the couch.

"To victory," Bill raised his bottle.

"To victory," Ann raised her wineglass.

"It's a relief," Charles said. "I don't understand these people."

"It's all politics," Ann said.

"We don't need much of that," Charles said.

"Well what now?" Bill asked. He assumed his assignment for Charles was at an end. He had already given the matter much thought. He would give up his position as editor of Computer News. If Charles did not object, Bill planned to continue to write his weekly column as Computer News' Washington correspondent. The column was now syndicated in fifty newspapers. This would give him time to begin work on his new political novel.

"Are we going to shut down our operation here?" Ann asked.

"I haven't given it much thought," Charles replied honestly. "I don't think it is all over. It might be too soon to shut down. In any case we're going to continue to need your lobbying effort."

"Great," Ann said. "I don't have to look for other clients."

Bill let his arm drop across Ann's shoulders, and she leaned affectionately into him.

Charles smiled. "I'm delighted you two have gotten together."

"So are we," Ann smiled. "What about you Charles?" Unstated was the fact that once Ann and Charles had been a couple, but Ann's refusal to give up her career or leave Washington had put an end to their relationship.

"You know, Ann, I'm ready to get married," Charles said.

Ann frowned, not quite understanding Charles' meaning. Bill sat up straight.

"Don't misunderstand me," Charles laughed. "I wish you two well. It's my loss."

"Do you have a prospect?" Ann asked.

"No but I'm looking. I need a traditional wife," Charles said. "Someone to make my nest for me and to raise our children. I've got too much yet to do," Charles said.

"We wish you luck," Bill said, "but let me see if I understand you. Do you want me to get out of the information gathering business?"

"No way," Charles said immediately. "If anything we're going to have to expand it. This outfit is not going to give up. That Julie Winthrop has a gleam in her eye, and I'm afraid it's focused on me. Politicians, unlike old soldiers, don't fade away. Anyway, I'm afraid we might provoke them sometime soon."

"You're a big target," Ann said.

"Sometimes I wish I weren't," Charles said.

"How are you provoking them?" Bill asked. Charles' comment did not make sense. Why give the enemy ammunition to use against you.

"I've no choice. I've made a big mistake in not focusing on the Internet sooner."

"It is a big thing," Bill agreed.

"We're improving our browser and upgrading our network. Both Netscape and American Online will soon be crying foul."

"Why is that? Competition is legal," Ann said.

"I still intend to give the user the best possible setup that I can. We're integrating the browser into Panes so that it will work on the desktop just as it does on the Internet. Our users will be able to call up Internet and Web addresses right from their desktop."

"Linkage," Bill said.

"I call it progress," Charles said.

"The others will claim foul any way they can," Bill explained to Ann. "Since Panes dominates the PC operating system market, installing browsers on the desktop means that most users will use Charles' browser."

"It will destroy Netscape," Ann said.

"Competition is competition," Charles said. "What's unfair about it if we're giving our users the best of all worlds. Aren't we allowed to improve our product? That's what innovation is about."

"These Clintonites don't care about that. They're only interested in the headlines and percentages in the polls," Ann said.

"Tough shit," Charles said. "Can you continue to work for me, Bill?"

"Sure, Charles," Bill said. "But we'll need some things if we are going to stay at it."

"What? For example," Bill said.

"We need a young lawyer in the Antitrust Division."

"How do we get that?"

"We've bought some time," Bill said. "Could you get your lawyers to find us a young graduate from the University of Washington?"

"It's done."

"I know a tame Democratic Senator who needs money; we might him get to sponsor our him or her," Ann said.

"What else?" Charles asked.

"We need better media coverage."

"What do you suggest?"

"Washington lives for the political talk shows, Capitol Gang, that kind of stuff."

"We should start our own show?"

"No all we need is a sharp young journalist as a regular panelist. He could ask the right kind of questions," Bill said.

"This administration lives by the poll," Ann said. "We can influence them if we can influence opinion."

"I know the perfect candidate," Charles smiled, looking meaningfully at Bill.

"Exactly," Ann agreed.

"That's not what I meant," Bill said. "I wasn't volunteering." Privately, the idea appealed to him.

"OK," Charles said. "I got the idea. You guys work on it and let me know what you need from me."

"Do you mind if I move back to D.C?" Bill asked.

"What about Computer News?"

"It's time to turn it over to someone else."

Charles nodded agreement.

"If you don't object, I'll write a column for CN from Washington."

Charles smiled.

In Cupertino, Lawrence Starkey sat with his back to his desk and stared out the window with unseeing eyes. For some reason, the Board of Directors had asked that Starkey and his Chief Financial Officer, the two company directors, to leave the room. Starkey did not know what this meant and did not care. Starkey found it ironic that outside the confines of Astron Computers he was considered the elder statesman of the entire industry. Starkey's speaking calendar was full, and the television networks were always calling to elicit Starkey's expert commentary on the latest of technical developments. Starkey's fame was so large that many had expected him to be appointed Bill Clinton's Secretary of Commerce. Hillary had even insisted that Starkey sit beside her during Bill's first State of the Union message.

Starkey knew that his linkage to the Clintons irritated others, including members of Astron's own Board. Big Business was notoriously anti Democrat, and Starkey understood this. Privately, he considered himself an independent Republican. Publicly, he saw much to gain from association with the President of the United States. Starkey doubted the Board was considering his political connections. More likely, they were discussing Astron's dismal bottom line, as they always did.

Starkey knew that Astron Computers was in trouble. They had a new, more powerful processor and an attractive laptop. Unfortunately, they remained locked into a niche in the upscale user market while IBM's PC and its clones owned the low scale user and corporate markets. Stanley had focused on education users, and these were the loyal customers who continued to buy Astron computers. This was not enough, however, as Astron's market share had plummeted to a lowly seven percent.

Starkey after the departure of the difficult Pierre Dupris had tried to correct the problems. He recognized that it was too late to consider cloning the Asteroid. Allowing others to manufacture and market the Asteroid would not bring in new customers; Starkey knew they would nibble at the edges of Astron's existing base, taking away users Astron already owned. He had tried to lower prices, and this had worked somewhat. Sales had increased, but predictably the bottom line suffered; lower prices meant lower profits, and Astron had an incredibly high R & D overhead to carry.

Starkey had made himself Chief Technical officer to try to correct the deficiencies that Stanley and Dupris had left behind. Traditionally, the Astron engineers were an independent lot. Innovation required intellectual freedom, and thanks to its founders Astron prided itself on being the leaders of the industry. Despite Starkey's best efforts, the engineers continued to work on the projects that interested them. Starkey knew that behind his back, the engineers defamed him. Starkey was a sugar water salesman who married the CEO's daughter. They gave him credit for being a good front man; they just ignored his attempts to develop technical vision.

Starkey also had his problems at home. He had promised his eastern wife that he would linger in California no longer than five years. He had stayed ten, and his wife was now anxious to return to Connecticut. She felt so strongly that she was considering separate households. Starkey was ready to return home too. He wanted a climate where the seasons changed, and as he grew older he wanted to live nearer to his kids and grandchildren. Starkey had all the money and fame he needed. The time had come for him to think about family.

Two months previously, Starkey had almost attained it all. In response to overtures sent discreetly through a high-level talent search agency, Starkey had met secretly in New York with the President of IBM.

"We need someone with your skills and background, a marketing expert who has led a major technological company," the President had confided. "You are high on our list of candidates to be our next CEO."

Starkey had been flattered. He almost had the answer to his problems in the palm of his hand: to be CEO of IBM would cap off his career. Starkey could live in Connecticut and make Armonk his base. As a frontman for IBM, Starkey could rival even the President of the United States in worldwide stature. In his enthusiasm, Starkey had made a mistake. He felt obligated to share knowledge about his potential good fortune with Stuart Miller, a member of Astron's Board and one of the company's three founders. By doing so, Starkey believed he could meet his professional obligations.

Unfortunately, Miller had informed the Board of the offer at the very next meeting. It had caused a small uproar. Most had congratulated Starkey on his opportunity, then Miller without consulting Starkey had tabled a proposal.

"If Lawrence Starkey is leaving us to take on IBM," Stuart Miller had said. "He should take Astron Computers with him."

The majority of the Board, venture capitalists all, had immediately voiced agreement. Starkey, who honestly saw merit in the proposal, agreed to explore it. He phoned the President of IBM and unwisely made more of the idea than it deserved.

"As a condition of my acceptance, I want IBM to buy Astron Computers," Starkey had said.

Two days later, the talent search agency called and informed Starkey that his name had been withdrawn from consideration.

Stuart Miller and the Board had reacted negatively to the news. They felt that somehow Starkey had let them down again.

Starkey was wondering if possibly they were rehashing that old chestnut when Marjorie, his secretary, knocked and interrupted Lawrence's reverie.

"The Board wishes you to return," Marjorie said.

Starkey noticed she had an odd quiver in her voice and wondered why she was so nervous.

"Thank you, Marjorie," he smiled, trying to cheer her up.

Stuart Miller was seated in Starkey's place at the head of the table when Starkey reentered the Boardroom. Starkey smiled. "Thank you for coming, Laurence," Miller said, acting as if he were still in charge.

Starkey moved to the head of the table and was surprised when Miller did not surrender his chair.

"This won't take long, Laurence," Stuart Miller said. "We have decided to replace you effective immediately."

Starkey was shocked. He looked at Stuart to see if he was smiling at a lame joke.

Miller watched Starkey closely, almost as if he were expecting a violent outburst.

Laurence looked at the members of the Board, one by one. Most glared back. Starkey then realized he had committed the cardinal sin. He had permitted the bottom line to drop.

"Hans Wifler will serve as acting Chief Executive Officer until we identify a replacement," Miller said.

Starkey wondered if he were expected to reply.

"That will be all, thank you Lawrence," Miller coldly dismissed him.

Starkey left the room. He crossed the hall to his own office in a state of near shock. He paused near Marjorie's desk.

"They fired me," Starkey mumbled.

Marjorie broke into tears. Starkey could not tell if she were crying for him or herself for her career depended on his own.

"Don't worry, Marjorie, we'll come up with something better," Starkey said without conviction.

Starkey could not bring himself to re-enter his office. "I'm going home," Starkey said.

"I'll pack up your things," Marjorie gasped.

As he hurried from the office for the last time, Starkey imagined he could hear the word leaping from office to office: "They fired the old bastard without even a word of thanks."

Hans Wifler took over with a vengeance. His first day in office he announced a major reorganization and a reduction of ten thousand jobs.

Three weeks later Leonard Shepherd was sitting in his office when the intercom buzzed to announce a caller, the president of a major talent search agency. The interruption did not surprise Shepherd. He knew that as CEO of ZYTEC his name ranked high on many lists. As a forty-five year old engineer with twenty years of experience in Silicon Valley, Shepherd had been summoned to save a struggling company. In three years, Shepherd had turned the company around and the bottom line was showing higher profits than ever.

"Are you interested in Astron Computers?" The headhunter asked.

Shepherd hesitated. The question did not surprise him. He was a natural candidate to replace Lawrence Starkey. Shepherd had begun to believe his own media image: "Leonard Shepherd is a company saver, a miracle worker."

"It's an interesting little company," Shepherd had finally responded.

"Not so little," the headhunter replied. "And it has major problems. Just your cup of tea."

Again, Shepherd remained silent. Immediately, his mind began to run the figures. He would ask for five million up front, a forty- percent increase over his current satisfactory salary, and a firm golden parachute commitment.

"They have an opening on their Board," the headhunter spoke into Shepherd's silence. "Are you interested?"

"Why not?" A disappointed Shepherd replied.

"They'll be in touch."

The very next morning Stuart Miller phoned Shepherd and invited him to lunch at an exclusive San Francisco club, a preserve of the Valley's venture capitalists.

"Leonard, I'm very glad you could come on such short notice," Miller greeted Shepherd and escorted him into the darkly shaded room.

Shepherd admired the long bar, the richly paneled walls, the leather chairs, and the plush maroon carpet. Even more, he appreciated Stuart's choice of venues. The room was virtually deserted.

"A nice place for a private talk," Shepherd smiled.

"May I offer you a drink?" Stuart asked.

"Vodka martini on the rocks," Shepherd said. He knew it was acceptable for Board members to be congenial, even at noon.

"Two," Stuart turned to the elderly bartender. "We'll be at the corner table."

"Yes Mr. Miller," the bartender said.

"I assume you know why we're here?" Miller said as soon as they were seated and the drinks served.

"Yes Stuart. You have a vacancy on your Board."

"Yes. Interested?"

"Yes." Shepherd decided that if Miller wanted to play busy executive he certainly could match him.

"I'll speak frankly," Miller said.

"Please do. This conversation will remain confidential." Shepherd knew how prospective Board members were supposed to act. He also understood that Board memberships were a great deal. For lending the cover of a respected name and devoting very little time, companies rewarded Directors amply. Shepherd was always interested in easy money.

"Astron Computers lacks direction," Miller said.

"I understand you are searching for a new CEO."

"We've been drifting for some time. We discovered too late that Lawrence Starkey lacked vision."

"He had a great presence as company spokesman," Shepherd said. It did not cost him anything to appear charitable.

"Sometimes he lacked judgment, associated with the wrong sorts," Miller said.

"Like the President of the United States," Shepherd chuckled.

"Exactly."

"The general impression in the industry is that you treated Starkey poorly," Shepherd decided to show a little spine.

"The way we fired him?"

Shepherd nodded.

"The dumb bastard forced our hand. IBM offered him the job as CEO and Starkey screwed that up."

"IBM?" Shepherd was impressed. If he had been in Starkey's position, he too would have jumped at the chance. "How did he screw it up?"

"We can go into that latter. Suffice to say, he did. Now we have to find a new CEO for Astron Computers, one with vision."

"The vision to increase the bottom line?"

"Among other things. We're locked into a niche market with no prospect of escaping with our current product line. We've got to increase market share and profit."

"Lowering prices did not work?"

"Obviously."

"And it's too late for cloning?" Shepherd did not care about the answer. He was only demonstrating that he understood Astron's problems.

"We think so. However, our acting CEO plans to give it a try."

"Not a bad idea," Shepherd acknowledged. "If it doesn't work out of the gate, dump the clones and the acting CEO."

Stuart Miller smiled.

"We need to be upfront about one thing," Shepherd said.

"Only one thing?"

"Yes. ZYTEC sells chips to Astron Computers. This year, twenty million worth."

"You're afraid of conflict of interest?"

"No. I'm not afraid of it," Shepherd said confidently. "I just want the Board to know in advance about the situation."

"No problem. You can always recuse yourself when the subject of chips comes up. I'm not sure it ever has, not at a Board meeting."

"Just so we're straight," Shepherd said, pleased that that subject was out of the way.

The remainder of the lunch was devoted to getting to know you probing with Miller assuring himself that Shepherd would be a congenial addition to the Board.

One week later Astron Computers announced the addition of Leonard Shepherd to its Board. That very day, Stanley Pitts, owner and CEO of the Again Corporation, phoned ZYTEC to request an appointment with Shepherd. Assuming that the founder of Astron Computers simply wanted to meet the new Board member and to pass along some helpful advice, Shepherd juggled his busy calendar to work Stanley in.

Stanley appeared promptly at the agreed upon time. Shepherd studied Stanley closely as his secretary escorted the visitor into his office. Shepherd of course knew Stanley's reputation as a charismatic and difficult leader. Stanley's appearance impressed him. Stanley's attire lived up to his billing: long sleeve dark blue sports shirt open at the collar, freshly starched and pressed khaki slacks, and dirty, scuffed Nikes.

"Good of you to accept me on such short notice," Stanley said as he strode confidently across Shepherd's office.

Stanley offered his hand then sat down without invitation in the chair facing Shepherd's desk. Stanley acted like the man in charge.

Still standing, Shepherd admired the young man's bravado.

"I hope you know what you're doing," Stanley said, surprising Shepherd with his rudeness.

"Now?" Shepherd asked, making a joke of Stanley's question.

"Oh not here," Stanley smiled. "At Astron Computers."

"I'm only a new Board member. I've yet to attend a session, so I have a lot to learn," Shepherd tried to be congenial.

"I can tell you all about Astron Computers," Stanley said. "I speak from experience. Always cover your backside."

Shepherd did not comment. Silence was always best when you suspected someone was trying to set you up. Obviously, Stanley had his own agenda.

"Look at what they did to poor Lawrence Starkey."

"His termination did come across as rather heavy handed."

"But deserved," Stanley said. "Starkey was a great soda pop salesman. I'm the one who recruited him for Astron. Not many people remember that."

"You underestimate your fame."

"No I don't," Stanley said. "I make the best computers in the world."

"Asteroid and Starbright," Shepherd acknowledged.

Stanley smiled. "Astron lived off of Asteroid for ten years. Now it's junk, just like the Astron 2."

"The Starbright is a technical marvel," Shepherd said. "But it didn't sell very well," he showed his knife.

Stanley frowned. "Sometimes the market is stupid."

"This is all very interesting," Shepherd said, flipping his calendar to show Stanley that he too was a busy man.

"And you want to know why I'm here," Stanley said.

Shepherd nodded yes.

"Astron is in serious trouble and only one person can straighten the mess out."

"And that is?" Shepherd asked, not believing that Stanley was prepared to ask Shepherd to take over the company.

"Me," Stanley said. "The troops are demoralized, and I'm the only person who can straighten them out. Without me, they're out of business."

Shepherd almost laughed at Stanley' audacity. He had heard about Stanley arrogance, but this was almost too much.

"The world has changed out there, and Astron Computers and its narrow minded board don't realize it. The Asteroid as I said is dead. They need a new processor, something well ahead of its time."

"Like Starbright?"

"Exactly."

"You actually think a ten thousand dollar computer will capture market share?"

"Don't get hung up on price," Stanley said.

"Lower the price, keep costs up, and you eliminate profit, Stanley."

"Don't lecture me," Stanley said. "You sit in this second rate office," Stanley stared at the modest furniture that Shepherd considered appropriate to a CEO trying to turn a company around and shook his head disapprovingly.

"I'm not interested in lecturing you Stanley," Shepherd said. "You're my guest, and I'll treat you like one. I do have one word of advice."

"And that is?"

"Forget it. You're never coming back to Astron Computers. You had your day and fucked it up."

"Thank you for your time Mr. Busy Man," Stanley said. He stood up and marched triumphantly from Shepherd's office.

After Stanley's departure, Shepherd shook his head in wonderment. Stanley certainly lived up to his advance building. It was miraculous that the non-technical Stanley had played such an influential role in founding Astron Computers.

Shepherd did decide to keep one portion of Stanley's advice in mind: "Watch out for Stuart Miller."

Chapter 36

David, fully dressed, bow tie in place, patent leather shoes gleaming, sat on the side of the bed with legs wide and hands propped behind him. Terry was in the bathroom applying makeup.

"I'm not looking forward to this," David said. They were scheduled to attend a formal dinner at the IBM Country Club honoring a departing Chief Executive Officer, John Parsons, and an incoming one, Ernie Hendricks.

"Why is that, lover?" Terry asked. "I would think you would be happy to see Parsons go and cheering to welcome Ernie in his place."

"It's not that simple," David said. "Only Ernie and John would agree to a joint event under the circumstances. They're both weird."

"Maybe they wanted to get two events for the price of one."

"Maybe that's true," David laughed. "One's all the company can afford this year."

"Probably they both feel about parties the way you do."

"It's the circumstances. The Board fired John, and they considered everybody else in the country before they settled on Ernie. Certainly, John is not celebrating being forced out, and Ernie will be embarrassed by being put into the position of having to accept good wishes at being elevated."

"It's good for you, anyway," Terry joined David in the bedroom.

"You look great," David said, admiring the sleek silver cocktail dress that Terry wore.

"So do you," Terry said, adjusting David's tie.

David patted his wife's hips.

"Don't touch," Terry admonished, backing away.

"I don't know why you say it's good for me," David picked up on their conversation.

David watched as Terry sat down at her dressing table to put on her high heels.

"It's nothing to complain about, having your sponsor replace your enemy as the top man."

"John Parsons isn't my enemy," David laughed. "He's always treated me fairly."

"He's the one you almost killed at second base, if I recall the story correctly."

"I didn't almost kill him," David protested. "I just flattened him."

"And he liked it?"

"No. Who would?"

"And he talked mean to you ever after."

"No, it was just an act. Ernie explained all that."

"So enlighten me."

"Ernie said John's a competitor. He was embarrassed when he got knocked out. Ernie said John actually respects me because I am a competitor too. John signed off on every assignment that Ernie gave me."

"Then you're sorry to see John go?"

"No, I'm happy for him. The company is in terrible shape, and nothing that he tried worked. Fifteen years ago, we were one of the most profitable companies in the world. Last year John presided over the largest loss any company ever produced."

"Can Ernie do better?"

"I doubt it. I'm sorry he got the job. We don't have the products, and we don't have the organization to produce them."

"So why don't you change that? Aren't you the boy wonder who gave us the PC?"

"Come off it Terry. You were there. You know it was just an accident. The timing was right. That's all."

"And you are carrying the company now by selling PCs."

"Not quite. We're losing market share every year."

"That's not your fault."

"No. Thank you dear. The man responsible for that is no longer with us."

"DiGenova."

"You know the whole story."

"He quit before he was fired?"

"Yes. Everyone thought he would be CEO one day. The OS-2 and his personality did him in."

"But every one loves you."

"Hardly. I am still a lone duck. I've never fit into the organization, and I'll never be CEO, not that I want it."

"Ready dear," Terry turned out the bathroom light. "How do I look?"

"Let's just stay home," David said, meaning it.

"Sorry baby. I promised Margaret we would share a table with her."

"Why did you do that? You know I can't stand her husband. He's another DiGenova."

"Be nice," Terry said, taking David's hand. "Margaret's an old friend, of both of us."

David did not respond to that joust. Margaret was his old girl friend and Terry's current best friend. They both enjoyed provoking David.

David and Terry had difficulty finding a parking place among all the Lincolns and Cadillacs and were among the last to enter the huge dining room before the waiters began to serve the meals.

"Hi guys" Margaret called, waving her hand. "You look wonderful Terry. Too bad about your escort," Margaret smiled at David.

David kissed Margaret and offered his hand to Margaret's husband, Richard Adams. "Hi Richard."

"David," Adams replied.

Adams was a rising executive in the mainframe division. He was two years younger than David and for the sake of harmony muted his jealousy over the fact that David stood two tiers higher in the IBM hierarchy.

"Excuse me a minute," David said after seeing Terry into her chair.

Their table was a good one, located in the front, just to the right of the head table where Ernie and John Parsons and their wives and the other members of the Management Committee presided. Terry approached the head table from the front and offered his hand first to Parsons then to Ernie.

"John, Ernie," David said.

"David, so good of you to come," John Parsons spoke first, using the same lightly sarcastic tone that he always employed with David, a gentle reminder of their softball days.

David did not like the tone, but he tolerated it, understanding now that it was merely John's way of recognizing something special in David.

"David," Ernie greeted him.

Ernie also spoke with a humorous affectation when addressing David. David still wondered why that was. Ernie and David were now friends, at least as close as their respective positions in the company allowed.

"David," Ernie leaned across the table towards David. "Please come and see me first thing in the morning."

"Certainly Ernie," David said. David had a staff meeting followed by an important client meeting scheduled, but they would have to be rearranged. Ernie knew David would be booked up, so his request had to be important. David immediately started to worry. Ernie always liked to spring important decisions on David as surprises. David wondered where Ernie would send him this time.

David returned to their table and sat down facing Terry with Margaret on his left and Richard Adams on his right. David got the chair with his back to the main table.

"This is certainly a strange situation," Adams said, looking over David's shoulder at the head table. "It's not as if Parsons is retiring gracefully and is celebrating the appointment of his replacement."

David reacted negatively to Adams disapproving tone. Before he could reply, however, Terry kicked him from the left and Margaret from the right.

"Let's drink a toast to the departing and a hello to the new," Terry said raising her champagne glass.

"Hear, hear," Adams said, oblivious to the fact that he had troubled David. He raised his glass and waited for David to join them.

David lifted his glass, nodded to Margaret and Terry, then sipped the toast without saying a word.

David was tempted to ask Margaret what Ernie wanted to discuss so urgently but did not do so. Margaret, now Ernie's Executive Assistant, knew everything that went on in his office. Unfortunately, she did not often share that knowledge.

"I assume you will be changing offices," David said to Margaret.

"We move in tomorrow morning," Margaret smiled. "Early."

David could tell from the emphasis that she put on "early" that she knew David had been summoned.

"What's it all going to mean?" Adams said.

David knew Adams really meant: "What's it all going to mean to me."

"Doesn't your wife keep you informed?" David asked. David knew Margaret well enough to know that she would breach no confidence. When David had been a trainee sharing an apartment with Margaret, and she had worked as Ernie's secretary, she would never share office secrets.

"She doesn't tell me anything," Adams said.

"Poor boy, nobody does," Margaret leaned back in her chair, crossed her arms, and smiled. "You're a mushroom. They keep you in the dark and feed you..." Margaret did not finish the sentence.

"Cow shit," Adams said in a loud voice.

Terry patted Adams arm. "Poor boy. David has the same problem."

The two girls laughed.

After a lame meal of rice and rubber chicken seasoned by light female chatter, David leaned back to listen to the two featured speakers. John Parsons went first.

"I'm sure many of you are wondering why I'm still hanging around here," Parsons began. A few loyalists called "No, No," and the majority of the audience laughed politely. David responded with an embarrassed smile.

"I do so for two simple reasons," Parsons continued. "I want to publicly thank this marvelous company that put up with me for thirty glorious years, and I insist on the opportunity to applaud my successor, my friend and colleague Ernie Hendricks."

The nervous audience responded with an enthusiastic response.

"They're relieved that John is departing graciously," David smiled at Terry and Margaret.

"Times have changed," Parsons said, "and our industry has changed. Many of us older guys did not. Our loss. We thought our company was too big, too grand, too established to have to adapt. We were wrong. I was wrong. Now the burden has been passed to Ernie Hendricks. I wish him success where I have failed. I pray that all of you in this audience, the senior leadership of our company, understand what Ernie must now do, that you have the flexibility to adjust, to support him." Parsons paused.

The stunned audience sat silent.

"I thank all you heirs to the legacy of the two Tom Watsons for a wonderful career. I wish you luck." Suddenly, Parsons smiled wryly and sat down.

"That was short and sweet," Richard Adams said snidely. "I wonder if we can get another bottle of champagne."

"That was very touching," Terry spoke quickly before David could reply to Adams.

"I guess it's my turn now," Ernie Hendricks was now standing.

David wondered if he were the only one who appreciated the double meaning implicit in Ernie's words.

"I promise you I will follow my friend and colleague's example and keep this brief. I too thank our company's founders for the opportunity they have given me to work along side John Parsons for these many years. Most of you may not know nor care that John and I joined this company together during a troubled time. I won't bore you with the details but we began with the opportunity to watch Thomas Watson Jr. bet everything on the fabled 360, and we were there to see him win."

This time the audience responded with the obligatory applause.

"We again are facing a critical point in our company's history. I know that I, unlike Tom Jr., do not have the ability to face that crisis on my own. I need your help. With it, I am confident we will not let our past leaders like John Parsons down."

The audience waited for more, but Ernie abruptly sat down.

Again, David started the applause.

"What was that all about?" An indifferent Richard Adams said.

"Richard!" Margaret fired a warning shot across her husband's bow.

"I'll see if I can get find a waiter and get us some champagne," David said.

While most of the audience remained seated, David made his way to the back of the room where he found a group of waiters preparing to return.

"Two bottles of champagne," David said, grabbing the arm of a familiar face.

"One for my table, and one for the head table."

"Certainly Mr. Howard."

David did not immediately return to his table. He retreated to the bar and ordered a double scotch. The whole affair depressed David, the sad speeches that left so much unsaid, the indifferent audience, Richard Adams who unfortunately represented the new generation of employees.

"May I join you David?" A familiar voice said.

Surprised, David turned to find John Parsons standing next to him. Before David could respond, Parsons nodded at the bartender who was setting David's whiskey in front of him. "I'll have one of those if I may?" Parsons said.

"A nice speech," David said weakly.

Parsons laughed. "Don't be embarrassed David. I'm not. You've got to understand these things. We all get old."

"John, the way the Board treated you is indefensible."

"That's the way the game is played. But that is not what I want to talk with you about."

David, now curious, waited.

The bartender delivered Parsons' drink. He raised it and toasted David who replied in kind. David noticed the bar was filling up, but the crowd gave David and Parsons ample space. David wondered if they were afraid of contamination.

"David, don't you and Ernie make the mistakes I did. Act decisively. This grand old company must change if it is to survive. Its future is on your shoulders."

"But..." David started to protest, but Parsons placed one hand on David's wrist to stop him.

"Let me say a few things I could not out there," Parsons said. "You're one of our lone ducks."

David nodded, wondering if he should apologize for having been so independent.

"You probably don't know it, but every year we bring in one like you. Ernie and I started that. You were our first. We both recognized that the company had become one huge bureaucracy with committees running everything. We had our own culture, our own way. Ernie and I decided this was dangerous. Every company needs risk takers, but big organizations stifle them. We decided to experiment in our own small way. We tried to give you an opportunity to grow within the system without stifling your initiative or independence. It was no accident that you were sent to Boca Raton or Oshkosh. We protected you when we had to. Otherwise, the system would have devoured you. Do you think the DiGenova's would have allowed you to survive?"

"I never realized..."

Again, Parsons cut David off. "I'm telling you this not to flatter you. I'm warning you. It's almost your turn. Don't let us down. There are other lone ducks out there. Find them, and nourish them."

"But we still have Ernie," David protested. He was not sure he wanted the responsibility that Parsons was implying.

"Ernie will go down in flames. He knows that. If the company is to survive, Ernie must take drastic actions. Things will get worse before they get better, and the Board, the media, the public, the company will blame Ernie. Help him."

David was tempted to ask if things were so desperate why Parsons had not acted.

"I know," Parsons said. "You want to know why I did not act. The answer's simple. I thought I could take the evolutionary way. I failed."

"What happened to us?" David asked.

"We grew too big, and the committees took over. Look at Digital Software, at Astron Computers, at the upstarts in Silicon Valley. Do you see the difference between us and them?"

"Size?" David knew his answer was inadequate.

"Size yes, but more importantly culture. What do our IBM managers want?"

David hesitated, knowing that Parsons' question was rhetorical.

"They want to be important and safe. The system teaches them to work within it. That is the way to get a secretary, have big staffs, lots of employees waiting to carrying out your bidding. They want to live in Seniorsville, on Kings Row, to have big offices and to belong to this club and to play lots of golf and tennis with important people. They like to see their names in the newspapers and hear them on television. They like to see people's reactions when they let drop that they work for IBM, the biggest and the best. Most of all they want power, safe power, within the system. What could be better than power and security? It's like working for the government, only there the pay is less and the real power rests in the hands of the politicians who keep changing. What's the difference between us and the upstarts like Digital Software?"

"Money," David said. He knew that his colleagues envied the new companies their wealth. Digital Software had over 2,000 millionaires among its employees.

"Money. They have leaders who want to be rich. Look at Charles Swift."

"It's not money alone," David said.

"It's greed and power and the need to participate in a revolution. To be the best. Not safe and secure. The best. Your friend Charles prides himself on hiring the smartest, working them hard, and rewarding them. Note I said the best, not the most."

"How do we compete?" David had his own views, but he asked the question to give Parsons the opportunity to expound.

"Compete with greed? We can't. We've far too many employees. How do I give 100,000 employees stock options? The Board would faint. I tried in my little way with bonuses, but a couple thousand here and a couple thousand there is meaningless. We've got too many people to motivate with money."

"Then all we have is security," David said. IBM had always assured its employees that they had jobs for life.

"This is the era of downsizing. Lean and mean is the way to go," Parsons said.

"What do we do about management by committee?"

"Eliminate them. Downsize. Delegate. You can't imagine the kind of decisions that are kicked up for the CEO to decide. We have marketing staffs, sales staffs, engineering staffs, administrative staffs, armies of lawyers, all independent of our operating divisions. If the marketing people do not like an engineering proposal, they say they will not sell the new product. The decision gets kicked upwards."

"Then we must eliminate the staffs," David said. He had had this thought himself.

"Every new CEO reorganizes his company. Ernie must recreate it. To do so he will first have to destroy it. Do you understand what I'm saying? Remember when you went to Boca Raton and singlehandedly assembled the first PC. You liked the independence, being your own boss. Right?"

"Right."

"And the bureaucracy resented you. It wasn't the IBM way."

David nodded.

"We've got to build a company of independent bosses responsible for their units and products. With responsibility comes obligation. There's no one else to blame for failure. We're too centralized," Parsons waved to the bartender for two more drinks.

David was tempted to ask why Parsons did not give this speech to the assembled audience, but did not, recognizing why. They would not have understood. He would only have frightened them.

"Just putting the IBM logo on our products is no longer enough. Ten years ago it was. But no longer. But you don't want to hear about the old days. I just wanted to tell you, rely on your lone ducks."

Parsons downed his drink, patted David on the shoulder and left.

David returned to the restaurant to find an orchestra playing and Terry sitting alone at their table twiddling with an empty champagne glass.

"Have something going on the side, lover," Terry greeted him.

"I had a long talk with John Parsons in the bar. Sorry," David said.

"Tell me about it."

"Let me digest it, and I'll share later. I am worried for Ernie."

"Don't, David. Ernie can take care of himself. You worry about David Howard."

"And his wife."

"Don't worry about his little wife. She can look after herself," Terry smiled. "Now, do you think you could manage a little dance."

"With the prettiest girl on the floor?" David asked. "Certainly, I can. Where is she?"

"Try and keep up," Terry grabbed David's arm and pulled him to the center of the dance floor.

The next morning David woke promptly at six. Despite the fact that it was a Saturday, he prepared for work. Dressed informally in slacks, a long sleeved sports shirt, and tassel loafers, David arrived in the Armonk headquarters parking lot promptly at seven. To his surprise, he saw Ernie's Lincoln parked in the CEO slot. Two lanes back, he spotted Margaret's Mustang in its usual spot. David assumed that would change. David parked his Lincoln in its normal place in the second row.

He greeted the guard, stopped in the cafeteria, selected a cup of coffee and a glazed donut, and took the elevator to the fifth floor. He entered Ernie's old Sales Director office and found it dark and deserted. He was sure he had seen Ernie's car out front. Then, he realized that Ernie must already have moved into the CEO's office on the top floor.

Carrying his coffee in one hand and his glazed donut in the other, David took the elevator to the top floor. The doors slid open and David stepped out into a lighted reception room.

"Glad you could join us," Margaret greeted him.

"Nice digs," David said.

Margaret shrugged her shoulders.

"I had a late night," David said.

"I'll bet you did," Margaret smiled. Over her husband Richard's objections, Margaret had insisted that they leave early. "I saw you stumbling around the dance floor with the loveliest girl at the ball."

David sat down in the chair next to Margaret's desk and opened his coffee container. "Mind if I have breakfast while we chat?"

"Be quick about it," Margaret said. "He's waiting for you."

"How can he be waiting for me?" David protested. "It's seven o'clock on Saturday morning."

"You've got a lot of work to do," Margaret said, patting an airline ticket folder on her desk.

David pointed at the ticket. "Me?"

Margaret smiled.

"Give," David leaned forward.

Margaret slipped the folder into her desk drawer.

"Where to?" David whispered.

Margaret shook her head and pointed toward Ernie's office with her thumb.

"Is he alone?" David suddenly felt intimidated. He was about to enter the office of the CEO of IBM.

David set his coffee next to the donut and started towards the door.

"Wait," Margaret ordered. "Take this stuff with you."

David paused at the closed door with both hands filled.

"Oh for Christ's sake," Margaret complained.

She got up and knocked on the door before opening. She stuck her head inside. "He's here, finally."

"Come on in David," Ernie called.

David entered. As always, the grandeur of the office intimidated him. Ernie, sitting behind the huge desk with the heels of his crossed feet resting on the shiny surface, looked natural.

"Have a seat, kid. We've got work to do."

David decided that Ernie had taken his sudden promotion to the stratosphere in stride.

"Mind if I finish my breakfast?" David asked.

Ernie waved towards the chair that faced his desk.

"Want half?" David held up his donut.

Ernie laughed. "Eat and listen, David," Ernie said. "I don't have much time. I've got a Management Committee meeting scheduled for eight and then a news conference at nine."

"Something important happening?" David tried to make a joke.

"We're going to shake this place up David."

David wondered what Ernie meant by "we." He hoped Ernie was using the imperial we and did not mean he and David. The airplane ticket in Margaret's desk drawer worried him. David and Terry were scheduled to tee off at one.

"I don't have to tell you, David, that we have gone from being one of the most profitable companies it the world, the most respected, to a company that has suffered the biggest loss of any company ever."

David suddenly felt uncomfortable chewing on a glazed donut while Ernie discussed catastrophe. David placed the donut on its napkin on Ernie's desk and sat up straight. Then, he realized he had left his breakfast on the CEO's shining desktop.

"Just eat your donut and listen, David," Ernie said. "I need your help now more than ever. You're going to have to take a lot on yourself, however."

David wondered what Ernie had planned for the PC Sales Group. In David's opinion, they were doing as best they could.

"First, we're changing your office. Let Margaret take care of that. You're going to be busy."

David chewed the donut as quickly as he could.

"At eight I will tell the Management Committee that I am abolishing the Management Committee."

David who was sipping his coffee choked.

"And all of the division chiefs that they are to submit their resignations. Those who can will retire. Those who cannot, we'll set up an office to help them find new jobs. I want them out of their offices immediately."

The words shocked David. A revolution was truly underway. Word would spread like wildfire, and every IBM employee would begin worrying about what the changes would mean to him or her personally.

"That'll shake things up," David said. "What's going to replace the Management Committee? Who are the new division chiefs?"

"That's up to you," Ernie smiled.

"Me!" David almost dropped his coffee cup. He set it on Ernie's desk and forgot about it.

"For a start. I want to clean out about ten layers of management. The staffs have to go. Let each of your new division chiefs select their own support elements, sales, marketing, personnel. Make sure each understands that he will be held responsible for his own bottom line. Only R & D will remain separate. Note I said separate not independent. Reduce them by fifty percent for now, select someone who can manage the researchers, make him responsible for developing new technology that will produce hardware we can sell."

"If there is no Management Committee, who do the divisions report to?"

"You."

"Me?"

"I want the management of this place flattened, David. I mean it."

"One person can't make all those decisions," David protested.

"One man won't. Delegate authority and responsibility then monitor the hell out of them. You've heard of Blanchard and Davis?"

"The law firm?"

"Think man."

"You don't mean Mr. Inside and Mr. Outside?" David referred to the famous West Point backfield team from the mid forties. David had only read about them.

"That's right. I keep forgetting about the generation gap. You're Doc Blanchard, I'm Davis."

"I'm Mr. Inside?"

"That's right, and I'm Mr. Outside."

"Starkey tried that at Astron Computers and it didn't work," David said.

"He didn't have a Mr. Inside. Any questions?"

"What's my title?"

"I'll think of something."

"Do I get a raise?"

"We're going to institute greed here as a motivational force."

"I better get busy," David said, wanting to escape before Ernie unloaded more on him.

David was halfway to the door when Ernie stopped him.

"Three things before you get started," Ernie said.

David turned.

"One, you have an appointment for lunch with Charles Swift. It's all arranged. I talked with Charles last night. Margaret has the details."

"What do I tell him?"

"That IBM is back in business. We need Charles and Digital Software to help us develop vision and to build computers we can sell."

"I can handle that." David had always been able to work with Charles.

"Two, get to work on identifying your new division and group chiefs. I want them in place Monday morning. Don't forget the lone ducks. There are several of them out there, guys just like you were."

"I think I know who they are."

"Margaret has my list. Make your own decisions. The authority and responsibility are yours."

"The buck stops here," David joked.

Ernie did not smile. "That's right David."

David turned to go.

"And three," Ernie said. "Get your garbage off my desk."

David retrieved his cup and donut and retreated. He finished off the donut and approached a smiling Margaret with his hand out.

Margaret handed him the plane ticket and a large folder stuffed with paper. "A company car will leave here at nine to take you to Laguardia where you will catch the ten o'clock shuttle to Washington, D.C."

"Washington, D.C.?"

"That's where you are meeting Charles Swift. He'll be waiting in the Willard bar at noon."

Charles tossed his napkin into Margaret's trashcan. "Why don't I use the company plane?"

"What plane?" Margaret asked. "If you are talking about the old one, it's being sold."

David turned toward the elevator.

"Where are you going?" Margaret asked.

"My office," David said, then remembered that Ernie had implied he had a new office.

"First door on the right," Margaret said. "Have a nice trip."

David turned right and found a blue suited service employee busy installing a new bronze plaque on the wall next to the door of his new office. It read:

Associate Chief Executive Officer

ACEO. David liked the ring of that. David stepped around the worker and, balancing the coffee, his full envelope and his ticket folder, opened the door.

"Good morning sir," Jean, his smiling secretary, greeted him.

"Jean," David said. "What are you doing here on a Saturday morning?"

"Margaret thought it would be best. I understand you have much to do."

"This looks great," David examined the room. It was twice the size of his old office. Two desks, including the one where Jean sat, faced the door. Two sofas and several chairs were strategically aligned. Behind Jean to her right stood three large file safes. "Is that my desk?" Charles asked, nodding toward the empty one on her left.

Jean laughed. "That belongs to my assistant. I gave her the morning off. Your office is in there." Jean nodded toward a closed door to her immediate right.

"And that one?" David looked at the closed door to Jean's left.

"That's Mr. Hendrick's office."

"I don't think we'll be using that much."

"Oh it's not really to Mr. Hendrick's office. It opens in Margaret's office."

"I'll try to keep all this straight," David said.

He opened the door to his office and stopped. The room was about three quarters the size of Ernie's office. It was quite impressive however, far grander than any office David had occupied before. The desk, large enough to pass for an aircraft carrier, sat in front of ceiling to floor windows. David crossed the room, admiring the glistening furniture—it all looked brand new—and looked out the window. He had a panorama view of the IBM campus, acres of fresh cut grass.

"Do you like it?" Jean asked. "They said you could change it all if you don't."

"It's fine," David said. "When did they set this up?"

"Last night. Mr. Hendricks said you were going to be busy today."

"I need one thing, Jean," David said.

"Yes sir."

"I want a PC placed on that credenza," David pointed at the polished credenza to the left of his desk.

"One of the OS-2 models?"

"No. One of the good ones, the original PC."

Jean smiled. "I'll take care of it sir. Don't forget. The car leaves at nine."

David sat at his desk, examined the phone set up, pressed the button for a direct outside line and dialed home. When Terry answered, David tersely described the mountain that had just fallen on him, apologized for having to cancel their golf date, and promised to be home for dinner.

"Something light," David said.

"No problem. ACEO?" Terry asked.

"I'm afraid so," David said.

"Don't worry. You can handle it," Terry said. "Will we have to move?"

"No way," David said.

"Good. I'll check out King's Row anyway. Congratulations ACEO. I'll fix an omelet for dinner."

"Great. See you tonight." David started to hang up when Terry spoke.

"David?"

"Yes?"

"Kick ass!"

David, still dressed casually, went directly from Armonk headquarters to Laguardia where he caught the shuttle for Washington. He arrived at the Willard by taxi from Reagan National Airport without luggage and a little embarrassed to make an appearance in that stately hotel, a tribute to past generations of Washington power seekers, dressed so informally. To his surprise, he found Charles waiting alone in the bar.

Charles winked when David greeted him like an old friend and took the stool next to Charles.

"Good," Charles said. "Pretend we met by accident. I'm in the Presidential Suite. Stay here for ten minutes then come up and join me." Charles dropped a five dollar bill on the bar, shook David's hand and spoke loudly" "Good to see you again David. Sorry I must rush." With those words, Charles left the bar.

"It must be nice to be friends with the richest man in the world," the bartender smiled, retrieving Charles' half-empty bottle and glass.

"An old friend," David smiled. "I knew him when he was a mere millionaire."

The bartender chuckled, then moved on to serve other guests.

While he waited, David pondered Charles' strange behavior. Suddenly, he realized that Charles was protecting them both. With the Government in hot pursuit once again of Digital Software, neither Charles nor David should risk appearing to meet under circumstances that would give the government investigators ammunition to imply conspiracy. They had already tried that once. Erring on the side of caution, David waited a full half-hour before leaving the bar.

David tapped on the door, and Charles himself opened it.

"Come on in, David," Charles again enthusiastically shook David's hand. We were afraid the Feds may have rounded you up."

David laughed. "I hoped we haven't reached that stage yet." Over Charles' shoulder, David saw a smiling Ann and Bill Oldham.

After enthusiastic greetings, They seated themselves in the comfortable sitting room of the suite.

"Before we start, let's order something to eat," Charles said. "David?"

"What are you guys having?"

"I just want a salad," Ann said. "Something simple like a Chicken Caesar and an iced tea."

"Fine by me," Bill agreed with his wife.

"I'm hungry. I'll take a club sandwich, any kind, and iced tea," David said.

Charles called room service and placed the order including a cheeseburger and fries for himself.

"OK," Charles turned back to David. "Tell us what the hell is happening in Armonk. Ernie called but he said you would brief us."

"The story of Ernie's press conference was on the wire when I left the office," Bill said. "I didn't have time to read it."

"John Parsons is out. The Board dumped him," David said. "Ernie's replacing him as CEO."

"Wow," Charles said. "That was abrupt."

"Ernie's taking over with a bang," David continued. "He told the Management Committee this morning that he was abolishing the committee and replacing all of them as division managers."

"Good move," Charles said.

"Is this all off the record," Bill asked, gently reminding his friends that he was a working journalist.

"I would appreciate it if you would keep this as deep background," David said. "Certainly don't quote me."

Bill nodded agreement.

"Ernie plans a severe downsizing," David said.

"About time," Charles agreed.

"He's going to eliminate about fifteen levels of management and delegate authority and responsibility to the lowest possible levels."

"That's pretty hard to do," Ann said.

"If we're going to survive as a company, we have no choice," David said. "The fat days were over years ago."

"Ernie mentioned something cryptic like Mr. Inside and Mr. Outside," Charles smiled.

David immediately realized that Ernie had briefed Charles in more detail than he pretended.

"Ernie says he will be Mr. Outside," David hesitated. "And he's making me Mr. Inside. We've got to work all of that out," David said.

"Congratulations," Charles, Ann and Bill took turns shaking David's hand.

"Maybe."

"What's Mr. Outside and Mr. Inside mean?" Ann asked.

"Like the football players, Davis and Blanchard," Bill said.

Ann shrugged. The names meant nothing to her.

"I'll explain later," Bill promised.

"You know what he is doing, don't you?" Charles asked.

"I'm not sure," David said honestly. "I'm just frightened by the responsibility he has thrust upon me. If he means what I think he does, I have to tear the company apart and rebuild it."

"I couldn't think of a better man for the job," Charles said.

"What do you think he's doing?" David asked. "I only learned about all this three hours ago, most of which I spent getting here."

"He's going to take the heat and let you work behind the scenes to do the job," Charles said.

David had not thought about that. It made sense and was perfectly consistent with Ernie's management style. Ernie would keep as much pressure off David's back as he could, freeing David to decide what needed to be done and to do it.

"I'm not sure we will ever get rid of enough bureaucracy to let us become competitive in the industry again," David said. "Let along dominate it like we once did before some people I know changed everything." David smiled at Charles.

"IBM still has a quality reputation," Bill said.

"That's about all we have," David said.

"You have a lot of talent," Charles corrected him. "You have to free them to create. Your job is to cover their backsides when they make mistakes."

"Astron Computers never recovered," David said, letting his worry show.

"That's because they brought in a bunch of managers to lead them," Bill said. "They need a Charles on top, a man with vision who can lead, and let the doers do."

"What do you think of Shepherd?" David asked. Astron had elevated Leonard Shepherd, the company saver from ZYTEC and Astron's Board, to the CEO position."

Charles shrugged. His response told David eons. "How can we help you?" Charles asked.

"We're not going to conspire to fix prices and divide up the market," David laughed.

"We're not going to even come close to discussing anything like that," Charles laughed. "I meant," Charles spoke loudly as if he assumed the walls had ears, "what can we sell you."

"Have you thought about software?" David laughed.

Charles smiled.

"I haven't had time to think about where we should go," David said. "Clearly, we've got to continue to ride our mainframes as far as the diminishing market permits. We can write own software there. As far as everything else goes, we're wide open. We're going to have to do more with less."

"It makes sense that you continue to concentrate on the corporate market," Charles said.

"And we'll stay in the microprocessor business too," David said. "Our PCs are good machines. They're just not price competitive with the clones."

"You can cut costs," Charles said.

"And we have the engineering talent to move the technology forward."

"You don't have a chance in hell of taking over the PC market again, no matter how poorly Astron does," Bill said.

"We know that," David smiled. "We can fight for our fair share. We're going to need a good operating system," David said.

"I think I might know where you can find one," Charles smiled back.

"It wouldn't have a name like Panes or Screens or something like that would it?" Ann asked.

Charles did not answer Ann. "What do you think about the Internet?" Charles asked David.

"It's here to stay," David said. "We can go in on the server end," David said, "but we can't compete with you and Netscape and won't waste resources trying."

"Working hard on servers might be a good tactic for you," Charles said. What he did not say is that server hardware would in no way compete with his products.

"We recognize that big business will continue to need mainframes, not as many as previously, and the smaller companies might go for desktop stations from the clones rather than buy our more expensive products, but they will need servers," David said. "We can still capitalize on our reputation for quality and service. Tell me Charles, how are you coming with your Internet plans?"

"We're pleased with our browser but running into political problems. We keep picking up signals," Charles looked at Ann, "that the Feds are displeased with our technical progress."

"What does that mean?" David asked.

"They don't like innovation. We're integrating our desktop totally for our next version of Panes. I think its great progress, and our users will too. How would you like to go direct to an Internet address from your desktop?"

"If I were an Internet user, I would," David deflected the question.

"I don't understand why the government thinks it should regulate technical innovation. They don't try to control scientific research or engineering in other fields," Bill said.

"That's a good line, Bill," Charles approved. All three listeners understood exactly what he was saying.

"Congratulations on your talk show," David said.

"It's fun. I'm enjoying it more than I thought I would," Bill said.

"And Ann," David said. "You've been quiet. One of the main reasons for this trip, aside from spending time with you three of course, was to ask you a question?"

"Me?" Ann feigned surprise.

"Yes. What do you think of our Washington office?"

Ann hesitated, looking at Charles, who nodded affirmatively.

"Honest talk?" Ann asked.

"Yes."

"They're losers. You don't have a good man among them."

"Could you be more specific?"

"They all are long time senior managers put out to pasture. They're having a great time with a liberal expense allowance. They spend more time at the Congressional Country Club than they do on the Hill. Is that enough?"

"Tell me more."

"They don't even know why they're here. They think it is public relations, putting on a good front for IBM."

"When they should be doing what?"

"Mobilizing political support for IBM objectives and establishing defenses against unwarranted government intrusion."

"Aren't you speaking to Digital Software objectives?"

"To the entire industry," Ann said. "Thanks to the Clinton mercenaries, even the courts are contaminated with the idea that it is their job to manage our industry."

"Why is that?"

"Money and votes."

"What should I do?"

"Sack the whole bunch. Send down some eager young men with clear-cut marching orders. Delegate then audit the hell out of them."

"I think I've heard that philosophy someplace else," David smiled.

Chapter 37

Leonard Shepherd sat in his office on the eighth floor of the Astron Computers headquarters building. He found his office drab with its white plaster walls, tan carpet and black furniture with brown tops, but he liked the view of the mountains. Shepherd had schemed and maneuvered to get here. Sometimes, like this moment, he wondered why. A realistic Brooklyn boy with a Ph.D. in Physics and a sharp analytical mind, Shepherd did not delude himself. He had had a good deal at ZYTEC, twenty-seven million in salary, stock options and performance bonuses, and he had left that affluence behind to move to Astron for a meager thirty percent salary increase that took him up to $990,000 a year, plus a five million dollar signing bonus, and a million in stock shares plus options. He calculated he would have to stay at Astron Computers five years to come close to his former prospects at ZYTEC.

Money had not been the reason Leonard had moved to Astron Computers, despite all the criticism his emoluments drew in the media; some even accused him of having held up Astron for ransom like a common criminal. Others just called him greedy. No, money had not been his motivation. Leonard frankly admitted to himself that it had been pride. He enjoyed his reputation as a company saver—look what he had done at ZYTEC—and Astron Computers had appeared like the answer to a dream. Astron was a troubled company, and a media darling to boot. Astron CEOs commanded national attention; they were media stars of the first order. Leonard had craved that attention and that was why he had jumped.

First, he had joined the Astron Board as a common director. There, however, Leonard had aggressively challenged Acting CEO Hans Wifler's inept leadership. It had taken less than a year of declining market share, lowered profit lines, Saturday morning quarterbacking and considerable backstabbing for Leonard to achieve his objective.

When he became CEO, Leonard inherited a company with no corporate strategy, an antique financial system, a major cash flow problem, an aging product line, an out of control engineering department, a manufacturing line that produced products with inadequate testing, a declining market share, defecting users, and a shrinking bottom line. These things Leonard knew how to handle.

In this moment of retrospection and review, Leonard thought about his predecessors: Willard Temple, Lawrence Starkey, Hans Wifler, even Stuart Miller who served as acting president at odd moments. Leonard wondered if they too had felt confidence in their respective abilities to manage the maverick corporation called Astron. Over them all, the shadow of Stanley Pitts lingered. Leonard had difficulty comprehending this phenomenon. Stanley had limited technical skills; he was arrogant, and he treated people with disdain; yet, somehow, Stanley had unquestioned star power. Stanley's choice of names for his company had been prophetic.

Unfortunately, the past year had not unfolded as Leonard had planned, not from the very beginning. The media had shocked Leonard with its mindless attacks on his employment package. He had really underestimated the fascination that the world held for Astron Computers. His first press conference had attracted the major television networks and over two hundred reporters. Unfortunately, he had slipped and referred to his first hundred days.

A reporter had asked: "What do you intend to do about the slipping deadlines for your next generation operating system?"

Leonard had glibly replied: "This is my second day in the job. Ask that question in one hundred days."

To Leonard's dismay, that simple comment unleashed a media frenzy. Pundits cited John F Kennedy's first hundred days and speculated about what Leonard might do that would let him re-create that magnificent period for Astron. Many criticized Leonard for daring to position himself in the same league as their hero.

It did not take Leonard long to conclude that he had been wrong to thirst for greater public recognition. He had tried to correct the slip. Leonard truly believed that he could save Astron Computers. He was an experienced manager, and he knew what a successful company needed. He turned his attention inward. He commissioned a mission statement. After several false starts, Leonard had been satisfied with the product of his assigned drafters. His subordinates praised the statement, then promptly filed it. Leonard expected action. At ZYTEC, Leonard issued orders, and his managers and their subordinates marched. At Astron, there existed a different climate. The division chiefs considered themselves autonomous, sovereign in their commands. They expected to make their own decisions while the CEO played at media star. Leonard tried to change that. His division chiefs responded with a brisk "yes sir," then stubbornly marched in different directions to the beats of their individual drummers.

Leonard's intercom buzzed:

"Ms. Stuart is here," Leonard's secretary announced.

Chris Stuart, Astron's Vice President for Technology, had been an early Leonard Shepherd appointment. The heart of a technology company is its engineering department. They are the ones that conceive and create the marvelous new products that drive the technological revolution, excite the customer base, and fatten the bottom line. On Leonard's arrival, he had discovered that this company with the reputation of being the innovation leader was paralyzed by a drifting engineering department that resembled a university faculty; each of the engineers pursued his own dream, oblivious to the market and the desires of management. To give discipline to this chaotic division, Lawrence had hired an experienced professional engineering manager he knew and trusted. He had imported Chris Stuart from ZYTEC.

Chris, despite being female, had the credentials. She had earned her spurs at IBM, moving from a position as a working engineer into management. From IBM, she had migrated to ZYTEC where Leonard learned to admire her skills. At Astron, Leonard assigned Chris two primary tasks: establish the engineering division on a professional footing, and get control of Beethoven. Chris had attacked the merely difficult first. Now, the engineering department had discipline. It had priorities and deadlines. Marginal studies had been abandoned, and those engineers and scientists who resisted direction had been encouraged to seek employment elsewhere.

Beethoven, though, remained a problem. Stanley's Asteroid was on its last legs. Panes 95 had moved the PC ahead of Asteroid. Even the most ardent of the long time Asteroidphiles were buying PCs. It was no consolation that the clones had moved ahead of a faltering IBM in market place. Astron now faced the very real risk of sinking to a five- percent share in a market they once dominated. They still had twenty million users clamoring for more, but that was not enough. They needed a new operating system to survive, and Hans Wifler had committed them to Beethoven.

"Leonard," Chris Stuart said as she marched into the room and took the chair of honor facing Shepherd's desk.

It's too bad she's not better looking, Leonard thought to himself for probably the hundredth time. Leonard, like most modern managers, was sexually liberated. He knew that equal employment opportunity suits could demoralize a company. He was still enough of a chauvinist, however, to admire good-looking women. Chris was stocky. She weighed at least one hundred and eighty pounds, had melon breasts that drooped over a barrel shaped body supported by telephone pole legs.

"Chris," Leonard returned her terse greeting. "Any progress?"

Chris shook her head negatively, and Leonard was not sure he wanted to know more.

"I don't have to tell you the problems," Chris said. "I'm sick of talking about them. I haven't solved them, and I'm not sure I ever will."

"We're working as a team, now, aren't we?" Leonard referred to the fact that they had five hundred employees urgently pursuing Beethoven.

Leonard had been shocked to learn that Wifler's Beethoven team was not really a team at all. It had been a conglomeration of teams, all working on different facets of Beethoven, all oblivious to the fits and starts of the other teams. Such an organization, or disorganization, if left to its own devices, would have produced nothing more than full employment for Astron's independent engineers.

"They don't like my new organizational structure," Chris said, "but they have no fucking choice."

"I didn't promise you it would be easy," Leonard said.

He was dismayed to find Chris in this mood. He needed Beethoven. They were already a year behind and a victim of Astron's high profile. Anything dealing with Astron's new operating system made headlines, and disgruntled employees leaked to a hungry media faster that than national security secrets left the White House and State Department. Leonard's press office estimated that over one thousand stories a week featuring Astron appeared in the world press. Leaks about Beethoven had become the chief line of defense of disgruntled employees fearing for the jobs. Leonard had recently been forced to release another 3,000 employees, making six thousand out of their sixteen thousand-man workforce that had departed since he had taken over.

"I don't know how Stanley Pitts produced Asteroid in three years or Harold created Astron 2 in months. It took Digital Software nine years to design Panes, and they still don't have it right."

"Unfortunately, Panes now has ninety percent of the market," Leonard said.

"We need memory protection and multitasking," Chris said.

Memory protection involved establishing the operating system in one protected memory and putting the word processing, spreadsheet programs and other applications in separate areas. With memory protection, one application could crash while the others would continue to run. Otherwise, the whole computer would crash and have to be restarted, a time delay that irritated busy users. Multitasking gave the users of an operating system the option of doing more than one task at a time. For example, while the printer was printing, the user could draft new documents in the word processor.

"We need both," Leonard said.

"I know we can solve our problems," Chris said with a frown, "but I can't guarantee when."

"We don't have time to wait," Leonard said. "We almost had a user riot when we missed the last deadline."

"Then stop setting deadlines, Leonard," Chris said.

"I thought this company was supposed to set the cutting edge for technology," Shepherd said.

Chris shrugged.

"Following your last projections, the Software Division stopped writing for Asteroid," Shepherd complained.

"Why write new applications for a dead operating system?" Chris countered.

"Right. But we're a year behind our timetable now. We have no new applications. Our competition does. Even worse, the independent software companies aren't writing for Asteroid either. I'm on my knees with Digital Software, and Charles keeps saying: 'OK. You bring out your new operating system, and we'll write a lot of jazzy new applications for it.' Of course, in return all we have to do is integrate his damned browser on our machines. We're dying without Beethoven."

"I'm working on it. I guarantee you, our engineers will do the best they can," Chris said. "They understand what's at stake, but they can only work so fast. This stuff takes time."

"Time we don't have," Shepherd said.

"Want me to see what's available elsewhere?"

"We already know that Chris. There's Panes, there's Dupris' Future, Stanley's "Starbright" and that's about it."

"I could take a sounding, find out what they want," Chris offered.

"No! You concentrate on engineering, on Beethoven, and I'll handle the outside stuff."

After Chris departed, a dispirited Shepherd buzzed his secretary.

"I'm going to make some phone calls. I don't want to be interrupted."

Leonard took a pad and a pencil out of his desk drawer and started making notes. He set up a page for Dupris and another for Charles Swift. He thought about a third page for Stanley and his Starbright, then dismissed it. He refused to bring Stanley back into Astron. The man was an uncontrollable flying object who destroyed everything in his path.

Without thinking, Leonard wrote Licensees at the top of the third page. Immediately, he scribbled through the word. Licensing was another of his many major problems. Wifler had left him two abominable legacies, Beethoven and licensing. Wifler had done the unforgivable. He had authorized Dallas Computers, a small organization based in Dallas, to build and sell Asteroid computers. This was a desperate attempt to broaden market share through licensing clones.

When they signed their agreement with Astron Computers, Dallas Computers had promised not to go after Astron's already dwindling base of high end educational users, and they had lived up to their word. Dallas Computers produced a good machine, however, and its quality was luring customers away from Astron's base anyway. Other licensees had proved to be not so trustworthy. Two had immediately targeted the education user community, and this threatened to take huge chunks out of Astron's high-end base.

Since the licensees were authorized only to manufacture Asteroids, Leonard needed a new operating system. If Astron Computers refused to license Beethoven or whatever, the cloners with their outdated Asteroid machines would die a natural death.

Leonard decided to call Dupris first. To his surprise, he was immediately patched through.

"Leonard!" Dupris, who Shepherd had never met, greeted him like a long lost friend.

"Pierre," Shepherd responded in kind.

"I've been hoping to get out to Cupertino and talk with your personally," Dupris said, beginning his sales pitch. "We've had some breakthroughs that should intrigue you. I know they do me."

"You're welcome anytime," Shepherd said. "I would like to learn a lot more about your operating system."

"I bet you would," Dupris laughed, allowing some of his old arrogance to leak through.

"Oh?" Leonard decided he did not like the Frenchman. He had heard many Dupris stories, and few of them had been favorable.

"I hear things from old friends," Dupris said. "Nothing important. How is Beethoven coming along? Do you have the piano tuned yet?"

Leonard did not react. He wished they had selected another code name. Beethoven invited too many jokes. He owed the unimaginative German for that too.

"I'll be in the Valley next week. Could we make an appointment for, say, Wednesday?"

"Sounds great," Leonard said, wondering if he had made a mistake. "Have your assistant call mine."

Leonard then dialed Seattle. Charles picked up his private line, and they resumed their ongoing conversation. Charles reiterated his desire to write software for Beethoven and stressed repeatedly his interest in getting his browser on all Astron machines.

"I'm not asking for exclusivity," Charles said. "If anyone else is interested, please stress that. I only want parity."

"Do you really think your phones are tapped Charles?"

"Yours and mine."

"I find that hard to believe," Leonard said.

Charles grated on Shepherd's nerves. He found the aggressive young man pushy. Usually, Leonard discounted Charles' hard line. He assumed that came with phenomenal success. Adding paranoia, however, pressed Leonard's tolerance. He decided he could not broach the subject of putting Panes on Astron hardware. He knew that Charles would jump at the chance. His operating systems already were used by the IBM and clone PC machines—ninety percent of the microprocessor market—and to add Astron's seven- percent market share would make his dominance complete. Charles' crown would be too big for one man to wear. Leonard rang off without broaching the fatal subject.

After two personal meetings and several lengthy phone conversations, Shepherd lost confidence in the elusive Dupris who constantly agreed then sidestepped to the left and right. One time he convincingly stated: I don't care about money. Anything we work out is fine with me." Two conversations later he offered to sell his company and all of its assets for $500 million dollars.

Leonard estimated the top value for the property Dupris was offering was no more than $50 million. After two more weeks of negotiating on the price issue, Dupris dropped to $300 million and in disgust Leonard decided to contact Stanley. Given the sour outcome of their last meeting, Leonard was reluctant to take the direct approach. Leonard briefed his management team on his interest in talking with Stanley, then waited for the word to slip out, as he knew it would.

Leonard with Chris Stuart for moral support met Stanley in the Astron conference room.

Stanley arrived with a broad smile and dressed in a dapper mode, a dark blue long sleeved sports shirt buttoned at the neck, clean khakis, and new running shoes. Leonard, wearing his normal dark blue suit, polished black loafers, white shirt and maroon power tie had been prepared for anything including bare feet and shorts. He interpreted Stanley's attire as a signal that he was serious and interested in what Leonard had to say. Leonard assumed that Stanley knew Astron Computers was shopping for a new generation operating system.

Stanley greeted Leonard and Chris like old friends. "You can't imagine what it means to me to be here in this room."

Immediately, Leonard wondered if Stanley was putting him on. The words were certainly ambiguous. For all Leonard knew, Stanley was referring to having been fired here.

"Astron certainly remembers you," Leonard said, using a little ambiguity of his own.

"I'll bet," Stanley laughed, then immediately got down to business. "I think I might be in a position to help you," Stanley spoke directly to Chris Stuart.

Chris smiled.

"Tell me if you have no interest in what I have to say. I'm probably making another big mistake," Stanley hesitated.

"Stanley, let's talk straight. That's my style," Shepherd said. "I'm sure you've heard that we need a new operating system."

Stanley smiled again at Chris. She had the feeling that he was challenging her to admit that she had failed with Beethoven. That was not something she was prepared to do. All she needed was time. Chris smiled, sweetly.

"I'm not unfamiliar with the computer hardware business," Stanley began again, "but please stop me if I step over the line."

Leonard doubted that anything he might say would deter Stanley at this point. Stanley under full steam was an interesting phenomenon. If he did not need the operating system so badly, Leonard might have leaned back and enjoyed the performance.

"If you think we might have something useful for you, I'll go along with any kind of deal you want. We'll license our software or sell you the whole damned company. Pardon the language, Chris, it just slipped out."

"That's all right Stanley. I've heard fucking worse," Chris said.

Chris as instructed asked the necessary technical questions. Stanley answered each to her satisfaction. For most of the meeting, Leonard sat back and let Chris and Stanley spar. Finally, Chris suggested that engineers from her shop meet with Stanley's people so that they could assess the Starbright software thoroughly. Stanley agreed, and they parted with Stanley taking one final shot: "You'll want to buy the company and take all my people. They're great."

"I'm sure they're insanely great," Leonard said, unable to resist the opportunity to use one of Stanley famous phrases.

The engineering meetings proved positive, and Chris submitted her recommendations along with the technical appraisals. The Astron team ranked Starbright as their first choice and Dupris' Future second. Leonard then scheduled a meeting at which each of the two contenders would separately make formal presentations to Leonard and his Management Committee. Both Dupris and Stanley accepted the venue and the format without protest.

Leonard's people booked the neutral meeting room at the Hilton and scheduled the presentations, Stanley first, and Dupris second. Leonard was amused by the fact that he, an outsider, would sit in judgment over sales pitches being made by one of Astron's founders and by one of the company's former chief technical officers.

Stanley turned in an awesome performance. Leonard was dazzled by his presentation and was ready to make his decision before hearing from Dupris. He of course waited. Dupris disappointed him. Although Chris had briefed him thoroughly on what was expected, Dupris did not bother to make a formal presentation. He acted as if he had already won the competition. He summed up his position with one brief statement:

"Your people know what we have. There's no need for me to repeat the technical details. We've made a fair offer. I understand your situation, I doubt that anyone is in a position to say that better than I, and believe me, all pretense of modesty aside, you have but one choice, buy our system."

Leonard was so impressed with Stanley's performance that he decided to have Stanley repeat it for the full Astron Board. He knew Stanley would enjoy the opportunity to confront the institution that had caused him so much personal pain, and Leonard wanted to have the Board participate in the delicate decision he was about to make. There were times, he knew, that a good executive chose to share responsibility, particularly when there was every likelihood of an adverse reaction. Leonard could imagine the media response when they learned that the once vaunted innovator had to turn to others, particularly one who carried so much adverse baggage as Stanley, for its next generation operating system.

Stanley entered the room like he owned it, confident and in complete mastery of his emotions. He immediately approached Stuart Miller, the man who had fired him without a thread of ego protecting cover, and shook his hand warmly.

Leonard could tell that a nervous Stuart Miller was flattered.

"It's good to see you again," Stuart said.

Leonard was surprised when Stanley did not reply with sarcasm.

"It feels good to be here," Stanley said.

Stanley, acting like a political candidate, deliberately shook the hand of each Board member, profusely thanking each for giving him the honor of appearing before them.

"It means so much to me," Stanley said when he took his place at the head of the table.

He took out his handkerchief and wiped one eye. Leonard could not tell if Stanley faked the tear or not, but it looked real. He gave Stanley the benefit of the doubt and waited a few seconds, giving Stanley the opportunity to regroup before beginning.

Stanley, milking the moment, turned his back on the Board and stared at the blank back wall. Finally, he turned back, and Leonard announced: "The floor is yours, Stanley."

Again, Stanley gave a mesmerizing performance. He described his product in terms that would have flattered the emperor's jewels. When he finished, the Board surprised itself by applauding, even Stuart Miller. As one they rose to their feet and chatted with Stanley as Leonard escorted him slowly to the door. After Stanley departed, Leonard returned to conduct the vote. The Board supported the decision to purchase Stanley's company without dissent. Not one Director voiced a concern that Leonard was certain they all shared: "Can they trust Stanley?"

Two days later, Leonard met with Stanley at his home to work out the details of the purchase. There were several key points that worried Leonard:

—Did Stanley and his two key partners want dollars or stock?

—What would happen to Stanley's employees?

—What role would Stanley play in the enlarged Astron Computers?

—Would Stanley and his partners accept a fair offer?

—How would they handle Stanley's employees vested stock options?

There were a thousand other details to be handled, but these were the deal breakers that worried Leonard.

Stanley, playing host, brewed them tea, and then the two CEO's sat down in the kitchen to negotiate.

Leonard, still preferring the direct approach, sipped his tea then started:

"Stanley we all want you to come back as an employee of Astron Computers, as a member of the management team."

"I understand," Stanley said, noncommittally.

"When our users think of Astron Computers, they think of Stanley Pitts and the Asteroid," Leonard tried again.

Again, Stanley failed to respond.

Leonard immediately began to worry that he had made a fatal mistake giving Stanley an opportunity to return. Leonard, of course, had heard all the Stanley stories, but he reassured himself that Stanley had matured. Leonard, recognizing his company's need, moved on, leaving the decision about a role for Stanley in abeyance.

"Tell me, Stanley, before we set a price for your company, are you thinking in terms of stock or money?"

"Money," Stanley said.

Good, Leonard thought. One hurdle crossed successfully. "And your partners?"

"Money."

Leonard began to worry again. This was too easy. Did Stanley and his partners know something he didn't? If they did not want stock, did that mean they had no faith in their operating system's ability to solve Astron's problems?

"Agreed," Leonard said.

"What are you thinking about in terms of price?"

"Twelve."

Leonard disliked Stanley's terse answers. Where was Mr. Charm now? Setting a price was not, however, a key part of the deal for Leonard. He knew other factors were more important. The purchase would give Astron its much-needed operating system, at least $50 million a year in income from the sales of Starbright software, the system itself, and a team of talented people. One hundred million dollars expended would represent three days of Astron sales. Stanley's $12 a share made this a $500 million dollar deal, the equivalent of fifteen days of Astron revenue. Leonard knew the company could manage that.

"That's too high, Stanley. It's more than we can handle. The Board won't buy it," Leonard countered.

"What number are you looking at?"

"Not a penny more than ten."

"Agreed," Stanley said, surprising Leonard who had expected to haggle for some time over price.

"Settled then," Leonard said. "Let's go back to the cash and shares issue." Leonard had given this subject much thought. Astron was buying not only an operating system, but Stanley the showman's talent as well. It would not do for Stanley to emerge with cash only. The media would immediately interpret this as an indication that Stanley had no confidence in his old company. Leonard decided to strike while Stanley was in an agreeable mood.

Stanley nodded and poured more tea.

"Cash for your partners is no problem, but we want to give you a motivation to work for our company." Leonard paused and waited for Stanley to react. He said nothing so Leonard assumed Stanley understood where he was going. "We ask that you take two million shares in stock and the rest in cash."

"One million."

"Million and a half."

"Agreed."

"We'll register the shares for you, but we must agree on one more thing," Leonard pressed on.

Stanley sighed and smiled to show that his reaction was good-natured.

"We must ask that you agree not to sell the stock immediately. If you sell it, you will send the wrong message to the market."

"Three months," Stanley said.

"You must hold it for longer than that. Selling the stock would undermine everything we're trying to accomplish here."

"Three months," Stanley repeated. "I'll hold it for three months, then I'm free to do with it as I please. After all, it's my money."

"A year," Leonard offered.

"Six months," Stanley said.

"Agreed."

"What about my people," Stanley said. "Their futures are key to this deal for me."

Leonard doubted that but did not say it. "We'll take everybody but your Chief Financial Officer and your main lawyer. Our CFO is my man and I trust him. The same applies to our lead lawyer."

"All right, but I want some say when it comes to assigning positions."

"Agreed." Once the deal was in place, and Stanley's company belonged to Astron, Leonard as CEO would decide who would form his management team.

Leonard knew how that would work out but decided to defer any debate over the subject until later.

At Stanley's suggestion, they took a walk around the block. Leonard hated the idea but complied. He knew that rumors would fly if the media or competitors or even their own employees got word of the meeting. By the time they got back to the house, they had a tentative deal. Astron would buy Again Corporation for $380 million, plus one and half million Astron shares for Stanley. Stanley was smiling when they parted; he had just pocketed close to $160 million.

Leonard and Stanley turned the deal over to the lawyers. Word leaked, and the media frenzy began. Even Charles Swift phoned. Leonard expected Charles to react positively. Whatever Charles really thought, he predictably put on the face that he thought would best benefit Digital Software. Instead, Charles surprised him.

"Leonard," Charles said. "I think you people at Astron have lost your collective minds. Why are you letting Stanley back in?" Before Leonard could think of a diplomatic response, Charles continued. "I know Stanley's technology, and it's nothing."

"Charles," Leonard said. "Our engineers have closely studied Starbright and we're convinced it's our system of the future. It would take us another five years to build, and that will not happen. Our company would die first."

"Don't be that desperate," Charles said. "There are other alternatives."

Leonard then realized what he should have seen from the beginning. Charles was upset that he had lost the opportunity to get Panes on Astron hardware.

"Starbright won't work on your machines," Charles insisted.

"Charles," Leonard addressed the real reason for Charles' call. "We considered Panes and Dupris' Future and the prospects for our own Beethoven and decided, quite calmly and coolly, that Starbright is best for us."

"Then you made a mistake," Charles insisted. "Stanley's not an engineer. He's a showman that's all."

"He's a great salesman."

"He's a great snake oil salesman," Charles countered. "You can't trust him. You'll learn that soon enough."

"You may be right," Leonard tried to calm Charles. "But the deal is done. Can we continue to count on you for software support?" Leonard thought it best to remind Charles that he too had something to lose if he went too far and started threatening.

"Certainly," Charles said. "And you will feature our browser?"

"The best I can promise you is parity," Leonard said.

"Will you ship with ours as the installed browser?"

"We haven't reached that decision yet, Charles. I promise you Digital Software will get a fair deal."

When they finally hung up, Leonard was of the impression that he had calmed the excitable Charles.

Finally, the lawyers ironed out the wrinkles. With Stanley's concurrence, Leonard had his public relations people schedule a joint news conference. Leonard and Stanley met in the conference room with the lawyers to sign the papers. Sitting with pen in hand, preparing to sign for Astron, Leonard deliberately raised the subject of Stanley's future role.

"Stanley," Leonard asked. "Have you decided what role you want? Come on the payroll? Be an adviser?"

Stanley shrugged.

"I have to know Stanley. That will be the first question that I'm asked after I make the formal announcement."

Sensing real trouble with Stanley in this kind of mood and no agreement, Leonard recapped his pen without signing.

"It's best you and I go into my office and work this out."

A strangely silent and passive Stanley followed Leonard back to his office.

"What's up Stanley?" Leonard asked as soon as they were seated. "Are you having second thoughts?"

"I've been having trouble sleeping," Stanley said.

"If this deal is causing you problems, maybe we better not go through with it." Leonard was bluffing. He was sure that Stanley knew as he did that forces were in motion that would make it almost impossible to cancel the deal.

"I've been thinking about everything that needs to be done, and about the deal, and everything. I tried and can't think clearly."

Leonard wondered if Stanley was on some kind of drugs.

"I don't want to be asked any more questions," Stanley said.

Leonard was not about to let him get away with that. "Stanley. You know me. I'm an up front guy. If there's something really bothering you, put it on the table."

Stanley irritated Leonard by not responding.

"Do you want to take over our engineering group?" Leonard could put Stanley in over Chris Stuart. He was sure she would understand even though it would mean a downgrading of her prestige.

"No."

"Do you want to head up our Sales and marketing? It's a natural for you."

"No."

"Is there any division you want?"

"No."

"Do you want to be an adviser?"

"No."

"Stanley, this puts me in a difficult position."

"Leonard, I told you I didn't want to answer any questions."

Leonard thought about letting the matter ride but decided against it. If he and Stanley were going to have a battle, now was the time for it to happen before it was too late to back out. "Stanley," Leonard let his tone rebuke the stubborn young man sitting before him. "I have to say something out there. I'll say you are going to be Adviser to the Chairman of the Board."

"OK," Stanley said indifferently.

Chapter 38

Bill Oldham's Washington operation alerted Charles to the brewing trouble in Washington. Julie Winthrop, the ambitious Chief of the Department of Justice's AntiTrust Division, had departed government service after acquiring another fine entry for her resume. Julie had done nothing wrong; she merely followed a time honored Washington tradition: members of the legal professions hopped in and out of ever more influential government positions as they worked their way up the ranks of the predatory law firms that thrived on their New York and Washington practices.

Herman Shehee, another ambitious young lawyer, had been Deputy Counsel in President Bill Clinton's White House when Julie compromised the Digital Software anti-trust investigation by negotiating the consent decree which most had interpreted as a Charles Swift victory. As soon as he moved in as Julie's replacement in the shabby office on the third floor of the Department of Justice, Shehee had ordered that the Digital Software investigation be reopened.

According to Bill's sources, primarily a young lawyer from the State of Washington who owed his position to the intervention of a senior Democratic Senator from California, Shehee was determined to show that the Antitrust Division had teeth. Shehee planned to prove that the Digital Software was violating the consent decree which forbid predatory licensing and product linking. Panes 95 featured Charles' new browser and DS' proprietary internet network which hoped to challenge America Online as the country's number one internet access and content provider. Shehee was determined to prove that Digital Software was illegally forcing the hardware manufacturers to license his browser. This linkage, Shehee argued, constituted predatory practice designed to drive Netscape out of business.

What the government labeled predatory product integration, Charles described as innovation. His media campaign, orchestrated by Bill and Ann Oldham, accused the government of partisan politics, ignorance of technical innovation, obliviousness to the way the computer industry did business, and general hostility to big business. Charles had many allies, but his enemies lined up at the Antitrust Division door, competing to see who could produce the most damaging stories about Digital Software's predatory practices.

Even the Attorney General got into the act. She and Assistant Attorney General Herman Shehee held a joint press conference that the media immediately dubbed a resounding success. Herman tediously outlined the perceived consent decree transgressions, and the Attorney General attempted to refurbish her tarnished image by announcing that she intended to ask the courts for a one million dollars a day fine which would continue ad infinitum until Digital Software complied to the letter and spirit of the consent decree.

Charles reacted by visiting Bill and Ann Oldham in Washington for a strategy meeting. The three old friends met in Bill and Ann's Crystal City apartment.

"It's the same old shit," Charles said. This time he was truly angry. "I'm losing my faith in the American people. They are the ones that gave us Clinton and Gore and this lousy administration."

Bill fought back a smile. Their old roommate, Mike Hamilton, now Charles' long time right hand, had openly supported Bill Clinton. He wondered what Charles had thought of that.

"And they've given them another four years to pander to minorities and to base every decision on the polls," Charles continued. "The American people are getting exactly what they ask for. I don't blame Clinton. We all knew what he was before we elected him president."

"I don't know why anybody wants to be president," Ann said. "The media selects our candidates now." When she used the word media, Ann stared at her husband. She tended to blame him for every distortion that his print and television colleagues perpetrated.

Bill did not respond. He waited for Charles to vent before trying to turn their conversation towards deciding how to react to the Department of Justice.

"Most of our candidates are good men," Charles started again. "At least when they start out. But in order to get nominated they have to compromise every principle they once held. By the time they get elected, they are nothing but weather vane men idly turning in response to every breeze."

"That's pretty good," Bill laughed. "Mind if I use it."

"Be my guest," Charles glared at his friend. "I'm tired of being the target."

"People are jealous of your fabulous success," Ann said.

"I earned it," Charles said defensively.

"Even your most ardent supporters are jealous," Ann persisted. She felt it important that Charles understand the deep source of the sentiment against him. "You're so young, and you have all that money."

"Every time one of your software applications freezes a computer," Bill joined his wife, recognizing what she was trying to do, "the user blames you personally. Digital Software has a reputation for rushing its applications to market without adequate testing."

"We test the damned stuff. We rush to market because the users and the media demand it. Everyone wants the latest stuff."

"We know that, but it's like Ann said. You have wealth beyond everyone's ability to comprehend and when the applications freeze they blame you personally. You're greedy."

"Like hell I am." Charles in his anger turned on Bill.

Bill smiled and held up his hands in surrender. "I'm not saying you're greedy. That's what the frustrated average user says. They're jealous, but they just want their computers to work."

"They don't understand the industry. Where do they think technical innovation comes from? It's not easy bringing out a new product every three months. It takes brains and lots of hard work."

"Calm down, Charles," Ann said. "We're your friends. We above all understand what a marvelous thing you have accomplished."

"I like the word innovation," Bill said, determined to focus his friend. "Innovation sounds good. Nobody can argue against it."

"It's true. Integrating our browser is innovation. It's not easy. It's a very complicated technology. We're improving the user's experience. The Internet and browser integration is as an important development as the graphical operating system was."

"And the Department of Justice calls it product integration evilly designed to support predatory practices. You use one application to force your customer companies, ninety percent of the market for operating systems, to buy your other applications. This unfair competition is designed to promote a monopoly," Bill summarized the government's case.

"And I still say its innovation," Charles glared at Bill.

"Bill is just trying to channel our discussion," Ann defended her husband. "We're on your side. If you say its innovation, it's innovation. Right Bill?"

"Damned right," Bill said. "Now we know what they are up to, what are we going to do about it?"

"All we can do is defend ourselves. We turn our lawyers loose, and we put as much political pressure on these guys as we can. We know they listen to public opinion, polls, etc. I'll do the best I can with public appearances, and you guys will have to generate as much negative publicity for the government's action as you can. And, Ann, everything your lobbying effort can get out of Congress will be useful."

"Then we're supposed to go all out," Ann said.

"Yes," Charles said. "This time let's see if we can bury this monopoly nonsense once and for all."

"After it gets to the courts, it'll be a different story," Bill cautioned.

"I know," Bill said. "But let's see if we can stop them before it gets that far. Clinton is a chameleon."

The three sat in silence for a while, then Ann smiled. "Charles. How was your golf game this morning?"

"Thought you would never ask," Charles said.

Both Ann and Bill knew that Charles had played golf with the President that morning at the Congressional Country Club.

"How do you do it?" Bill asked. "Playing golf with Clinton makes you as big a hypocrite as he is."

"I know. Greg Norman, David Howard, Clinton and myself made up a foursome."

"Did the lawsuit come up?" Ann asked.

"Not a word. Clinton acted as if the three of us were his best friends. I waited for the opportunity to lambaste him, but I didn't want to be rude and raise the subject myself."

"Only Greg Norman would not have known what was going on," Bill said.

"How did the match go?" Ann asked.

"He had the nerve to suggest that he and Norman play against David and myself for a hundred bucks a hole," Charles said.

"Did you accept it?" Ann asked.

"No way. I'm a lousy golfer, and David is a natural athlete, but he doesn't get to play much, and he's no match for Greg Norman anyway. Greg is a delight to watch, and a nice guy too."

"What about Clinton?"

"He's a phony, and he cheats," Charles said. "I shot a hundred and ten, David had a eighty-five, Norman had a seventy, and who knows what Clinton really had. He claimed to have an eighty-eight, but I know he used as many strokes as I did."

"You didn't call him on it?" Bill asked.

Charles laughed. "Not with all those secret service men following us around."

"Too bad we didn't arrange to get your picture taken with him," Bill said. "We could have gotten mileage out of a story describing how your golfing buddy is busy stabbing you in the back."

"How do you like your new house?" Ann asked, referring to the fifty million-dollar state of the art home Charles was building on Lake Washington.

"We're still camping out waiting for the workmen to finish it," Charles said.

"I assumed you were already in it," Bill said.

Charles laughed. "After Penny and I got married, I turned the house over to her. She's made so many changes it might be a year before we get in. She thinks some bathrooms are too small, and we need a nursery and a room for a governess, and that sort of thing."

"It shouldn't take too long to move a wall here or there," Bill, who had never owned a home, said.

"It does when the walls are two foot thick and made of poured, reinforced cement," Charles said.

The Government moved faster than Charles, his lawyers and his friends anticipated, so fast in fact that Charles suspected the golf match had been a set up with Clinton wanting to see first hand how worried his opponent might be. Since Charles had neither appealed to the President nor attacked the President's men, Clinton decided that he was free to act.

In October Herman Shehee filed suit in District Court claiming that Digital Software had violated the consent decree by forcing other equipment manufacturers to license DS's browser as a condition for using Panes.

Jeffrey A. Potter, Charles' lead lawyer, flew in from Seattle and established his defense team in an adjunct to Ann's H Street office.

With Ann listening, Potter phoned Charles to report the details. "Are you sure you don't want to brief Charles?" Potter said as he waited for the call to go through.

Ann laughed. "I'll let you have all the fun." She knew Charles would be livid over the government's latest transgression.

"Yes?" Charles came on the line.

"Charles, it's Jeffrey Potter."

"I'm busy," Charles said.

"Aren't we all," Potter replied. He knew Charles liked his employees to stand up to him whatever his mood.

"All right, give me the good news," Charles said.

"It unfolded just as..." Potter hesitated when Ann waved her finger back and forth. Potter had been about to say "just as Bill anticipated." "Shehee's case is just what we anticipated," Potter corrected himself.

Charles did not reply.

"Do you want the long or the short of it?"

"Just tell me the bad news," Charles ordered.

"Shehee argues that Digital Software has a monopoly."

"That's not true," Charles erupted.

"Digital Software maintains that monopoly using anti-competitive contracts which force hardware manufactures who wish to distribute processors using the Panes operating system to also license other DS software, specifically the DS browser. This anticompetitive practice is harmful to other manufacturers and denied the consumer the right to choose its own software."

"That's bullshit. We gave the damned browser away free just like Netscape."

"Shehee has offered to settle if we agree to change our predatory practices."

"We can't stop what we're not doing," Charles said.

"I naturally declined," Potter continued, ignoring Charles' outbursts.

"I thought we settled this issue two years ago," Charles said.

"The Government has asked the Judge to issue a cease and desist order and asks for a penalty of a million dollars a day until we do."

"Shit, that's nothing. I make a million dollars every two hours," Charles said.

"The Judge took the motion under advisement."

"What happens now?" Charles demanded.

"I'll file a rebuttal."

"Don't file anything until you clear it with me," Charles ordered then hung up.

Potter smiled thinly as he returned the phone to its cradle.

"That wasn't too bad," Ann said. "He didn't fire you."

"Not yet," Potter said glumly. "The government's just getting started."

"That's what we told you," Ann said. Privately, she had advised Charles to beef up his legal representation. She suspected that Potter was in over his head.

"Does Bill have anything on what the government plans next?"

"Nothing. They're waiting to see how you react to this suit."

"It's just the beginning," Potter said.

Potter flew to Seattle where he met with Charles and drafted a curt counter-motion. In simple terms, Digital Software noted that it retained the right to define the parameters of its products. In this instance Digital considered the browser an essential feature of Panes 95 and had designed the interrelationship of system features accordingly. Digital Software had the right to require any other equipment manufacturers who wished to license Panes 95 for installation on their processors to install all of Panes 95. Digital Software could not allow its licensees to pick and choose elements of Panes 95 because the features are all interrelated. Such selective actions would be harmful to the product and ultimately to Digital Software's reputation. The other equipment manufacturers are free to license other products if they so choose. Therefore, Digital Software's business practices are neither predatory nor monopolistic.

The Judge replied to Digital Software's counter motion by issuing a temporary injunction that required Digital Software to unbundle Panes 95 and the browser until the issue was resolved by a special master to be appointed by the court.

Charles in a fit of pique ordered that his company comply with the temporary injunction by offering two stripped down versions of Panes. One, without the browser, would not boot, and the other was stripped of essential features. Charles was determined to show that essential elements could not be removed from his operating system without negative impact. Potter filed a motion asking that the Judge rescind his order calling for a special master.

This ploy angered the Judge who immediately ordered a demonstration in his courtroom in which an independent contractor showed that the browser could easily be removed using Pane's add/remove function. Following the demonstration, an infuriated Judge denied Digital's motion.

Digital Software responded by appealing to the U.S. Court of Appeals for the District of Columbia.

Chapter 39

For a while after the purchase of Again Corporation and its Starbright operating system, Leonard Shepherd was confident that things were back on track. He even tried to work with Stanley as an adviser, talking with him at least once a week. Leonard assigned Stanley an office, not in the Headquarters building, but nearby. As best as Leonard could tell, Stanley did not spend much time there. Stanley appeared to be preoccupied with his animation company.

This calm hiatus did not mean that everything ran smoothly. Stanley's talented employees began their move to Astron. Most fit in nicely. One problem developed when Leonard tried to fit Stanley's senior management people into Astron's structure. He had difficulty deciding what to do with Misak Ayvasian, Again Computers' Chief Technical Officer. Stanley argued that Misak should be made senior to Chris Stuart whatever the titles might be. Stanley hinted that Misak might seek other employment if his wishes were not honored. Leonard capitulated and named Misak Chief of New Product Development. In this capacity, Misak reported directly to Leonard, and Chris, retaining nominal charge of engineering, reported to Misak. Chris protested her loss of prestige, and Leonard calmed her by confiding that his door remained open to her. It was only later that Leonard learned from Misak that he had not insisted on the new title.

Chris was not the only senior management chief to find unhappiness in the revamped Astron Computers. The still declining bottom line forced Leonard to seek out scapegoats. He terminated four key managers that he blamed for his company's underperformance. Morale sagged.

Then, rumors began to appear in the media. This did not surprise Leonard; disgruntled Astron employees had always used the media to vent their wrath, but one vicious cycle unsettled Leonard. Reliable journalists quoted Stanley as saying that he "lacked confidence" in Shepherd's leadership. Leonard confronted Stanley, and Stanley denied being the source, ardently claiming that the allegations did not represent his true attitude. Stanley promised to issue a denial, but he never did. At least none appeared in print, and the stories became even more vicious. One even hinted that Stanley was maneuvering to replace Shepherd.

Four months after Astron Computers acquired Stanley's company and operating system, an event occurred that truly angered Leonard. The <u>Wall Street Journal</u> reported that the previous day one million five hundred thousand shares of Astron Computers stock had traded as a block. Leonard immediately thought of the million five hundred shares that Stanley had acquired as part of the deal. At Leonard's insistence, Stanley had promised that he would hold the shares for six months and would inform Leonard before selling them. Leonard firmly believed that Stanley should own a substantial block of Astron stock to demonstrate his confidence in the company's future prospects.

Leonard immediately picked up the phone and dialed Stanley's private number.

"Stanley, I thought we had an agreement," Leonard started as soon as Stanley came on line.

"Hi, Leonard, what's the problem?"

"Have you seen the <u>Journal</u>?"

"<u>Wall Street</u>?"

"Of course."

"No. I don't follow it regularly."

"You should," Leonard did not try to hide his irritation. "A block of Astron stock traded yesterday. A million and a half shares. I thought we had an agreement."

"We did," Stanley said.

Leonard chose to interpret that Stanley's answer constituted a denial. "Good," Leonard said.

Stanley said nothing.

"It's important that we stand together and offer a united front to the public," Leonard said.

"That's right," Stanley said. "Anything I can do for you?"

After he hung up, Leonard began to worry. When he reviewed the conversation, he realized that Stanley had not specifically denied having sold the stock. Since he did not want to worsen relations with Stanley by accusing him of lying, Leonard decided to let the matter ride until he got the next quarterly report which would identify the seller.

Hoping to take advantage of Stanley's love affair with the media, Leonard decided to mount an extravaganza at the next Astron Computers fair scheduled to take place in Boston. Leonard planned to personally introduce an upgrade to the Asteroid operating system, then to have Stanley present the operating system of the future based on Stanley's Starbright. Leonard remembered the dramatic impact of a black clad Stanley standing on the darkened stage in San Francisco and presenting Starbright for the first time. If they could together duplicate that feat, they could jointly calm Astron's user and stockholder concerns.

Leonard checked with Stanley, got his agreement to participate, and delegated the details to his staff. Leonard personally decided to stage the consecutive appearances of himself and Stanley at the end of a two-hour show.

Leonard, now confident of his public persona, did not review the ghost written speech before traveling to Boston. Upon arrival, his staff assured him that all was in order. Leonard assumed this applied to Stanley's role as well as his own. He spent his time before appearing studying his speech. He was not informed that Stanley was indulging in one of his not uncommon temper tantrums. Stanley, always a headliner, expected to appear at the opening, and he exploded when he learned that he had been relegated to a closing role. Things were made even worse by a couple of the early speakers who rambled, extending the length of the program by almost a full half-hour.

When his turn came, Leonard, still unsuspecting of the tumult around him, took center stage. From the beginning, the teleprompter malfunctioned, forcing Leonard to ad lib a speech that he had not written; he fumbled badly. When the end came, a relieved Leonard turned the audience over to Stanley ardently hoping that Stanley would be so brilliant that his own presentation would be forgotten. Stanley lived up to expectations, even running overtime. On camera Stanley always performed. Then came the finale that Leonard had carefully orchestrated. Leonard planned to join Stanley in front of the audience and appear to surprise Stanley by calling Harold Dumbroski to the stage for a dramatic reunion of old partners. Leonard anticipated that this impromptu staging would bring down the curtain to thunderous applause.

Unfortunately, Leonard's staff, anxious to implement their superior's wishes, ignored the fact that Harold and Stanley had not spoken for years and did not consult Stanley.

Leonard marched back on to the stage and stood next to Stanley. He leaned into the microphone: "Ladies and Gentlemen, I have a special treat for you and Stanley. May I call our next guest who needs no introduction: Harold Dumbroski!" Leonard shouted the name.

The surprised audience cheered.

Harold walked in from the right, and to everyone's surprise a frowning Stanley stamped off to the left.

Leonard had planned that the three of them stand together with arms linked, representing the past, present and the future of Astron Computers. He could only stare at Stanley's back. Stanley had turned as soon as he saw Harold make his appearance, and had departed without a welcoming gesture or uttering a single word to his old friend.

As time passed, the Astron market share eroded. Leonard was not concerned; this was predictable. Nobody was buying the Asteroid, not even with the most recent update. Alerted by media anticipation, the loyal Astron base kept their wallets in their pockets while they waited for the new operating system. The loyalists grew impatient. The second quarter statement disclosed that Astron revenues had fallen twenty-seven percent from the previous year's dismal return. Even worse, for Leonard, the statement confirmed that Stanley had misled him about the million and a half shares. Stanley, despite his promise and half denial, had sold every share of Astron stock he owned but one. Leonard considered this a shocking betrayal. Stanley had lied, and the media immediately concluded that Astron Computers was on the rocks; why else would Stanley have sold all his stock?

Recognizing that he could do nothing at this point about the public reaction, Leonard called Stanley.

"Stanley, I asked you about these shares and you denied they were yours." Leonard knew this was not quite accurate. Stanley could rightly argue that he had been told they had had an understanding and Stanley had answered evasively, that's right. Leonard had accepted it as a denial.

"Yes, that's right," Stanley replied brazenly. "I was embarrassed."

"Why did you do it? It places the company in an impossible position."

"I guess I was depressed at the time and felt the company was hopeless. It was a spontaneous thing."

Leonard hung up, too angry to respond.

This conversation and the company's report so depressed Leonard he decided to do something he rarely did. Take a full two-week vacation with his family. Leonard gathered his extended clan around him at his Lake Tahoe retreat and prepared to celebrate the Fourth of July. Ignoring Astron and its problems, Leonard drove the family boat while his offspring waterskied behind. Leonard enjoyed the kids and the warm comfort of his family. They celebrated the Fourth with their own fireworks. Leonard who watched from the porch thought they were rather unspectacular, but the young kids loved them.

The next morning Leonard was relaxing in his study with a copy of the Los Angeles Times, taking a respite from the kids, when the phone rang.

"Leonard, sorry to bother you," Ed Brown, the Deputy Chairman of the Astron Board said. "Something's come up."

"What's the problem?"

"Not a problem exactly," Ed said. "The board has reached a decision. It's not good news."

The words were like a sharp blow to the head. Leonard realized immediately that he was being fired.

"Why?"

"It's obvious, Leonard. Don't fight us. Sales are down. The stock is in the pits. Morale at the company is at an all time low."

"That's unfair. You know very well why sales are down. As soon as we bring out Starbright, we'll go through the roof."

"I doubt that," Ed said. "We need a new face to drive sales."

"Are you sure it's not an old face?" Leonard asked, bitterly suspecting that Stanley was behind this.

"Sales and marketing aren't your thing, Leonard."

"When I came here I told the Board it would take at least three years. I'm only half way there."

"The decision is made," Ed said coldly. "The board is not open to further discussion."

"Then I have no choice."

"No. We'll try to make the transition as gracious and cordial as possible," Ed said.

"Who else knows at this point?" Leonard asked. He had to find out about Stanley.

"The board of course, and Stanley. The board wanted his views."

"Why involve Stanley? He's just an adviser, not a Director."

"The decision is effective immediately. You are no longer associated with Astron Computers," Ed said and hung up.

Monday morning Stanley reported to work at Astron Computers at eight o'clock sharp wearing shorts, a wrinkled T-shirt, and bare feet in sandals. Employees who saw him immediately assumed he was passing a message. He was starting again just as he had when he and Harold had founded the company in Stanley's father's garage. While Leonard cleared out his desk, the Board met and offered Stanley the CEO position. He declined three times, then gracefully agreed to serve as acting CEO with no salary.

Among Stanley's earliest actions from his temporary office, Stanley canceled the cloning licenses, forced Stuart Miller off the Board, took credit for Starbright, announced that priorities were being altered to allow for dramatic new products, let it be known that Harold Dumbroski was no longer an adviser nor welcome on company property, and took long walks through his domain delivering on the spot decisions.

Overhead, the pirate flag reappeared on every roof of every building on the Astron campus, and the word bozo reappeared in the Astron lexicon.

Chapter 40

Before the District Court of Appeals ruled on Digital Software's appeal of the lower court judge's decision sending the consent decree violation charge to a special master, Assistant Attorney General for AntiTrust matters Herman Shehee filed a new lawsuit against Digital Software. The government charged that Digital Software illegally used its monopoly to block potential competition from Netscape Communications' Internet browser. Attorney generals in twenty states joined the suit.

Thoroughly bored with travel, particularly to Washington D.C. where he did nothing fun, merely consult with lawyers about his legal problems, Charles invited Bill and Ann to visit him. This would be their first chance to view the fifty million-dollar structure on the lake that Charles now called home. When they arrived, Charles gave them the grand tour, demonstrating his electronic toys controlled by a massive computer that adjusted the house to the idiosyncrasies of Charles guests. The computer adjusted the climate, the environment and the temperature, changed the pictures on the ever present screens, provided electronic books, and games and movies and music, all to the guest's taste. Each guest was assigned a wand programmed to their personalities, giving them the power to command their individual rooms.

Consisting of five separate pavilions connected by underground corridors, the complex had a central ballroom/entry hall, a guest house with bedrooms, a library and a huge dining room; a reception hall for large parties; a beach house with pool, hot tub and dock, a twenty car underground garage, and a separate house for the caretaker. On top was Charles private residence. Viewed from the lake, the five connected structures appeared like a Frank Lloyd Wright creation, sunk into its environment, formidable, interesting, but not particularly grand.

"What do you think?" Ann asked when she and Bill retreated to their room to change into swimming suits prior to their meeting with Charles and Jeffrey A. Potter by the pool.

Bill who stood with his wand before a Vincent Van Gogh screen hesitated before replying. "I'm overwhelmed, but these gadgets are neat."

"I wish he had shown us the family pavilion," Ann said.

"He will. He said Penny was not feeling well."

"You wouldn't either if you were seven months pregnant."

"How do you know? You've never been pregnant."

"Who told you?" Ann smiled.

Bill waved the wand at the curtains and commanded them to close.

"I just know," he said. He pointed the wand at the bed and it began to vibrate. "What about that?"

"We don't have time so don't start anything," Ann said, retreating to the bathroom.

"I'm tired of talking about Charles' lawsuits," Bill said.

"So quit."

"Where would I find an easier job that pays so well?" Bill asked.

"I wonder how they knew what size suit I wear or what color I prefer," Ann emerged from the bathroom. She held up a bright red suit emblazoned with the initials A.P.O. for Ann Page Oldham. As soon as she spoke, Ann realized she had made a mistake.

"I'm sure Charles remembers," Bill said, referring to a delicate subject they usually avoided, Ann's past relationship with Charles.

"There's one here for you too," Ann said. "It looks like your size," she held up a blue man's suit with large white initials B.O. "I wonder what that means? Were you two really close? Did you sleep together in the same room?"

"Don't get started," Bill laughed. "I surrender. A.P.O. Yours sounds like a military postal address."

"And B.O. What does that sound like?"

After inadvertently taking side passages to the garage and the kitchen, Bill and Ann joined Charles and his lead lawyer at the pool. Charles, like Bill and Ann, wore a bathing suit with his initials, but Jeffrey A. Potter was still dressed in a dark blue suit and tie.

"Jeffrey has news," Charles said sourly. "He'll brief us and then take off. He has to go sue an orphan or something."

"Bill, Ann," Potter greeted them. "Sorry I can't stay and chat, but I do have some good news. The Appeals Court has just ruled in our favor. The grand master business is dead."

"Great," Bill said. "That's two cases we've won."

"But the next one has just started," Charles said.

A white coated Filipino appeared and gave each of them a drink, a screwdriver for Charles, a Virgin Mary for Ann, a Budweiser for Bill and a glass of Perrier for Potter.

"I really like these wands," Bill waved his in the air.

"Careful," Charles cautioned as a cover began to unroll over the swimming pool and the sound of loud rock music burst from hidden speakers.

Bill put his wand in the pocket of his robe. "Sorry."

"No sweat," Charles said. "Volume down," he shouted and the music subsided. "Pool open," he shouted, and the cover rolled back.

"God it must be nice to be a King," Ann said.

Charles smiled. He turned to Potter. "Jeffrey, tell us the rest, and then you can go chase your orphans."

Potter smiled. "It's the same old stuff," the lawyer began. "Shehee thinks he can get us on the browser issue."

"Bullshit," Charles said. "America Online is getting ready to buy out Netscape. There goes their poor little Netscape crap. America Online wants to monopolize the Internet. How can they complain about our little browser?"

"You know as well as I do, Charles," Potter said. "Improper linkage."

"How can they prove that?"

"We'll see. The state suits are going to focus on the allegation that Digital Software illegally linked licenses for our Office application to Panes. They probably have Lotus and Word Perfect lined up to testify." Potter looked at Bill for affirmation.

Bill nodded affirmatively.

"And Shehee is going after Panes 98. They're going to try and force you to agree to ship Netscape's browser along with ours, so the user has a choice."

"Such garbage," Charles sipped his screwdriver. "We both have our browsers available on the net. Any newby can download either or both for free."

"I didn't say that Justice knows what it is talking about."

"Somehow Shehee has gotten your latest status reports," Bill said. "He knows that DS now has more than sixty percent of the browser market." All four knew that two years previously DS had only a five- percent browser market share.

"Some market. Both browsers are free," Charles repeated himself. "Any market gain comes from user choice."

"That's not what the government thinks. They claim its because you have been leveraging your monopoly power and illegally pressuring customers."

"They are also taking aim at our innovation defense," Potter said. Charles's media campaign had stressed the fact the government's lawsuit would stifle technical innovation.

"How are they doing that?" Charles asked.

"By demanding that we show how linking the browser and the desktop constitutes innovation."

"I'm tired of this shit," Charles said. "We keep going around and around."

"That's what happens when you face a political problem," Bill said.

"What happens if we lose?" Charles asked.

"We're a long way from that," Potter said.

Both Bill and Ann waited. They could tell from the lawyer's body language that he really did not want to discuss this subject.

"Tell me please what are the Judge's options if he decides against us?" Charles insisted.

"He has four main options," Potter said. "Firstly, he could order that Digital Software be divided into two companies, one selling operating systems and one for the internet and other software."

"He could insist on more than two companies," Bill said.

"That's right," Potter continued. "He could decide that the browser and Internet applications be separated from the other software." Potter paused to give Charles an opportunity to comment, but he said nothing. "Secondly, he could order that we open our source code giving competitors the ability to modify Panes to use their own software. Thirdly, he could order that Digital Software refrain from giving discounts to companies that use DS software. Fourthly, he could go the AT&T route and create a number of smaller but identical versions of Digital Software."

"And what do you think is most likely?"

Potter hesitated. "I think we can win this one. If we don't, we can string the case out with appeals for years. If we lose, that's what I would recommend."

"Of the four options, which are the most likely?" Charles insisted.

"I doubt he would go with open source code or impose pricing controls. Each of those has obvious pitfalls. If I had to chose between the first or fourth options I would say number one."

"He will break us into two companies keeping the browser separate from the operating system," Charles said.

"Yes."

"Thank you Jeffrey," Charles dismissed the lawyer. "I'm tired of this legal talk. You fight our case, and I'll sit here and talk with my friends."

As soon as the lawyer had departed, Charles turned to Bill and Ann. "OK guys, you heard what he said. The worst that is going to happen will be the Judge will force me to divide up my company. We'll fight the case for years if we have to, but in the end I might lose."

"We of course have a hole card," Ann said.

"The 2000 elections," Charles agreed.

"We've got to work on and for the Republicans as hard as we can," Ann said.

"Right," Charles said.

"We have one problem there," Bill said. "The case is now in the courts. It may get settled before a new administration gets into office, or it may pass beyond political influence's reach."

"All the same, we must work the political side every bit as hard as we do the legal battle. Even if they split me up, a new law could put me back in business."

"Agreed," Bill said. "But our best bet will be a strong legal defense."

"OK, Bill, I won't debate the point. Its not either or. Its full speed ahead on both fronts. I asked you out here to discuss something extremely confidential with you. I need your opinions."

"Shoot," Bill said.

"Take the worst case scenario. Say, the courts do order that Digital Software be split into two companies. What do I do?"

Recognizing that Charles' question was rhetorical, both Ann and Bill waited for him to continue.

"I'm preparing for that day. Just in case. I've already taken one step back from Digital Software management." Charles referred to the fact that he had recently appointed Mike Hamilton President of Digital Software and made himself Chief of Product Development.

"If they decide to divide me into a software/internet company and an operating system company, Mike will be in position to take over the latter. I could then take control of the software/internet Company."

"How would you handle ownership?" Bill asked.

"Good question. I assume the Court can't prevent me from investing in both companies. It's still a free country. I can buy any stock I want on Wall Street. I could keep a forty-nine percent share of the operating system company and own as much as I want of the software company. Mike could buy as much stock in the operating system company as he has the money to handle."

"The more shares he owns the better," Bill said.

"You still should control as much of the stock in Mike's company as you can," Ann said.

"Right. That's where you two come in. How much can you buy on your own?"

"Between us," Ann said looking at Bill who nodded agreement, "I would say maybe a million dollars worth at the outside."

"I think I can help you leverage that," Charles said. "You are both employees of Digital Software now. Ann as our Washington representative and Bill as a correspondent of <u>Computer News</u>. I think we can work out a stock option deal that will let you two own enough to guarantee that Mike has voting control of the operating system company. Can you live with that?"

"Certainly Charles," Bill easily recognized that Charles was offering them a chance to acquire a reasonable fortune. "But it is not necessary."

"You would be doing me a favor."

"Do you plan on managing the software/internet company yourself?" Ann asked.

"Good point Ann. As long as I am CEO, I would be a target. I have reached the point where I attract too much attention. As a stockholder, major or minor, I should be out of the firing line."

"Do you have someone in mind, someone you trust implicitly, someone with an unimpeachable management record who could be CEO?" Bill asked.

Charles laughed. "Don't worry. I'm not thinking of you." Charles knew that because of Ann's career Bill would not consider leaving Washington. "It's yours if you want it."

"No thank you," Bill said. "I don't need that kind of stress."

"I hope you're not thinking of someone like Starkey or DiGenova," Ann said.

"No," Charles hesitated. "Except for Mike, you are the only two who know about this deal. Remember, we're just thinking what if at this point."

"You've got our attention Charles. Your secret's safe with us," Bill said.

"I've been thinking about David Howard," Charles said.

"A perfect candidate," Ann and Bill said together.

At that point, Charles' wife Penny joined them. They sat beside the pool chatting, then retreated to their respective pavilions to prepare for dinner. Bill and Ann chatted about what they had heard while dressing. Both recognized the opportunity they had just been offered.

Several times during the evening telephone calls interrupted their relaxing conversation. Three seemed particularly important. All were intended to commiserate with Charles about the court case. During the first conversation, Charles twice responded "Yes Governor." Charles during the conversation deliberately referred to Texas. Ann winked at Bill and mouthed: "George W." During the second, Charles said, "I agree Mr. President." "Daddy." Ann whispered. The last call was from Bob Dole. Charles clearly had started his political campaign.

The next morning Charles and Penny drove Bill and Ann to the Seattle airport in Charles' big black Mercedes limousine. Bill chided his old roommate on his sedate choice of cars.

"I still have my old red Mustang," Charles said proudly.

"I saw it in the garage," Bill said, remembering the good times they had had as young men in that car.

After the lumbering 747 fought its way into the air, Ann turned and took Bill's hand.

"Why are you so pensive?" She asked.

"Just think," Bill said. "You could have lived in that house. Are you sorry you made the wrong choice."

Ann laughed. She squeezed her husband's hand. "I made the right choice. I think that's a big ugly old house."

"No so old," Bill smiled.

"I wonder if Penny's happy?"

"I think she is. Bill's busy, but Penny has everything she wants. She's handled the media just right."

"By refusing to admit they exist," Bill said. "Poor Charles."

"Why Poor Charles? He's anything but poor."

"He works all the time and has the Government after him constantly."

"Charles thrives on the competition and power," Ann said. "Poor Government."

"Poor Government?"

"They don't know what they're in for."

"Poor Government," Bill agreed.

Robert L. Skidmore

Author Biography

A graduate of West Virginia University and a teaching assistant at the University of Wisconsin where he worked on his doctorate in American History, Robert L Skidmore spent thirty-five years in the foreign service of the United States Government. Now, long retired he devotes himself to two lifelong passions, researching history and writing, both of which permit him to play with his computers.

www.ingramcontent.com/pod-product-compliance
Lightning Source LLC
Chambersburg PA
CBHW020719180526
45163CB00001B/28